AMERICAN MILITARY GOVERNMENT
OF
OCCUPIED GERMANY

1918–1920

REPORT OF

THE OFFICER IN CHARGE OF CIVIL AFFAIRS

THIRD ARMY

AND

AMERICAN FORCES IN GERMANY

AF797408

HEADQUARTERS

AMERICAN FORCES IN GERMANY

OFFICE OF CIVIL AFFAIRS,

COBLENZ, GERMANY.

March 4th, 1920.

From: Officer in Charge of Civil Affairs, A. F. G.
To: Chief of Staff, A. F. G.
Subject: Report on Military Government.

1. I herewith submit to the Commanding General A. F. G., a report on the activities of the Office of Civil Affairs, from December 1st, 1918 to January 10th, 1920.

2. While principally treating the activities of the Staff section, which during this period has been under my direction, I have nevertheless endeavored to give the report a somewhat more ambitious scope, and to discuss therein the entire subject of the American Military Government in Germany. It is hoped thereby, that a complete record of our administration in the Rhineland may be preserved, as a basis for technical study of Military Government by the General Staff of our Army as well as for future historians.

3. Volume I contains a narrative account of the Military Government, while the remaining volumes are devoted to collections of the more important documents. These documents are arranged by subjects, chronologically, in such a way that it is hoped they may be readily accessible to lay student or historian. A number of treatises on various phases of the German body politic, together with essays on the organization for Military Government in the French, Belgian, British and German Armies, have also been incorporated into this portion of the Report.

4. It is requested, that if the approval of the Commanding General is secured for this report, it be forwarded to the War Department, for either publication or record, as the Secretary of War may desire.

(Signed) I. L. HUNT,
(Typed) I. L. HUNT,
Colonel, Infantry, Officer in Charge Civil Affairs.

"Finally, I pay the supreme tribute to our officers and soldiers of the line. When I think of their heroism, their patience under hardship, their unflinching spirit of offensive action, I am filled with emotion which I am unable to express. Their deeds are immortal and they have earned the eternal gratitude of our country."

Preliminary report of General Pershing on operations of the American Expeditionary Forces to the Secretary of War, November 20th, 1918.

INTRODUCTION

At five o'clock A. M. Sunday, December 1st, 1918, more than a million men extending from the frontier of Holland to the mountains of Switzerland, under the supreme command of Ferdinand Foch, Marshal of France, moved forward in the closing scene of the greatest drama of history. In the center, a quarter of a million men of a new race from overseas began their march down the valley of the Moselle to the Rhine—that historic highway along which for centuries have tramped multitudes of men from Roman to American, now downstream, now upstream as the pendulum of civilization swung toward the east or towards the west. For a century and a half the men of this new race had been shaping in the new world an idea based upon the immutable principle that all men are created equal and are endowed by their Creator with certain inalienable rights, among which are life, liberty and the pursuit of happiness. Their lives had been challenged, their pursuit of happiness had been halted by a false political philosophy, teaching as one of its fundamental principles, that economic development can take place only when preceded by a battering ram of military force. They had seen the vision and heard the cry of Democracy: "Come over into Macedonia and help us" and had responded. Their motives and their objects in the war had been set forth by their President: "Let us be very clear and make clear to all the world what our motives and our objects are. We are glad to fight for the ultimate peace of the world, and for the liberation of its peoples, the German peoples included; for the rights of nations great and small, and the privilege of men everywhere to choose their way of life and obedience. The world must be made safe for democracy. We have no selfish ends to serve, we desire no conquest, no dominion. We fight without rancour, seeking nothing but what we shall wish to share with all free peoples. We enter this war only where we are clearly forced into it, because there are no other means of defending our rights, in armed opposition to an irresponsible government which has thrown aside all considerations of humanity and of right, and is running amuck. The day has come when America is privileged to speed her blood and her might for the principles that gave her birth and happiness and the peace which she has treasured, God helping her she can do no other".

So the American came to keep the watch on the Rhine, unfurled his flag from its highest fortress and from his outposts looked down on the hills and valleys of Hesse-Nassau whence came men, for hire, in other days, to deny him freedom, while at Versailles statesmen were gathering from the furthermost corners of the earth to redraw the map of the world.

This report on the Military Government of Occupied Germany by the American forces covers the period from December 1st, 1918, when our forces entered Germany, to January 10th, 1920, when the Inter-Allied Rhineland High Commission became the supreme Allied authority in the occupied territory. During this period the Third Army was commanded in succession by Major General J. T. Dickman and Lieutenant General H. Liggett and, upon the dissolution of the Third Army, the American Forces in Germany were commanded by Major General H. T. Allen. The Senate of the United States had not, on January 10th, 1920, ratified the Treaty of Versailles. The American forces therefore had still to retain their war powers under the Armistice of November 11th, 1918. By a *modus vivendi* proposed by the Commanding General, American Forces in Germany, and accepted by the High Commissioners of Belgium, France and Great Britain, the ordinances of the High Commission were placed in effect in the American zone as orders of the Commanding General. To a large extent therefore in a practical, though perhaps not in a technical sense, the *modus vivendi* terminated the Military Government.

An effort has been made to supplement the narrative part of the report by appendices in which the more valuable documents are arranged in logical order so as to show the development of American policies in the occupied area. Chapters of the report have been prepared by various officers, as follows:

Chapters 1, 2, 4, 6, 14, 15, 16 and 18—by Major Truman Smith, Infantry.

Chapters 3, 10, 11, 12 and 13—by Capt. Thomas H. Barber, Infantry.

Chapters 7 and 8—by Lt. Col. Walter L. Bensel, Medical Corps.

Chapter 5—by Colonel William O. Gilbert, Judge Advocate's Dept.

Chapter 9—by Major W. L. Post and Captain F. R. Lyons, Corps of Engineers.

Chapter 17—by Colonel I. L. Hunt, Infantry.

The whole has been edited by Major Truman Smith, Infantry, to whom the service at large is indebted for a report which it is hoped may be of value in presenting the subject of military government for the first time in our history in such form as to be of real value to military students.

A list of the officers connected at various times with this office is given in Appendix No. 52. Collectively they represent a cross-section of the college and university life of America. To them and to the numerous officers who have administered affairs in the Corps, Division and Kreis areas our country is indebted for whatever impressions of American Military government have been left on the minds of the defeated enemy. They have carried out sympathetically the policies laid down by the Army Commanders. Mistakes of execution have no doubt been many, especially in the early days of the occupation when we had nearly three hundred thousand troops billeted on less than a million Germans, but we leave to the governed the decision whether wrongs have been promptly redressed when brought to the official notice of those in authority, in the confident assurance that Germans in the American zone of occupation have gained a new opinion of America. In all fairness it must be recorded that German officials with very few exceptions have cooperated to the full extent of their authority in making the occupation bear as lightly upon the civil population as the exigencies of the time would permit.

In Chapter 6 will be found a recommendation that the fines levied on Germans by military tribunals be created into a trust fund by act of Congress and used for some educational purpose designed to bring to Germany a better understanding of America. It is believed that this can be best accomplished by assisting the Amerika Institute in Berlin (in coperation with the Smithsonian Institute in Washington) in disseminating American scientific literature and by establishing in Coblence, (the seat of government of the Rhine Province and from the beginning of the occupation the headquarters of the American forces, a library designed to bring to Germans a better knowledge of American democracy and American ideals. It is not in keeping with those ideals that we should profit by so much as a pfennig on account of their minor disobediences of military orders. Such use of these funds would demonstrate to Germans that America is really what she claims to be—a friend of humanity. When their new freedom has demonstrated to the world that they have rejected forever the philosophy of blood and iron and intend to remodel their political structure so as to give their primitive virtues of industry, thrift, order, loyalty, love for home and children and strong religious instincts an opportunity again to develop for them the prestige they enjoyed in Heine, Schiller and Goethe, and in Mozart, Beethoven and Wagner, America should be the first to extend the helping hand. In the philosophy of war there is no principle more sound than this: that the permanence of peace depends, in a large degree, upon the magnaminity of the victor.

TABLE OF CONTENTS

Chapters

Chapter No. 1

GERMANY ON THE EVE OF OCCUPATION

Geographical Data of the Rhineland; Population and Industry; History of the Rhineland; Political Institutions of the Rhineland; Local Self-Government; Economic Conditions; Food Conditions; Political Unrest Antedating the Armistice.

Geographical data of the Rhineland.—Geographically and historically, the river Rhine is one of the important landmarks of western Europe. Broad, swift, and flanked by rugged hills, it has always presented a strong barrier against movements of barbarian hordes. Rome, conqueror of Gaul, saw the Rhine as the bastion of her empire, and on its banks erected those fortified camps and military roads around which, medieval and modern Germany grew up. In the first centuries of the Christian era, the Rhine became the bulwark of civilization against which, for two hundred years, the Teutonic tribes hurled themselves in vain. Later, when, in the Autumn of the year 1918, at the moment of Germany's decline, the Allied military leaders looked to the Rhine as the assurance of their safety, it became once more the symbol of victory to the western world. With the establishment of Allied bridgeheads on the right bank of the stream, the struggle closed, and the last line of defense of central Europe passed into the hands of the victorious nations.

Rising in the Swiss Alps, the Rhine flows in a general northwesterly direction, alternately through fertile fields and lofty mountains, until below Remagen it enters the great plain of northern Europe. Just within the Dutch frontier, it divides into several branches and forms the Netherlands delta, finally merging its waters with those of the North Sea. Between Coblence and Bingen, lie the slate-mountains of the middle Rhine. In the Devonian age, impelled by some subterranean disturbance, the ocean-floor emerged from the sea and formed a highland here, in which the Rhine, the Moselle and the Lahn, then but riverlets, entrenched themselves in their course to the sea. The Rhine of that period bore little resemblance to the mighty stream of to-day. Its source was near Bacharach, and its valley even more rugged and tortuous than now. The highland was formed of gray wacke, quartz and slate. The river wore a bed for itself along the seams of slate formations, which were most subject to erosion, the quartz remaining as hills. In the tertiary age, a great inland sea, fed by such streams as the Main and the Neckar, filled the Mayence basin. Towards the close of that age, this sea finally wore down the slate formations of the Taunus near Bingen, and united its waters with those of the Rhine. Flowing north, the pressure of this inland sea widened and deepened the valley of the original rivulet as it wound through the quartz hills. In this period also, the waters of Switzerland, which had heretofore found an outlet into the Danube basin, between the Alps and the Jura, combined with the Rhine. This is explained by the fact that during the middle diluvial age, the glaciers of Switzerland formed terminal moraines, which gradually grew until they became a watershed, barring the way to the Danube and forcing the waters to seek a new outlet. The Rhine of to-day was therefore formed successively by the uniting of the Moselle, the Lahn and the Rhine, then mere mountain streams, by the inland sea in the Mayence basin, fed by the rivers of the Black Forest, and, finally, by the glacial rivers of the Alps.

Below Coblence, where the Moselle joins the Rhine, the river enters the Neuwied basin. Here the mountains to the west partake of a predominantly volcanic character. The center of activity of this region, known as the Eifel, appears to have been around the Laacher-Sea. Here are found evidences of forty small volcanoes, owing their origin to the same disturbances which gave rise to the Laacher-Sea itself. Earth tremors are at present much more frequent here than in other parts of Germany, and the mineral springs at Ems, Rhens Bad Bertirch and Neuenahr are undisputable indications of former volcanic activity. The lava beds deposited here during the volcanic eruptions of this past age, form a valuable source of wealth to the district, and are extensively quarried for building material. The Rhine leaves the Neuwied basin at Andernach, and passes once more through a narrow valley. After forcing its way through the Siebengebirge (Seven Mountains), trachyte formations of great antiquity, it finally enters the northern European plain at Godesberg.

The Moselle, with its source in Eastern France, flows in a northeasterly direction. Near Treves it is joined by the Saar, another Lorraine stream, and just below the city enters the Middle Rhine highland. Here its course is even more tortuous than that of the Rhine, and the hills are as rugged and precipitous as those of the larger stream.

The geographical districts to the north and south of the Moselle are known respectively as the Eifel and the Hunsrück. Both are mountainous in character, partly cultivated, partly covered with forests. The Hunsrück is the wilder region, and at the Erbeskopf, the highest point in western Germany, attains an elevation of 875 meters. The Eifel is a rolling upland, spotted with craters of extinct volcanoes, and cut by deep water-courses. These gorges, through which streams like the Eltz and the Ahr flow, are wild and picturesque, bearing a striking resemblance to the canyons of Western America. The country on the right bank of the Rhine is divided by the Lahn, which joins the Rhine above Coblence into two districts known as the Westerwald and the

Taunus, the former to the north, the latter to the south, of the Lahn.

The theatre of American occupation lay principally in the Eifel, but its troops also garrisoned the Neuwied basin and a part of the Westerwald which lay within the outposts of the bridgehead.

Population and industry.—The American occupied territory is an integral part of the former kingdom of Prussia. The major part of it belongs to the Rhine Province, one of the twelve primary administrative divisions of Germany. A small portion of the bridgehead is however attached to the neighbouring province of Hesse-Cassel. The Rhine Province, popularly referred to as the Rhineland, is the richest and most populous section of all Germany. The Ruhr basin, with its coal deposits and industries and the great cities of Cologne and Düsseldorf, falls within its borders. These industrial sections lie almost entirely along the lower Rhine, in the northern half of the province. The American territory farther south is primarily agricultural, so that while the Rhine Province as a whole, with it 29,996.92 square kilometers and its 7,121,140 inhabitants (estimated in August, 1919), has a density of population per square kilometer of 213, the American territory, with 8,992.24 square kilometers and 884,009 inhabitants, has a density per unit of but 98.3.

By Kreise (governmental divisions approximately corresponding to American counties), the population was estimated in August, 1919, as follows:

Kreis	Daun	34,310	inhabitants
"	Prüm	38,380	"
"	Bitburg	48,180	"
"	Wittlich	44,850	"
"	Bernkastel	49,880	"
"	Treves (Urban Kreis)	55,190	"
"	Treves (Rural Kreis)	98,330	"
"	Saarburg	35,760	"
"	Adenau	26,482	"
"	Ahrweiler	50,539	"
"	Coblence (Urban Kreis)	65,434	"
"	Coblence (Rural Kreis)	69,074	"
"	Cochem	41,487	"
"	Mayen	82,558	"
"	Altenkirchen	2,550	"
"	Neuwied	69,930	"
"	Unterwesterwald	47,442	"
"	Westerburg	12,969	"
	Total	893,345	inhabitants

The Rhine is not only the most important geographical feature, but also one of the most valuable economic possessions of the Rhine Province. It has been for centuries the connecting link between northern and southern Europe, and still continues to be the avenue of exchange between the North Sea States, Italy and Switzerland. Great fleets of barges and steamers distribute the coal of the Ruhr basin to the manufacturing districts bordering the river, and to those of south Germany. Coblence, capital of the Rhine Province, situated at the junction of the Rhine and the Moselle was not only the political center of the American territory, but its largest city as well. Its population in 1919 was estimated as 65,434. Historical reasons and the fact that the Moselle has never been developed as a navigable waterway, have prevented it from becoming a city comparable to Cologne. It is a residential, governmental town, without large industries. Treves, near the Luxemburg frontier, also on the Moselle, has many of the same characteristics as Coblence. Its estimated population in August, 1919, was 55,190, and before the war it had a large imperial garrison. It is also the capital of the Regierungsbezirk bearing the same name. These are the only two cities in the area originally assigned to the American Army, although in the Neuwied basin there were several fairly large towns. The district has not a fertile soil but is carefully and extensively cultivated. In consequence, the persistent industry of its inhabitants has made it wealthy and, in times of peace, much of the food for the cities of the lower Rhine was supplied by it.

The Eifel, the Hunsrück and the Westerwald are primarily agricultural districts. The Neuwied basin, on the other hand, is industrial, with important factories and products. The Eifel comprises the larger part of the American territory. It is an upland of bald hills and narrow gorges. The greater part is bleak and forbidding, but yields a livelihood to the laboriousness and persistence of its inhabitants. Life there is primitive, the villages are small and scattered and trade is carried on by peddlars who travel from village to village. It was formerly a great forest, which has been in process of clearance since the middle ages. Because of its picturesque scenery, the many inns and hotels are much frequented by trampers and tourists. In striking contract to this bleak upland are the lands bordering the Moselle, known as the Vorder-Eifel. Here the soil is comparatively fertile, and large crops of wheat, rye and oats are raised. Before the war, this region also had herds of fine cattle, but the demands of the army were so heavy, that to-day few remain. It is principally known for its Moselle wines, which, of course, are famous the world over, and are by many preferred to the more celebrated Rhine wines. The vines are planted on the steep hills bordering the Moselle, which have been terraced and cultivated bit by bit for centuries. Vine-growing is not under the control of the large landowners, the peasants of each village having organized themselves into co-operative societies. In the Ahr valley, in the northern part of the Eifel, there is another wine district of smaller extent, which produces an excellent red wine.

The Neuwied basin and the volcanic region immediately to the west, in contrast to the agricultural Eifel, is the seat of numerous industrial enterprises. The prehistoric volcanoes have left important lava and tufa deposits, which are extensively quarried for building material. The quarries appear to have been worked as far back as Celtic times. Later, the Romans used them to obtain the materials with which to build their military roads. Today the quarrying industries employ over 9,000 laborers.

There are several large steel and chemical factories in the Neuwied basin proper. The many steel works around Engers and Bendorf are branches of the Krupp plant at Essen. The "Rasselstein" iron works at Neuwied employ over a thousand workmen. Another great factory of this region, the "Rhenania Chemische Fabrik" at Hönningen, manufactured poison gas during the war. It was formerly a small business, but the demands of the army quadrupled it. Beyond the Neuwied basin, the rolling Westerwald extends to the limits of the bridgehead. This region is agricultural, and also more fertile and wealthy than the Eifel.

In such rugged country, railway lines must necessarily follow the river valleys—the Rhine and the Moselle, and to a lesser extent, the Ahr. These lines became the channels of communication on which the army depended. The Moselle, perpendicular to the American front, became the artery of supply for the American bases in France; the Rhine, the connecting link between the Allied armies. Treves, with its network of railways, extending to Coblence, Cologne, Saarbrücken, Metz and Luxemburg, became in consequence an important supply point, and an advantageous site for Advanced General Headquarters.

The Rhine itself was little used until, in the Spring of 1919, Holland permitted shipment of supplies direct from America to her ports and thence by river barges to Coblence.

History of the Rhineland.—To the average American before the war, the Rhineland was known more for its antiquities, history, mythology and poetry than for its twentieth century civilization. Culture throve there when Prussia was still in darkness, and civilization had fallen and risen again on the Rhine before such comparatively modern cities as Berlin and Hamburg had outgrown the village state. Barbarian hordes and imperial armies have so often surged back and forth over its land that there has been a considerable admixture of foreign blood, which has produced a people quite different from either the Germans of Prussia or of Bavaria.

When Caesar entered Germany and built his military bridge across the Rhine at Engers near Coblence in 55 B. C., he found a people in the neighbourhood of mixed Celtic and Teutonic blood. Caesar was followed by Drusus, who erected a castle in Coblence and made the town, with the neighbouring hill later called Ehrenbreitstein, one of the strongest Roman positions. Rome's intention at this time seems to have been to accept the Rhine as her natural frontier. History, however, has repeatedly proved natural frontiers to be absurdities and the fact that their possession merely entails further conquests. Empires cease to expand only when confronted by people who cannot be conquered. The Rhine became the frontier of Rome not from choice, but because the Teutonic tribes of central Germany imposed it on her. The defeat of Varus and his legions in 17 A. D. by Arminius in the Teutoberg forest ended the period of Roman expansion and brought about a static condition in which all efforts of the conquerors were concentrated in strengthening their colonization on the left bank of the Rhine. The center of this civilization was at Treves (Augusta Treveri) on the Moselle. From this city led the military roads, arteries of the Empire, which today from the general outlines of the very roads over which American troops advanced to the Rhine in November and December, 1918. For adminstrative purposes, Germany was divided into two provinces, Germania Inferior, whose capital was Cologne (Colonia), and Germania Superior, in which were Treves and Coblence (Confluentia). Autunaun, now Andernach, also was in this period an important city, and the Eifel was studded with villas of wealthy citizens, who knew and used even in that day the medicinal hot springs of the region. Treves became the seat of the Western Empire in the latter half of the fourth century. Fragments of the baths, palaces and amphitheatre, which made the city the most important one north of the alps, may still be seen, and indicate a high degree of civilization. Under Domitian was built the Roman wall connecting the Rhine and the Danube. This started near Rheinbrohl, skirted the American bridgehead in the Westerwald, and thence crossed the Lahn river and the Taunus range to the river Main.

Christianity entered Germany probably about 200 A. D., and had gained a considerable foothold even before Constantine made it the religion of the empire. During the following centuries, in which the Rhine vanished as Roman barrier and the Teutonic tribes overran Western Europe, Christianity was yet able to survive. Out of the turmoil and destruction of the dark ages, Mayence, Treves and Cologne emerged as parts of the States of the Church, and became the bearers of the torch of civilization.

Charlemagne revived the Roman conception of the empire for a short period. At his death in 814, the territory under his rule was divided, and the present Rhineland was assigned to the kingdom of his son Lothar, together with Alsace and Lorraine. In 870, the province became permanently a part of the Holy Roman Empire. From the ninth to the thirteenth century, the present American occupied territory was under the sway of the Counts of the Palatinate, feudal henchman of the emperor, and a semblance of local unity was preserved. Even at this period, however, the authority of the emperor was more apparent than real. In the thirteenth century these counts disappeared, and the country dissolved into feudal anarchy. Knights built strongholds, where they could maintain themselves in security, prey on the neighboring trade-routes and wage war as their fancy dictated. At this time arose the castles whose ruins still dominate the landscape of the Rhine and Moselle. Among the best known are Eltz, Cochem, Monreal, Lahneck and the Marksburg. At the foot of the hill crowned by the castle usually lay a village, which was virtually the property of the knight or baron of the castle. The serfs offered their produce and labor in exchange for the protection which their lord's military force afforded them from neighboring barons. The district which is now Kreis Mayen affords an illustration of the dissolution into which the empire of Charlemagne had fallen at this period. While the greater part of it belonged to the Electorate of Treves, the Count of Bassenheim owned the towns of Brench, Galenburg and Oberweiler, the Count of Burresheim held sway over Glees and Burgbrohl, and the lords of Burresheim held the fiefs of Rieden, Waldesch, St. Johann and Nitz. Kruft belonged to Maria Laach; Niedermendig to the Cathedral chapter of Treves; Obermendig, Bell and Volkesfeld to the chapter of St. Florin at Coblence; Geisburgelhof, Ober and Neider-Lützingen were dependents of the Bishop of Essen; Wehr of the Abbey of Steinfeld; Buschholz and Neiderweiler of the Abbey of Munden, while Gladbach, Andernach, Miesenheim, Kell and Namedy all belonged to the Archbishopric of Cologne. Throughout the middle ages, the archbishop of Treves managed to maintain a dominant position in the Moselle valley, both the cities of Coblence and Treves being a part of his dominions. All these many princes and bishops, however, recognized the Germany Emperor as their sovereign.

In what is now the American bridgehead, the principalities of Sayn and Wied came into prominence in the sixteenth century and continued as states within the empire until Napoleon ended their existence in 1805. The larger part of the bridgehead, however, always belonged to the Archbishopric of Cologne.

Needless to say that the resulting conditions were indescribable. Each lord, temporal or ecclesiastical, made his own laws, levied taxes and imposed custom-restrictions as his interests dictated. Justice was very irregular and varied from village to village. That with all these heavy financial burdens, the country still prospered, is remarkable.

From the time of Charlemagne to the fall of Napoleon, the Rhineland suffered terribly from invaders,—imperial, French and Swedish. As it was a borderland of the empire, it became inevitably the seat of many wars. France had developed a stable government and a compact national spirit, while Germany was still in feudal chaos. The interminable wars were usually fought on the soil of the empire and the Rhineland consequently suffered from them.

The Thirty Years War, 1618–1648, proved the severest calamity of all. Statistics which have been preserved of individual communities lead to the conclusion that the territory now included in the American area lost two thirds of its population and three quarters of its live stock during that catastrophe. To the south of the Rhineland, the Palatinate was at a somewhat later period terribly devastated by Louis XIV, who believed that the only way to protect France was to create a desert beyond her borders.

The last elector of Treves, Clemens Wenzelaus, transferred his capital to Coblence just before the French revolution. There he built in 1787 the palace which later became a residence of the German emperors. Coblence incurred the animosity of the terrorists early in the revolution, because it harboured expatriated French nobles. It was attacked and captured by General Marceau in 1794, but was not destroyed. During the Napoleonic occupation, it became the capital of the Department of the Eifel, and was incorporated into France.

On the whole, this period proved extremely beneficial to the country. The Empire had become utterly outworn, impotent and decadent. Industrial and commercial development was impossible with each petty state establishing a customs frontier and enriching itself without regard to the empire as a whole. Napoleon appreciated the evils of the systems, and by withdrawing power from each little ruler, united the Rhineland into an organic unit which he governed wisely and efficiently. His work was continued by Prussia, who received the territory in 1815 by the treaty of Vienna in compensation for her sacrifices in the great struggle. Prussia has had a previous interest in the region by virtue of her possession of the Duchies of Cleves and Jülich on the lower Rhine. Since no attempt was made to revive the old empire, Prussian claims to the Rhineland appeared to the negotiators as valid as any others. The old conditions of anarchy and misgovernment therefore did not return. Prussia took up with energy the work begun by Napoleon, and her administrators, soldiers and teachers, proved as capable and efficient as their French predecessors. The astounding economic development of the Rhineland in the nineteenth century must be ascribed in no small degree to the wisdom and foresight of these foreign governors. If they never acquired the love of the natives, they yet gained their respect and admiration. The Rhineland became the Rhine Province and her fortunes were merged in the fortunes of Prussia. In the wars of the Bismarckian era, her young men followed the Prussian banners, but the enthusiasm and the loyalty which the Province displayed at this time was not offered to Prussia, but to a united Germany.

The era following these wars witnessed the transformation of Germany from an agricultural to an industrial nation. The Rhine Province, with its great coal deposits in the Ruhr basin, received much of the wealth which accrued to Germany from the transformation. In 1914, no sign of discord arose in this region, and she entered the war with every evidence of enthusiasm for the cause of the empire. While her troops never bore the reputation for aggressiveness which fell to the lot of true Prussians, Brandenburgers or Bavarians, yet more than a million citizens of the Province served with the colors, out of a total population of about seven millions.

Political institutions of the Rhineland.—The political institutions with which the American Army came in contact during the occupation were, without exception, those of the old *régime*. Ostensibly, the revolution of November, 1918, had converted imperial Germany into a republic. Practically, the transformation of an autocracy into a democracy can not be accomplished in a day. The prestige of institutions handed down from generation to generation is a force with which even the most ardent revolutionists must reckon. The recent experiences of Germany have proven how easily a new form of government may be proclaimed, but how difficult it is to put it in actual operation. Where the knowledge and practice of government have been for centuries the exclusive privilege of the aristocratic and landed classes, change becomes still more difficult. If the revolutionary leaders at heart desire the public welfare, reforms must be introduced gradually. State socialization and the wholesale replacement of officials of the old *régime*, however much they may be desired by certain elements of the German population, could have produced only anarchy in the winter of 1918–1919. Discredited as the old officials and old institutions were with the mass of the people, yet Ebert and his followers could not muster the courage to take a step which, in the disordered state of the country, would only make the fearful economic condition still worse. It was fortunate indeed for Germany that the plans of the independent wing of the Socialist party were foiled. The counsel of moderation won the day, and most of the ex-imperial officials remained at their posts, while their fate was leisurely debated at Weimar by their Bourgeois masters. In occupied territory their retention in office was made a condition of the Armistice. The Allies, while desiring the regeneration of Germany under democratic influences, had in the first instance to consider the practical application of the occupation. Trained and responsible officials, as agents of a temporary sovereignty, could transmit and carry out orders better than the creatures of a Workman's Council. Neither did the higher Allied authorities in occupied territory desire an atmosphere charged with excitement, which would have been the inevitable result if factions had begun to tamper with established institutions.

German civil administration and Prussian officialdom in particular have been described as an army in civilian clothes. Experiences in Germany have substantially established the truth of the simile. The primary duty of Presidents, Landräte, Burgomasters and Vorsteher is to punctiliously carry out the orders of their superiors. Orders are issued by the cabinet, pass from level to level of the bureaucracy, and are eventually issued, to reach the people as regulations for their guidance. To Americans who have served in the Army, such a system may be visualized, but the fact remains that it is utterly foreign to the spirit of American political life. Local self-government to a certain extent does exist in Germany, but its institutions are primitive and the wishes of the inhabitants, ever subject to disapproval of the local representative of the king or central authority.

If we compare Prussia to the United States, her provinces may be likened to our states, remembering, however, that the Prussian province has no attribute of sovereignty. Provinces, according to their size and population, are divided into a number of Regierungsbezirke. Our states have no similar governmental divisions. The Regierunsgbezirke in turn are subdivided into Kreise (Circles), about the size of our counties. These Kreise may be either urban (Stadtkreise) or rural (Landkreise). The Stadtkreis is the fundamental unit of government and is indivisible. Landkreise are on the other hand subdivided into Stadt and Landbürgermeistereien (urban and rural townships). The Stadtbürgermeisterei is another unit which, like the Stadtkreis, is not subdivided. Landbürgermeistereien are made up of a number of villages, termed Gemeinden. At the head of the Gemeinde is a presiding official called the Vorsteher, who represents the executive of the state and enforces the laws.

The conception of a chief official of a village is common to every territorial division of government. Each such official receives his appointment direct from a higher authority or, in the case of cities where Burgomasters are chosen by the city council, upon approval of representatives of the king. Much resembling the position of the colonel in the regiment is that of the Oberpräsident of the province. In neither instance does the position entail the support of those governed. The Oberpräsident is appointed by the central authority,—in the days before the revolution, by the king,—and he possesses very extensive powers. In wartime he may even assume the entire government of the province. Within the province, the Regierungsbezirke have directing officials, called Regierungspräsidenten. Stadt- and Landkreise are governed respectively by Oberbürgermeister and by Landräte. The former official is chosen by the board of aldermen, but the choice must be confirmed by the Oberpräsident of the province in which the city is situated. This exception to the rule, otherwise general, that directing officials are chosen by the government, is worthy of note.

Although kings and emperors zealously guarded their local prerogatives, the turbulence of the middle ages required them to seek support among their people. Cities offered an effective bulwark against the ravages of nobles and knights, and their allegiance to the royal cause was won by granting them special privileges. Today, these rights of self-government, zealously guarded for centuries, form one of the few bases of democracy within Germany.

Oberpräsidenten and Regierungspräsidenten are the officials in the higher spheres of government. They transmit the directions of the Ministers, co-ordinate the work of their subordinates, and work out the details of broad general policies. The actual operation of government as it affects the people is best found in the Kreis. Here the military simile may again be introduced by comparing the Landrat or the Oberbürgermeister to the captain of a company. Prussian officialdom succeeded or failed according to the success obtained by these all-powerful governors in winning the confidence and loyalty of their people.

Within the Landkreis, Stadt- and Landbürgermeistereien are presided over by Bürgermeister. These officials usually, but not always, belong to the civil service, but they may be prominent landowners or even professional men. The Gemeinde, the smallest unit of all, has as its executive the Vorsteher, usually a simple peasant, who serves without financial compensation, but is satisfied with the local prestige which his office gives him. As an army is dependent for success primarily on the ability and moral qualities of its officers, so it is obvious that the Prussian state requires much of its servants. Prussian and American theories of government differ widely in their attitude toward their public servants. Civil service is confined with us to the lowest grades alone. In Prussia, government service is a profession, and the servant of the state is promoted from position to position, theoretically according to his ability and achievements. In practice, this ideal was never achieved. Human nature is too strong a factor to be ignored, and Landräte became Regierungspräsidenten and Regierungspräsidenten Oberpräsidenten, quite as often through favor of some friend at court as for conspicuous success in promoting the prosperity of the people.

The governing officials are each assisted by councillors (Räte), who bear to the executive much the same relation as does the staff to a general. These councillors are of different grades, the highest in the province being the Oberpräsidialrat, who is the executive officer of the Oberpräsident. Oberregierungsräte and Regierungsräte are distributed between the offices of the province and the Bezirk. Below these grades are the Regierungs-Assessoren and Regierungsreferendare, young men who have just entered the service and have not passed their final civil service examinations. Under the tutelage of the Regierungspräsident they are expected to learn Prussian regulations and "red tape", and in due course of time, to qualify as a Landrat. Officials very seldom serve in their native districts. Such a natural matter, according to American theories, was abhorrent to the ideas of the Prussian kings. They felt that to permit a native to serve his own neighborhood paved the way to abuses and the growth of influences which might endanger the royal power. The king was to be the sole master, and the officials primarily and indisputably his servants. The people were never consulted as to their preferences for governors, and the officials only occasionally as to their choice of posts. In the Rhineland for a century nearly all important government posts have been held by Prussians. Herr von Groote, the Oberpräsident of the Rhine Province during the American occupation, was a notable exception to this rule; but he was the first Rhinelander to hold the position during the hundred years of Prussian rule. Here was a source of dissatisfaction which bit deep into the hearts of the Provincials, but especially of Rhinelanders. Without choice in governors, the Rhine Province had been ruled for generations by men who, if not foreigners, were at least alien in religion and temperament. For it was primarily the sons of the landed Prussian gentry who turned to the civil service as a profession, and who, thanks to the influence of their families, rose to high rank therein. Whatever reasons may be marshalled for or against such a form of government, it is self-evident that its autocratic nature peculiarly fits it to adapt itself to the wishes of an occupying military force.

Local self-government.—The right of a community to govern itself within its own sphere is one of the dogmas of American political thought. In Prussia, such local self-government must be regarded as a favor of the king, not as an inherent right. That it existed at all was because the Ministry was overloaded with work, and because it was considered advisable that the king should know the local viewpoint toward laws primarily affecting it.

In the province and Kreis, the existing organs of self-government are fairly well developed, but in Bezirke local representation is so slight that we must regard this political subdivision as entirely the sphere of the central government. Certain local representatives belong to the Bezirksausschuss (District Committee), but this body is in no sense a legislative assembly, but rather an advisory cabinet to the Regierungspräsident.

The province well illustrates the blending of central and local strains, each element having a separate local executive. In the Rhine Province, local self-government, represented by the Provinziallandtag (Provincial Assembly) and the Landesdirektor (Chief Officials), has its seat at Dusseldorf, while the Oberpräsident and the representatives of the central authority have their offices at Coblence. In the Kreis there is no such separation, the Landrat being the representative of both the central and local interests.

The Landtag and the Kreistag are the legislative bodies which represent local interests in the province and the Kreis respectively. Their jurisdiction is entirely local and confined to such matters as the budget, public works and distribution of taxes. The Landtag appoints the local members mentioned previously as serving on the Bezirksausschuss, as well as the Landesdirektor. The Kreistag appoints the Kreissausschuss (Circle Committee), a sort of advisory cabinet to the Landrat. In the Stadtkreis, we find the Stadtverordnetenversammlung (Board of Aldermen), the powers of which are somewhat greater than those of the Kreistag. It is this body which selects the Oberbürgermeister of the city, subject, under the old *régime*, to the approval of the king or Oberpräsident, depending on the size of the city. Towns and Bürgermeistereien have also their legislative bodies, and even the Gemeinde has a Gemeinderat (Communal Council), whose members are chosen directly by the people. In the towns and cities, the people also select their boards of aldermen direct, but in all other cases, the members of each legislative body are chosen by the next lowest assembly.

With these self-governing bodies, the United States military authorities were to have little to do. The transmission of orders and supervision of their execution fell naturally within the sphere of the centrally appointed officials. In fact no attention was paid to the locally elected bodies, except to sometimes forbid new elections, until near the close of the military government. Attendance of members from occupied territory at the Provinciallandtag at Düsseldorf was also forbidden by Marshal Foch.

The districts of western Germany allotted to the American Army belonged, excepting one Kreis, to the Rhine Province. As previously stated, this is the richest and most populous of Prussia's twelve provinces. It contains five Regierungsbezirke, known by the names of their chief cities, Düsseldorf, Cologne, Aix-la-Chapelle, Coblence and Treves. About half of about Coblence and Treves lay within American jurisdiction. Kreis Unterwesterwald, the single exception lying in the bridgehead, which is not a part of the Rhine Province, belongs to Regierungsbezirk Weisbaden, of the province of Hesse-Nassau. Early in the occupation, the German authorities temporarily transferred it to the Rhine Province in order to simplify relations with the American Army.

Economic conditions.—At the time of the armistice, very little was known of the enemy's economic condition. The press published certain articles of an alarming nature stating that Germany had continued to manufacture during the war and was preparing to dump vast quantities of goods on the markets of the world. That such stories received credence seems unbelievable in the light of what is now known. The Allied nations had seen the strain placed on their industrial organizations by the wastage of war. It should have been apparent to intelligent observers that the strain of the war on German resources must have been very much greater. For a period of some months following November 11th, 1918, neither England nor France was in a position to export manufactured goods in quantity. If the economic situation of these nations was unfavorable, that of Germany was worse, as the blockade still continued in effect. Raw materials had been utterly exhausted and the resumption of manufacturing depended on the availability of new supplies. Even if the necessary imports had been permitted, a considerable time must have elapsed before the actual process of manufacturing could actually begin. Probably these alarmist articles originated in the knowledge that the German industrial establishments had come out of the war practically intact. Although this was certainly an important factor in her favor, the inference does not follow that these factories were in any position to dump their goods in foreign countries. It was primarily fear of Germany which conjured up an imaginary situation, in which the enemy was preparing to conquer the world supremacy which he had lost in battle. In the winter of 1918–1919, Germany's economic situation portended the gravest consequences.

During four years of war, her establishments had one by one been placed at the disposal of the government. By 1918, this practice had proceeded so far that factories still engaged in their customary peacetime operations practically did not exist. New machinery had been installed, the old stored away, and the owners were employing new staffs and new personnel. Most of the old force had been scattered to the four winds on mobilization to be replaced by women, cripples, and men unfit for active service. This new personnel, however skilled they may have become in the manufacture of munitions, knew nothing of the former manufacturing processes of the establishment. Time was just as necessary for German factories to reorganize as for Allied ones. Without raw materials not even a start could be made.

Despite disabilities such as these, the German manufacturer was even further handicapped. During the war there has been created an imperial economic organization which had reduced the factories to a helpless subserviency to the state. This organization, known as the "Kriegsamt", was formed by Ludendorff, a man who keenly realized the necessity of intensive organization to offset the greater resources of the Allies. Methodical and far-reaching, the Kreigsamt had seized capitalist and laborer alike and diverted the money of the one and the labor of the other, to further the purpose of the High Command. Work unessential to the nation or the army became a crime against the state, and from coal deposit to the copper kettle in the peasant's dwelling, all resources of the Empire were placed at the disposal of the Kriegsamt. By 1917–1918 a condition had been created which, if not actual State Socialism, narrowly escaped being that. Officers at Berlin manipulated factories and raw materials as generals manoeuvered armies in the field. The manufacture of clothing, shoes and hats, was ruthlessly limited, so that the army might be properly equipped. To the people, the Kriegsamt said: "You may have just enough food to keep you from starving and just enough clothing to prevent freezing; the rest is ours, and patriotism demands that you give it without a murmur." Such an autocratic system, in which industry is subordinated to a central authority, becomes in time indispensible to the existence of the factory. Old conditions must come back gradually. An abrupt change to unrestricted free trade carries with it no small danger of absolute anarchy.

The Kriegsamt, being essentially temporary, had carefully prepared for the return of the factories to their former direction. The demobilization period for industry was intended to extend over many months and was to coincide with the demobilization of the army. This latter was also a factor which had received much thought, as it was realized that unemployment would carry with it the danger of Bolshevism. To avoid this, it had been planned to continue the Kriegsamt until conditions were normal, and to loosen imperial control bit by bit.

The revolution, however, made the execution of this plan impracticable. The Army could not be restrained and simply went home, without waiting for demobilization. The authority of the Kriegsamt vanished and the industrial fabric was shaken to the core. Unemployed soldiers filled the cities, a prey to soap-box anarchists. Government subsidy to the unemployed reached enormous proportions.

Disastrous as this situation was, it became even further complicated by disorganization of the railways. Up to the revolution, their operation had continued satisfactory despite war-wastage. The occupation of France, Belgium, Serbia, Poland and Roumania had given Germany additional engines and rolling stock. This had more than made up for the lack of construction during the war. The railways had always been government-owned, so that no problem existed of a return to peace conditions. Yet on the coming of the revolution, the fabric fell to pieces almost at once. Coal production decreased, strikes spread, paralyzing entire sections of the country. Cities and towns passed over bodily to the Sparticists. Through traffic from one section of the country to the other became almost impossible. On top of such conditions, the demands of the Allies for 5,000 engines and 150,000 cars, just though they were, further handicapped the railway authorities. In December, 1918, Bolshevism appeared certain to engulf the country, and the possibility of a return to normal conditions a fanciful dream.

The American occupied territory, primarily agricultural, felt the disorder less than did many cities in interior Germany or in Allied zones. The only cities, Coblence and Treves, were governmental and residential towns, without industries which employed large numbers of workmen. Even in Neuwied basin, the situation never became acute. Through the entire American area, the Catholic workmen, who were in the vast majority, were organized into Christian trade unions. These unions were dominated by the Church and were hence more conservative than ordinary unions. American occupation therefore never had to face a labor situation comparable to that which confronted the British at Cologne. It must be borne in mind, in following the development of American policies, that the people we governed were primarily farmers and peasants.

Food conditions.—If the public in Allied countries in December, 1918, knew little of the economic conditions in Germany, their ignorance of food conditions there was even greater. Germany had been pictured all the way from a land of milk and honey to a wilderness. Great curiosity was felt by the army as well as by home populations, as to which of these divergent statements was true, and most of the newspaper dispatches from Germany in the first days of the occupation contained rather personal descriptions of the food conditions as they appeared to the correspondents. As these stories were usually gathered from sources in a single locality, they were not entirely reliable. Even now it is difficult to obtain a satisfactory picture of food conditions as they existed in Germany during the war, so invariably is one's viewpoint prejudiced by stories heard in his own locality. Not until the Food Administration at Berlin gives up its secrets can any of the innumerable reports be regarded as accurate.

The system of rationing the population was introduced late in 1914 for certain foodstuffs, and later extended to all the so called "staple articles" of diet. These were meats, fats, bread, milk, eggs, sugar and coffee. At the end of the harvest each year, a survey was made of all available food supplies within the empire. Supplies required by the army were withdrawn from the total amount estimated to be available for the ensuing twelve months. These supplies were generous in quantity as compared with the total available resources, for it was realized that the soldiers must have sufficient rations to undergo the physical ordeal of the front. It was provided that troops on the line of communication or at depots should receive less than those in combat divisions. However, if the soldier in the rear of the fighting lines received little, the lot of the civilian was far worse.

Germany was usually in such dire straits that the ration, computed by dividing the available food, and after deducting the army's needs, was just sufficient to sustain life. In such a highly organized industrial country, which in peace-time had to import large quantities of foodstuffs, great economy and intensive organization was necessary to avert starvation. With so many of the agricultural laborers in the army, the problem became doubly acute. The situation was further aggravated by the fact that the great industrial centers had to be fed from the predominantly agricultural country bordering on Russia. This placed a severe burden on the railways, already taxed with the tremendous traffic of army supplies, and required that the system of equalization and distribution of food-supplies be highly elaborated and honestly and efficiently supervised. Every Kreis of the empire was classed, in respect to each article of the ration, as either a plus or a minus district, depending on whether it produced more or less than it required. The surplus above the amount which a plus district required for its own needs was taken over by the Food Administration and apportioned among minus Kreise.

The official ration was recognised in 1917 as insufficient to maintain the munition workers at a high efficiency, and special amounts were therefore withdrawn from the general stocks to increase their ration. This could be accomplished only by diminishing the ration of the civil population as a whole, and the action of the government in so doing instantly aroused discontent among wide classes of the population. The success of the system as a whole depended on the honesty of the individuals of the nation, and on their willingness to abide by both the letter and the spirit of the food regulations. Whether any nation can be honest and hungry at the same time is doubtful, but the German people certainly did not carry out the rules prescribed by the leaders, or even make a pretense of doing so. Secret traffic in foodstuffs and profiteering by clandestine dealers became the order of the day. Rich and poor indulged in the practice indiscriminately. Farmers ordered to sell their produce at stated prices to the government, disposed of it secretly for the highest price obtainable. That the wealthy and governing classes, who should have set an example to the country, never lacked for food, was a fact which added to the general discontent. Illegal traffic assumed enormous proportions.

Even the army of gendarmes and secret government agents made little headway in checking it.

With such a large clandestine traffic, it is not surprising that the ration at times fell short of the computed amount. There was never a lack of bread and potatoes, but meat and fat in many localities could not be issued for weeks at a time. The use of milk had to be finally restricted to mothers, babies and the aged. This lack of nourishment primarily affected the great industrial centers. Workmen, quite unable with their wages to pay to farmers the high prices which they received from the rich, willingly listened to the agitators and bolshevists, who took the opportunity offered by the discontent of the poorer classes to further their own ends. As the war continued, and the food situation showed no signs of improvement, the feelings of the entire country became depressed. This depression naturally reacted on the *morale* of the army.

The revolution of November 9th, 1918, and the armistice terms, brought about the first really critical situation. When hostilities closed, no real danger of actual starvation seems to have existed for Germany, although the ration would probably have had to be still further reduced in the near future. The revolution, however, lowered the public *morale* still more, and the disordered state of the country prevented enforcement of the food laws. Disorganization of the railways and surrender to the Allies of rolling stock still further tended to destroy the workings of the intricate machinery of the Food Administration. When the Americans, in December, 1918, entered the Rhineland, they found a predominantly agricultural country threatened in the near future with actual starvation.

Political unrest antedating the Armistice.—The occupation of the Rhineland offers one of the few instances in history in which a war has ended by an invasion in which the bulk of the hatred and resentment was on the side of the conquering armies. The astonishment felt in Allied countries when it was learned that their troops were received by the hated enemy in a friendly manner, was only equalled by the surprised of the troops themselves. The average soldier looked forward with curiosity to seeing Germany. In his first glimpses of his late enemy, he noted the peculiar discipline and the submissiveness of the German individual to foreign military rule. His surprise at the friendly reception given him was tempered by the natural assumption that it was merely an insidious form of propaganda. Of the course of events in Germany he knew little or nothing. That the German of 1918 was a different man from the German of 1914 did not strike his mind. The professions of democracy heard on every side in those first days sounded blatant and insincere, but a resident of Germany in pre-war days, returning with the invading armies, would have at once recognized the change in the psychology of the people. Hunger, disappointment, the collapse of idols, and the strain of four years of war, had done their work. To the peasants and workers, the armistice stood primarily for relief from the horrors of the front; to the wife and mother, it meant the end of uncertainty. These were major passions and beside them, victory or defeat meant little. All were desirous of forgetting the black past and of returning to the paths of peace and work which had been their lot in pre-war days.

Underneath this dominant passion, however, the casual observer soon became aware of divergent currents of thought. The expressions and comments heard on every side indicated an amazing welter of ideas. The theories of autocrats and radical revolutionists sprung up side by side, without any apparent perception by either of the incongruity of the partnership. One heard the burgher of the Rhineland vaunting at one minute his country's military prowess and achievements and in the next breath, pointing to the new-born Democracy as the symbol of her repentance. The last months of 1918 were a momentous epoch in Germany's history. It was a period in which century-old habits of thought were being profoundly revolutionized, and in which neither the past, present nor future could be perceived in its proper light.

The first conclusion of the casual observer probably would have been, that there did not exist any public sentiment common to the country as a whole, as it is understood in the United States. In those first days of the occupation, one began to sense the hold of the caste system on the country. One then saw that each person thought on the events of the past and the problems of the future, much as did every other member of his own caste, and the American soldiers, having once become aware of this fundamental fact, more easily understood the apparently hopeless confusion of ideas.

The psychology of the military and ruling classes had at this time already crystallized. There was no possibility of understanding the sentiments they expressed. Humiliated by their defeat, they hated their enemies who had come as conquerors among them, and even more bitterly hated and despised their Socialistic countrymen who had usurped the government at home. It was among the ruling and professional classes that one found the frankest expressions of undying loyalty to the emperor. The workers, also, had clearly defined opinions on the past and future. It was quite comprehensible that they should welcome the republic as a relief from oppression and misery. Having established a republic, however, they showed the most complete ignorance of the ordinary conceptions of Democracy. Politically untrained, enthusiastic for the new *régime* in the erection of which they have taken a leading part, they became the prey of every soap-box

521572°—43——3

orator and the victim of the communistic theories which were being propagated on every side. Between these two extremes, the psychology of the Bourgeois and the peasant classes appeared as a rather more complex problem. It was in conversations of the different class of society that one most often heard amazingly contradictory comments on the war and politics. In the first days of the occupation, the ideas of the great middle class were being revolutionized under the pressure of internal disorder and the opinion of the outside world. This was unfortunate, since it was to this body of public opinion that Germany would have to turn if she was to erect a stable democratic government.

The welter of ideas abroad in November, 1918, was in marked contrast to the uniform, made-to-order psychology of 1914. It is a striking instance of the futility and shallowness of the German policy of attempting to make the people think by imperial decree. When the system broke down with the flight of the emperor, patriotism disappeared on the rush of each class to insure its own special interests. This policy of selfishness had gradually been developing in the four years of war; but so strong was the iron hand of the military that even up to the last eventful days, the industrials kept up a pretense of loyalty to the government, while actually losing no opportunity to profiteer and cheat at its expense.

To understand the psychology of 1918 one should examine briefly the point of view of the average German at the outbreak of the war. In those days, the opinions of individuals of all castes on national subjects were moulded as a unit by their rulers. As a young man the German was entertained with stories of the glorious days of Blücher and Bismarck. Either his father or grandfather had probably served in one of the struggles of those days and was intensely proud of the successes achieved by German arms. At school, his teachers lost no opportunity to instill reverence in him for the Kaiser. At church, on Sundays, the pastor or priest preached on the goodness of the emperor and prayed for the continued health of his majesty. If in his pursuit of learning he began to feel doubts on the desirability of an aristocracy or the emperor, the mere expression of his concern was sufficient to bring down on his head all the angry arguments of those he respected in family, church and school. His attention was summarily directed to the wonderful prosperity which the existing form of government had insured his country. As the arguments were impressive and the future a closed book, the young German usually stifled his doubts and became like his father before him, a disciple of the emperor and of the institutions of his country, in their accepted form. His military training further influenced him to accept the established order. He heard again in the barracks the stories of German prowess. He was filled by his officers with tales of the envy felt by the western democracies towards his country's wealth, of the French desire for revenge, of the jealousy of Britain, and of the enmity of Russia. In the newspapers he read of the iron bands which Germany's commercial rivals were forging to throttle her trade. His knowledge of the actual workings of government was very slight unless he was a member of the privileged classes and destined for a position in the civil service. He rather vaguely believed his country to be a constitutional monarchy and felt that his vote for a member of the Reichstag satisfied the behests of twentieth century democracy.

His attitude was definitely formed when the war came, for his psychology was ready-made. He was told by the papers and the speeches of men whom he had been taught to revere that war had been forced on his country. Papers and statesmen played especially on the popular fear of Russia, a fear which they themselves had consistently propagated. The German, consequently, accepted war in the spirit of 1870, and believed his country to be fighting for her existence. Nothing illustrates the success of government propaganda better than the tremendous rejoicings over Hindenburg's victory at the Mazurian Lakes. The people feared Russia and her defeat was considered far more important than the vastly more vital battles on the western front.

No soldier of any nation loves war. If he is a good soldier he makes the best of it. Early enthusiasm soon wears off and he settles down to hoping that the horrors will soon be over. This the German soldier did, and his hopes were constantly buoyed up by propaganda depicting alternately the approaching victory, or a peace negotiated on the basis of a mutual understanding. The first note of disillusionment came when the promises made him one after another proved illusory. Not even pride of his country's victorious marches in Russia, Italy, Roumania and Serbia, could assuage the weary routine of the trenches. His letters from home and occasional furloughs informed him of the hunger, the unhappiness, the death of kin. By 1917 there can be no question that these tales had seriously affected his *morale*. While still loyal to Germany, his confidence was shaken in the judgment of his country's leaders who, while promising him an end of war, found themselves in no position to fulfill their promises.

There followed in the Spring of 1917 a heated discussion in the German press on the value and advisability of the submarine campaign. For the first time during the war, the people managed to get a peep behind the scenes and perceive the discord and wrangling which was going on in high government circles. Neither was the government's final decision to continue the war popular, and the soldier saw the entrance of America into the struggle with misgivings, despite the government's assurance of that nation's military impotency.

The hunger did not abate, the burden of war grew ever heavier, and in sheer weariness the people began to seek a way to escape from the struggle. Then followed President Wilson's fourteen points, which were widely discussed and received with favor in many liberal and radical circles. Clever propaganda, emanating from Allied sources, harped on the friendship of the Allies for the German people and stated that the war was being fought solely against autocracy. This propaganda attained a considerable success, and many individuals came to believe that America, at least, was at heart a friend of the German people. The government, however, was still too strongly entrenched for revolt, and even if the discontent was general, it had not yet shaped itself into tangible form.

The astonishing victories of the Spring of 1918 aroused the drooping spirits of the country, and furnished material for the Pan-Germans with which to combat the discontent at home. However, the *morale* of the army, even in the midst of its victories, was giving the high command cause for concern. Certain influences had been weakening the spirits of the soldiers since early in 1917, and there had been considerable discussion among the staff before the Somme battle as to whether the army would bear up under the heavy sacrifices which certainly would still have to be made.

These influences were two in number, and existed quite apart from the general war-weariness of the army and the nation.

The first weakening factor was the static condition on the eastern front which had existed since early in 1917. Germany had employed the lull in hostilities to weaken the enemy, by permitting the soldiers of both armies to fraternize. The results of this policy, while immediately successful, in the long run proved disastrous to her own military fabric. Bolshevism obtained many supporters, in such numbers, in fact, that divisions thereafter transferred from the eastern to the western front proved unreliable, and had to be broken up.

A more important cause of the decline in *morale* came with the failure of the young classes of recruits to adopt a patriotic spirit. These young boys had been removed from school to the factories at a time when they ordinarily would have been receiving instruction in the infallibility of their emperor and the country's institutions. The manufacturing towns were nests of immorality. In consequence, the boys were subject to the worst influences, and drifted into every kind of crime. They therefore received without enthusiasm their orders to exchange the comparative luxury in which they were living for the mud of the trenches. The army had suffered an irretrievable loss of *morale* even before March, 1918, and only prompt victory could have averted eventual dissolution. The Allied military position was too strong, and as soon as the tide turned, Bolshevistic propaganda more and more asserted itself. In July and August, 1918, the defeats on the western front brought to a head the various streams of discontent and dissatisfaction which had been circulating everywhere in the army and the interior.

With each successive retreat, the position of the government weakened. There was no personality strong enough to take the responsibility of suing for peace or demanding internal reorganization. When finally conditions at the front became so threatening that some action was inevitable, the attempt to liberalize the government was half-hearted and unavailing. The period of constitutional monarchy, with Prince Max of Baden as Chancellor, proved short-lived. Then the government, having yielded to the liberals, drifted further from conservative councils and the spirit of revolution, once afield, proved too powerful to be held in check.

It is probable that the Bourgeois class looked at this time on a constitutional monarchy as the solution of their troubles. There was a real reverence for the Kaiser and a general conservatism among all classes which feared above all extreme revolutionary doctrines. If peace could have been attained, success of the moderates was certain; but as day by day passed, marked in the field by defeat after defeat, their position rapidly became intolerable.

The revolution of November 9th proved the culminating stroke. The Majority Socialists accepted the responsibility of government, and the abdication and flight of the emperor destroyed at a stroke the respect of the Bourgeoisie for the monarchy.

The great middle class therefore found itself in a predicament. It despised the emperor for running away, it feared the Spartacists who had already begun to organize their Soldiers and Workmen's Councils, and it accepted the personalities of the new government without confidence in their ability to guide the ship of state. There was little of the political horizon which could command the respect of the young German. He either saw or heard of the crimes of the Bolshevists throughout the country and believed Allied intervention preferable to the excesses of such fanatics. Patriotism to Germany had previously meant love for the emperor. Now that his faith in the Kaiser was gone, he could not associate patriotism with loyalty to Ebert or Scheidemann.

Confronted with all this political turmoil, the American soldier appeared to the German almost as a friend, rather than an enemy, and a friend who would end the Bolshevist menace once and for all. So in the late Autumn of 1918, did the German meet the American soldiers. He was curious and friendly, and seemed quite unable to realize the stern frame of mind which animated the victorious army.

CHAPTER No. 2

MILITARY GOVERNMENT DURING THE MARCH TO THE RHINE

Formation of the Third United States Army; The Third Army in Luxemburg; Proclamations and Orders Issued by the Armies during the Advance from the Luxemburg Frontier to the Rhine; Tactical Survey of the Advance to the Rhine; Development of American Policies during the Advance from the Luxemburg Frontier to the Rhine; Arms and Ammunition; Rate of Exchange; Alcohol; Billets; Requisitions; Relations between Army and Civil Population.

Formation of the Third United States Army.—The Third United States Army came into existence on November 7th, 1918, four days before the signing of the armistice. This was a period of great activity throughout the American Expeditionary Force, as the pursuit of the German troops was still being continued on that date, despite knowledge that negotiations had been begun for an armistice. The orders (G. O. 198, G. H. Q.) directing the formation of a Third Army, while issued on November 7th, were not put into effect until some days later.

Organization of an army staff is under any circumstances a difficult matter, but when it must be done in the presence of ever-changing conditions of battle, it becomes a serious problem. It became known almost as soon as the armistice was signed, that the Third Army was to be entrusted with the task of following the retreating German troops and enforcing the terms of the armistice. Composed of over 200,000 men it began its forward movement puntually at 5.30 A. M. on November 17th, 1918. Its mission required great effort on the part of every officer, if the high command, in the short time at its disposal, was to have necessary machinery to direct the complex movements of hundreds of units, together with their constant supply. The Army was placed under command of Major General J. T. Dickman, who had formerly commanded the Third Division and later the First Corps. The following General Staff officers were assigned:

CHIEF OF STAFF SECTION

Chief of Staff	Brigadier General Malin Craig.
Deputy Chief of Staff	Lt. Col. George Grunert.
Secretary General Staff	First Lt. C. C. Colburn.

CHIEFS OF SECTIONS

G-1 Section............ Col. Jas. A. Logan, Jr., A. C. of S.
G-2 Section............ Col. R. H. Williams, A. C. of S.
G-3 Section............ Col. John C. Montgomery, A. C. of S.
G-4 Section............ Lt. Col. J. H. Perkins, A. C. of S.
G-5 Section............ Col. Walter C. Short, A. C. of S.

While, at this time, few people believed that Germany desired a resumption of hostilities, there was no intention on the part of the Commanding General to permit any unit of the army to relax proper military precautions. The armistice terms provided that Germany must surrender large quantities of military equipment. If these terms were to be rigorously enforced, the Allied armies had to be ready to resume hostilities in case the German army proved refractory and refused to abide by the signatures of their representatives. The problem confronting the army commander was therefore primarily a military one. The American Army at this time, composed of eight divisions, had to be moved forward to the Rhine, and be ready during the movement for battle at an instants notice. Having finally occupied the bridgehead, it was to take up a defensive position and await the signing of peace. Yet, if the immediate problem was formidable enough, the future held other problems equally serious, which, were they to be properly solved, required a large amount of immediate preparatory work.

The armistice provided that the supreme governmental power in occupied Germany should be lodged in the military authorities of the occupying army. Occupation by the American army of a part of the Rhineland placed therefore a heavy burden on the American High Command, for it was necessary to at once commence the drafting of regulations, necessary for the security of the army, and preparation for the many military tribunals which would be required to enforce our laws, when published. Furthermore, the intricate machinery of civil government had to be ready to move the instant the army set foot on German soil. Incoherence of policy and vacillation of purpose in the first days of the occupation would have forever after placed our military government at a disadvantage, by lowering the respect of the civil population,—a respect which, once lost, would have been difficult to regain. While reports of a contradictory nature often portrayed Germany as a nation in the throes of anarchy, the requirement of the armistice that the civil government should continue to be conducted by German officials assured a continuance in office of experienced and competent men.

Unfortunately, the unexpected close of the war with its occupation of enemy territory, left little time for study of the problems confronting the military government. The time between the signing of the armistice and the occupation of Germany (less than three weeks) was too short to organize completely a civil administration with formulated policies. G. H. Q. was therefore confronted with the necessity of adopting such temporary expedients as would tide over the situation until some permanent organization could be devised.

In accordance with historic military precedents, it was decided to publish proclamations to the civil population in both Luxemburg and occupied Germany. As it was General Pershing's intention from the beginning to retain a large amount of the power of civil government in occupied territory in his own hands and in those of the staff at Chaumont, the drafting of both these proclamations was left to G. H. Q. Work was begun simultaneuosly on a fundamental plan for the American military government of Germany, which was to be put in operation as soon as the army reached Coblence and settled down in its permanent billets. To cope with the immediate situation, incident to the presence of an American army in Germany, and until a policy was outlined by G. H. Q., authority ordinarily given to the Commander of an occupying army on enemy soil, was lodged by the Commander in Chief with Major General Dickman.

At 5.30 A. M., on the morning of November 17th, 1918, the Third Army crossed the old front-line positions and began its advance toward the Luxemburg and German frontiers on the way to the Rhine. The army was composed at this time of two Corps, the III and IV, each consisting of three Divisions. Later the VII Corps, composed of the 89th and 90th Divisions, was added to General Dickman's command. However, the advance to the Rhine was made by the troops of the two first mentioned Corps only, as the VII Corps went into garrison shortly after crossing the frontier near Treves. Four of the six forward Divisions of the army stood in first line when the advance commenced from left to right, the 2nd, 32nd, 1st and 3rd. The 42nd was to march in support of the 2nd and 32nd, these three Divisions forming the III Corps on the left wing. On the right wing, the 4th Division was to follow the other two Divisions—the 1st and 3rd—of the IV Corps.

The Third Army in Luxemburg.—The American army entered Luxemburg on November 20, 1918. The Commander in Chief took this occasion to issue the following proclamation, which sought to make clear what the attitude of the occupying army would be to the civil population of this neutral county.

PROCLAMATION TO THE PEOPLE OF LUXEMBURG

After four years of invasion Luxemburg is now happily liberated. Your freedom from German occupation was exacted of the invaders by the American and Allied armies as a condition of the present armistice. It now becomes necessary for American troops to pass through Luxemburg and to establish and maintain for a time their communications upon your territory.

The American troops have come to Luxemburg as friends and will conduct themselves here strictly in accordance with international law. No hardship need be expected from their presence, which will not extend beyond what is deemed strictly necessary. Your government and institutions will nowhere be interfered with. You will be undisturbed in the peaceful pursuit of your lives and occupations. Your persons and your property will be fully respected. It will be necessary for the American army to utilize certain buildings, railroads, telegraph and telephone lines, and possible other public works, for shelter, transportation or communication; but beyond that, whatever supplies may be required, will be paid for at a just valuation.

It is assumed that you will commit no aggressions upon the American army and will not give any information, aid or comfort to the enemies. You are expected cheerfully to observe such rules as the American military commanders may find it necessary to make for the safety of their troops and your own protection.

(Signed) JOHN J. PERSHING,
Commander in Chief of the American Expeditionary Forces.

As Luxemburg was a disarmed neutral and had sharply protested against Germany's invasion of her rights in the first days of the war, there was no intention on the part of the American army of treating her as an enemy country. All relations with the civil population, and particularly dealings with the Luxemburg civil authorities, required, as may be imagined, constant tact on the part of the military authorities responsible for carrying on the occupation. The invasion and occupation of any country by a foreign army, even if it comes as a liberator, is apt to give rise to constant sources of friction. During the stay of the Third Army in Luxemburg, the deep appreciation of the Luxemburgers for their liberation, together with the just policy of the American authorities in all civil matters, prevented any unfortunate incidents. The stay of the army in Luxemburg was very brief, a fortunate matter, as the billeting of 200,000 men within the frontier taxed the accomodations of the tiny Grand Duchy to the utmost. Although Luxemburg continued to be occupied by American troops until the Summer of 1919, it never, except for a few weeks in the Spring of 1919, was attached to the territory under the jurisdiction of the Third Army. The powers of military government exercised by the American army in Luxemburg will therefore not be treated in this report.

The cordial and even enthusiastic reception of our troops by the civil population of the Grand Duchy made a deep impression on every soldier of the army. The parade of units of the Thirty-second Division through the streets of the Luxemburg city on November 21st, 1918, seemed to have been made the occasion for a national festival. Bands, local societies and boy-scouts escorted the parading troops. A banner bearing the inscription "To the saviours of our dear country", made a particularly striking impression on the soldiers. The reception of the army in other parts of the Grand Duchy seems to have been no less hearty. Observers in the First and Thirty-second Divisions comment strongly on the apparent genuiness of the joy of the Luxemburgers in bidding the Americans welcome. Though the American soldiers seemed somewhat surprised at first to find that German was the language generally spoken by the people, their attitude toward the Luxemburgers was friendly, and they showed their appreciation of the reception given them in no uncertain way. While the Luxemburgers evinced deep resentment toward the Germans, there did not seem to be any of the marked causes of complaint which were so common in other countries invaded by the enemy. Requisitions appeared to have been partly paid for in cash, and no evidence of wanton destruction could be observed. Luxemburg suffered great economic loss, however, by having a large amount of German marks floated on her population at the rate of one mark to 125 centimes, whereas the mark had at the time of the occupation, shrunk to approximately 80 centimes, representing a capital loss of more than 35%.

Proclamations and orders issued by the armies during the advance from the Luxemburg frontier to the Rhine.—While the Headquarters of the Third Army was still in Hollerich, a suburb of Luxemburg City, Brigadier General H. A. Smith, formerly director of the military schools at Langres, arrived from G. H. Q. to consult with the army commander on questions of civil policy in occupied territory. General Smith had been designated as deputy chief of staff for civil affairs at G. H. Q. His title was changed a few days later to that of Officer in Charge of Civil Affairs in Occupied Territory, which designation became permanent. (The Commander in Chief had by this time decided that as soon as practicable an Advance General Headquarters should be established at Treves, and that the Office of Civil Affairs should form a part of that organization which was to operate directly under him). In the meantime, Colonel I. L. Hunt arrived at Luxemburg for duty in connection with the Civil Affairs of the military government. In view of the desirability of the Office of Civil Affairs having a representative with the Third Army, General Smith direct Col. Hunt to accompany that body as an advisor to the Commanding General in civil matters.

In order to aid the personnel which would be eventually designated to administer German territory, a pamphlet entitled "Notes on German Local Government", was prepared by the Second Section of the General Staff at G. H. Q. This pamphlet contained important notes on the organization of the German governmental system and in addition devoted several pages to German laws on billeting and requisitions. It further contained notes on the German administration of justice. This pamphlet was at that time the only source of information distributed among officers of the army on the organization, government and laws of the country which they were called upon to govern.

During the advance to the Rhine, three important documents were published by the Headquarters of the Third Army. Two were proclamations for the guidance of the civil population, which the Army Commander had received from higher authority; the other, a set of instructions for officers and men, which latter appeared as an army memorandum. One of the proclamations originated with Marshal Foch and was published simultaneously by all the Allied armies.

Marshal Foch's proclamation follows:

PROCLAMATION

of the Commander in Chief of the Allied Armies, to the Inhabitants of Occupied Territory.

The Allied military authority herewith assumes command of the country.

It demands strictest obedience from all.

The laws and regulations in force at the moment of occupation will be continued inasfar as they do not affect our rights or our safety.

Public services will continue to operate under the direction and control of the military authorities.

Public officials will be held responsible for the conscientious and honest discharge of the duties with which they are entrusted. The courts will continue to dispense justice.

The inhabitants must abstain in word and deed from any act of hostility, direct or indirect, toward the Allied authorities. They must obey the requisitions which may be made of them in conformity with law.

Every person guilty of a crime or misdemeanor, whether as a principal or an accomplice, will be immediately arrested and brought before a court martial.

Every infraction of the regulations issued, as well as every refusal of obedience to orders, will be severely punished.

The present proclamation declares the occupation of the country by the Allied armies. It points out to each one his duty which is to aid in the re-establishment of civil life by work, calm and discipline. Let all persons diligently discharge such duty.

FOCH, MARSHAL OF FRANCE,
Commander in Chief of Allied Armies.

As may be inferred from a casual perusal of this proclamation, it was essentially a formal document whose phrases permitted a wide interpretation by each of the several Allied commanders. While bringing home to the civil population the essential truth that the Allies were a unit in occupation as well as war, it very wisely permitted each nation to treat the defeated enemy according to the dictates of its natural policy and characteristics.

The American proclamation, signed by General Pershing, was issued by the American army alone. It was as follows:

PROCLAMATION TO THE INHABITANTS

The army of the United States, in conjunction with the Allied military authorities, has taken possession of and garrisoned with troops in the district of Treves (Rhenish Prussia), the entire localities of Daun, Prüm, Bitburg, Wittlich, Berncastel, Treves (City) and Treves (Suburbs); in the district of Coblence (Rhenish Prussia), the entire localities of Adenau, Ahrweiler, Coblence (City), Coblence (Suburbs), Cochem, Mayen, Neuwied, St. Goar, Simmern and Zell, and the entire district of Alterkirchen, with the exception of those parts lying further to the east than thirty kilometers from the eastern end of the bridge over the Rhine at Coblence; in the district of Wiesbaden (Province of Hesse-Nassau), the entire localities of St. Goarshausen, Unterlahn, Unterwesterwald, with the exception of those parts that are more than thirty kilometers distant from the eastern end of the bridge over the Rhine at Coblence.

The above mentioned areas and their inhabitants are under the regulations and authority of the American army. These orders are succinct, and strict observance is expected of all. Those who observe these regulations have nothing to fear. The American army is not bringing war against the civilian population. All that lawfully and peacefully abide by the regulations laid down by the military authorities may count on protection for their persons, homes, property and belief. The others will be arrested at once and brought strictly to account.

The American army on their part will adhere strictly to the laws of nations as well as the laws of civilized warfare. The population, on their part, must offer no signs of enmity towards the American troops nor place obstruction in their path, either by word or deed.

It is the duty of the population to regain their normal mode of life and to re-establish the schools, churches, hospitals and charitable institutions, and to continue in their regular local activities. Therein they will not be disturbed, but rather assisted and protected. In so far as their scope and bearing permit, the courts, city departments and civil establishments will be continued under the supervision of the American army. The existing laws and regulations, in so far as they do not interfere with the duty and security of the American troops, shall remain in force.

All violations of the regulations, inimical conduct or overt hostility against the orders of the military authorities will be severely punished.

(Signed) JOHN J. PERSHING,
Commander in Chief of the American Expeditionary Forces.

The first paragraph of the proclamation concerns itself with geographically delimiting the area of American control. The boundaries therein established for the Third Army did not prove to be permanent. Even while the Allies were still advancing to the Rhine, a modification of this area was found to be essential, and the Kreise (termed "localities" in the proclamation) of Simmern, Zell and St. Goar, St. Goarshausen, and Unterlahn, were transferred to the jurisdiction of the Tenth French Army on our right. The other paragraphs of the proclamation are notable for the repeated declarations that the occupation would be conducted in accord with the laws of nations and of civilized warfare. While warning the civil population that any word or act inimical to the occupying army would be severely punished, the proclamation breathes throughout a spirit of humanity and justice, in marked contrast to the proclamations of the German army in France and Belgium.

Memorandum No. 4, Third Army, was published on November 28th, 1918. It is notable as the first act of specific regulations issued by the American army for observance by the German civil population. The authorities realized that regulations issued without first-hand information of actual political and economic conditions, must naturally be temporary in nature, and subject to either revision or complete revocation, as the occupation progressed. In a force of over 200,000 men, it is inevitable that some individuals, even officers, will have a false conception of the proper treatment to be accorded the defeated enemy. If regulations for their guidance are lacking, they will follow the dictates of their personal feelings, and will either err in undue severity or go to the other extreme and, by marked leniency, weaken the respect due the occupying troops. Definite regulations by the high command were therefore necessary, in order to impress on the civil population the fact that they were dealing with a disciplined, united army.

The opening sentence of the second paragraph of Memorandum No. 4 strikes the keynote of the whole American occupation.

"In stopping in a town where a halt for the night or a longer period has been ordered, the commanding officer will *send for* the Burgomaster, chief of police and other prominent officials."

This sentence informed all officers that the civil population was to be at the start impressed with the power and authority in the hands of the occupying army, and that respect and courtesy must be accorded its representatives on every occasion. Civil officials were to be made to realize at once that they were subordinate to the American commander of their city or town, and were to obey, promptly and cheerfully, the slightest order given. The sentence implies further that American officers when dealing with civil officials in territory under their control, should order and never request.

The sub-paragraphs of Memorandum No. 4, which follow, are designated to serve as an outline of the proper directions which town commanders should give to civil officials. Only those sub-paragraphs of the memorandum are quoted which pertain to purely civil matters.

> "He will hand them copies of the proclamation of Marshal Foch and the Commander in Chief. He will inform them that military government has been established in the town and surrounding district; that the principal object of this government is to provide for the security and efficiency of the United States Army; that so long as the inhabitants conduct themselves peacefully and quietly, the ordinary civil and criminal laws will be continued in force, and will be administered by the local officials, and the private property and personal rights will be respected. Should an official decline to serve, the commanding officer will direct the Burgomaster to name his successor. Should the Burgomaster be absent, the commanding officer will direct the official next in rank or a suitable civilian to act as such.
>
> "The commanding officer will obtain from the Burgomaster a map of the town showing the locations of all banks, hospitals, railroad stations, libraries and all public buildings. He will inquire if added police protection is needed. If in his judgment such added protection is needed, extra guards will be posted.
>
> "The commanding officer will direct the Burgomaster to inform his people to avoid assembling in crowds and to go quietly about their ordinary affairs of life.
>
> "He will direct the Burgomaster to prohibit the sale of liquors, except beer and light wines. The sale of beer and light wines will be prohibited between the hours of 9 A. M. and 5 P. M.
>
> "He will direct the Burgomaster to forbid the sale or carrying of deadly weapons."

Although under the terms of the armistice, civil officials might very properly have been required to remain in office, there was no intention on the part of the American authorities to compel them to serve. As a matter of fact, no case of an official declining to continue his duties, was brought to the attention of the Third Army Headquarters. In many cases, our arrival was welcomed by the officials, who regarded our coming as being of great assistance in enforcing their authority, which had been weakened by the revolution. During the month of November, Soldiers' and Workmen's Councils had been created in many cities, and these organizations had usurped much of the power of former officials. Our coming therefore was a matter of prime importance to them, as it restored their authority and power.

In the concluding paragraphs quoted above, three restrictions are imposed on the civil population. These were believed necessary in order to prevent friction between the civil population and the troops.

The first of these restrictions, forbidding the gathering of assemblies or crowds, had ample precedent in history and was undoubtedly necessary under the circumstances. It is gratifying to note that no infractions of this order occurred during the march to the Rhine. The restriction on alcohol, also proved to be a wise measure, in the best interests of both the army and the civil population. With large numbers of demobilized German soldiers already in the area to be occupied by the army, the possibility of brawls with our troops, which might well assume serious proportions could not be lightly dismissed. It seemed probable that the chance of such disorder would be reduced by entirely forbidding the sale of strong alcoholic drinks and by limiting the hours in which wine and beer might be sold in inns and cafes. Arms and deadly weapons were definitely forbidden to be sold or carried by civilians, this being the third restrictive measure.

The final paragraphs of Memorandum No. 4 read as follows:

> "He will direct the Burgomaster to furnish billets for so many officers, so many men, and so many animals and such supplies (fuel, forage, straw, etc.) as may be needed for the troops under his command. *Food will not be requisitioned.*
>
> "(*The attention of all officers is called to the fact that requisitions shall only be demanded on the authority of the commanding officer in the locality occupied.*)
>
> "He will inform the Burgomaster that these requisitions will not be paid in cash but that receipts for all billets and supplies furnished will be given. Requisitions should generally be made upon the municipal authorities, but may be made upon individuals, if necessary. The receipts given should be filed with the Burgomaster.
>
> "All receipts should be signed in duplicate by a supply officer and approved by the commanding officer. One copy will be given to the municipal officer or individual furnishing the supplies and one will be forwarded to the Military Secretary, Army of Occupation, through military channels."

The right of the occupying armies to billet and requisition was specifically provided for in the armistice terms. The declaration that food would under no circumstances be requisitioned, is a striking commentary on the policy of humanity which actuated the army. So many stories had come to the notice of the army commander of starvation and misery within the German territory already occupied, that he believed it wise to abstain from requisitioning foodstuffs until the true food situation could be ascertained by investigations conducted on the spot.

Tactical survey of the advance to the Rhine.—At half past five on the morning of December 1st, the American Army crossed the

German frontier and began its long march to the Rhine and the Coblence bridgehead.

A brief account of the tactical phases of the advance is essential to an understanding of the difficulties which confronted the army during this period, in its exercise of the powers of military government.

When the armistice was signed, the Allied armies at the front were drawn up in the following order from north-west to south-east: the Belgian army, the British armies, a group of French armies, the American armies and, finally, another group of French armies on the Alsace-Lorraine frontier. From each of these groups of armies, a force was detailed for the occupation of Germany. These were:

A part of the Belgian Army;
The Second British Army;
The Fifth French Army;
The Third American Army;
The Tenth French Army;
The Eighth French Army.

At five o'clock in the morning of November 17, these six vast armies, under command of Marshal Foch, started to move in a north-easterly direction, the whole battle-line from the Swiss border to the North Sea surging forward in one great wave. At first it was deemed not improbable that hostilities might again break out, and therefore it was essential that this great force in its advance should be so disposed, both laterally and in depth, that battle-order might be resumed at any time.

On November 18th, the Third American Army reached the 1914 frontier of France opposite Luxemburg. Fear now began to be felt that the population of Luxemburg might become disorderly, once the restraint imposed by four years of German occupation was lifted. Accordingly, by mutual agreement between the American Army and the German Command opposed to us, the terms of the armistice were sufficiently modified to permit the advance guards of our army to follow the last detachments of the retiring Germans at a distance of ten kilometers. This change of plan resulted in the American army completing the occupation of Luxemburg sooner than was originally intended, and consequently made a halt of several days necessary before the old German frontier might be crossed under the terms of the armistice.

Division of zones.—The boundaries of the armies' zones during this halt were extended to the Rhine, and the Belgian and British armies were assigned the territory which they eventually occupied as shown on map II accompanying this report. [Map not printed.] The southern boundary of the Third American Army began at Schengen, where the right then rested, and extended along the southern

521572°—43——4

limits of Kreis Saarburg, thence along the northern and eastern boundaries of Kreis Merzig, thence along the northwestern boundary of the principality of Birkenfeld, thence along the northwestern boundary of Kreis Kreuznach to the Rhine. The line was approximately Schengen-Saarburg-Sitzerath-Mühlfeld-Zusch-Kempfeld-Kellenbach-Dichtelbach-Trachtingshausen. The left boundary of the American zone was extended northward through Belgium until it coincided with the right or southern boundary of the British zone, thus absorbing the front previously occupied by the Fifth French Army. The line of this northern American boundary ran from Carignan to Merzig, thence to Oberfeulen-Niederfeulen-Burscheid-Horscheid-Consthum-Bochholz Marburg-Roder-Heinerscheid-Kalborn-Luxemburg frontier-administrative boundary separating Bezirke of Treves and Coblence from those of Aix-la-Chapelle and Cologne. This latter is approximately the line Lommersweiler - Manderfeld - Kronenburg - Waldorf - Rohr-Hilberath-Oberwinter.

Following this reorganization of the old army front, Marshal Foch decided to attach certain French Divisions to the other Allied armies. This plan contemplated the assignment of two French Divisions to the Belgian army, one Division to the British, and two to the Third American Army. Further, one American Division, the Ninetieth, was to have been detached from the Third Army for duty with the Tenth French Army in the Mayence bridgehead. Information of this organization was furnished to American G. H. Q. informally, and was never confirmed in orders. The French Divisions were therefore left by the American and British Armies on the German frontier and the Ninetieth Division was not assigned to the Tenth French Army.

On December 12th, Marshal Foch, who evidently thought it expedient to alter the plans already commenced, issued orders directing that the dividing line between the French and American armies, north and east of the general line Budenbach-Trarbach be changed in such a way as to make the Moselle river north of Trarbach the general boundary between the two armies. The French Army was assigned one of the Coblence bridges and the roads of approach, and the Ehrenbreitstein-Hamn road was to be the boundary between the American and French armies in the Coblence bridgehead. At a conference between representatives of the Third American, and Tenth French army, to exactly establish the dividing line between their respective areas, it was decided to leave the boundary from Schengen to near Schneppenbach as it was. Thence it was to follow the administrative boundaries separating Kreise Berncastel, Wittlich, Cochem, Mayen, Coblence, Montabaur and Westerburg from those of Simmern, Zell, St. Goarshausen, Diez and Limburg. The French armies under General Fayolle were grouped as follows: the Tenth army in the southern half of the Coblence bridgehead and in the Mayence bridgehead;

the Eighth army in the Palatinate. The English troops garrisoned the Cologne region, their line of supplies being westward to Aix-la-Chapelle and Malmedy. The Belgians were further north around Crefeld and München-Gladbach, with their headquarters at Aix-la-Chappele. A French Corps (the Thirty-third), was assigned to the Belgian army.

Development of American policies during the advance from the Luxemburg frontier to the Rhine.—Army Headquarters were moved on December 3rd from Hollerich to Bitburg, chief town of the Kreis of the same name. Investigations of the local situation were begun at Bitburg by the Office of Civil Affairs, in order to frame intelligently an American policy in many pressing matters.

Arms and ammunition.—The German people are accustomed to having among their population a much larger proportion of firearms than ordinarily is the case in an average American community. This is particularly true of sporting weapons, hunting retaining many of its ancient feudal characteristics in the Rhineland. The hunting privileges for a district are let out in public auction by the town or village in whose territory it lies, to the highest bidder, provided the purchaser holds a hunting license. Hunting by other than persons so authorized is rigorously forbidden. The government employs a large corps of foresters, in addition to the regular police, to prevent poaching.

Carrying of arms by all German citizens had been strictly forbidden in Memorandum No. 4. In the first conference held with any civil official by the Officer in Charge of Civil Affairs, the Landrat of Kreis Bitburg, Count Adelmann von Adelmannsfelden, took occasion to inform the representative of the Commanding General of the disposition and strength of the police under his direction and of their unquestioned reliability. He emphasized that it would be to our interests to permit them to carry firearms. This request was carefully considered and finally granted, as it was believed proper to permit the German police a certain measure of authority over their own citizens. Directions were therefore given the Landrat, permitting the police to retain their pistols and bayonets, but at the same time requiring their carbines to be turned in to the Burgomasters. Although the Landrat further begged that foresters might retain their arms to check the depredations of wild boars, this request was not granted at this time. However, on December 8th, the matter having been investigated in the meantime, the Commanding General authorized foresters temporarily to retain their hunting weapons. The question of carrying of arms by police and others again arose when Headquarters of the Third Army were moved to Mayen. It appeared from the statement of the Landrat of that Kreis, Herr Wilhelm von Loeser, that all units of the army had not strictly abided by the requirements of Memorandum

No. 4, and that certain commanding officers had even issued instructions of their own on the subject of firearms. This had naturally produced considerable confusion in the minds of German officials. However, while the army authorities were still considering how far the privilege of carrying of arms might be extended to civil officials without endangering the safety of the army, General Pershing issued his *Anordnungen* governing the subject.

Rate of exchange.—Germany had during the war established and enforced an official rate of exchange in both Belgium and Luxemburg. The value of her currency at home had been maintained by the diminution of imports. This had been caused largely by the British blockade, but also by government restrictions on such goods as still found their way into Germany through neutral countries. Imports were restricted to food, clothing and raw materials for manufacture of military equipment. In consequence, there had been little demand in Germany for foreign money and the value of the mark had not fallen in neutral countries as much as might have been expected. With the signing of the armistice, restrictions on German trade with neutral countries were practically removed. However, as the political and economic fabric of the nation was already badly shaken and the temper of the Allies indicated that a demand would be made in the peace treaty for a huge indemnity, the value of the mark began to fall rapidly. After the evacuation of Belgium and Luxemburg in November, 1918, a vast amount of German marks still remained in circulation. When the army entered Luxemburg, the soldiers with their French currency found that the natives demanded a franc and a quarter for the German mark. This was manifestly far more than it was worth in neutral countries and was, in fact, the official rate formerly enforced in the German armies. As soon as the army entered Germany, the value of the mark began to fall, at first being rated on a par with the franco and gradually going lower. The people seemed perfectly willing to accept francs at this rate, as they evidently anticipated a still further decline in the value of the mark. Much dissatisfaction was however felt by the troops who found themselves the prey of tradespeople and speculators who took every opportunity to obtain a favorable rate for themselves at the expense of the army. An observer with the Thirty-second Division during the advance reports that nearly every officer and enlisted man in that unit made a considerable personal sacrifice in changing his francs to marks. This officer's statement is substantiated from other sources in the First and Second Divisions, and may be regarded as applying generally in the whole army. No attempt was made by the military authorities during the advance at setting a definite rate of exchange. Such action was impossible, largely owing to the lack of information as to the commercial rates in neutral countries. As soon, however, as troops settled in the permanent areas, Advance

General Headquarters in Orders No. 1, announced an official rate of exchange of 142.85 marks for one hundred francs, or, one mark equals seventy centimes.

Alcohol.—The provisions of Memorandum No. 4, forbidding the sale of alcoholic drinks and restricting the hours in which light wines and beer might be sold, were not modified during the march to the Rhine. Although elements of the civil population resented the American standpoint in the matter and looked askance at a measure which they believed an attempt to transfer American ideas to Europe, the better element soon appreciated the fact that it operated to reduce friction between soldiers and civilians. The order was very generally obeyed by the civil population, although secret traffic in Cognac and *Schnaps* could never be wholly prevented. When the attention of the military authorities was called to the continued sale of illegal liquor, strict measures were taken to punish all offenders, including not only fine and imprisonment by Provost Courts, but confiscation of liquor and closing of the place of business of the offender on orders of the Commanding Officer.

Billets.—The provisions of Memorandum No. 4 were so explicit in the directions given commanding officers as to billeting, that very little trouble was encountered in obtaining satisfactory accommodations for officers and men. The German people were accustomed to having troops of their own army quartered on them, and as billeting is provided for in their own laws, they accepted the demands of a foreign army as a matter of course.

Two German attempts, however, to take advantage of our American good nature were speedily recognized and checked. Observers in the Fourth and Eighty-ninth Divisions reported that during the advance, well-to-do families generally attempted to conceal the number of rooms at their disposal. Such action naturally threw the burden of billeting the occupying army on to the poorer people. So general did the practice become in some districts, that the billeting officers of the Eighty-ninth Division found it necessary to assign definite rooms to the civil population and take the rest of dwellings for the troops. In districts occupied by the First Division, Burgomasters occasionally demurred at furnishing the amount of billet-space demanded. However, a slight show of firmness by the billeting officer was generally amply sufficient to bring local officials to terms. The Forty-second Division reported that even less difficulty was encountered in Germany in billeting their troops than in friendly countries. The Thirty-second and Ninetieth Divisions procured necessary billeting facilities without special trouble and their reports state that Burgomasters apparently did all in their power to satisfy the wishes of the occupying army.

Requisitions.—As Memorandum No. 4 absolutely forbade units to requisition food, demands on the civil population were limited to fuel and forage alone. During the enemy's retreat, his high command had utilized its requisitioning power to the fullest extent. The German lines of supply had completely gone to pieces and the army, stranded as it was, had either to live on the country or starve. In consequence, many parts of the occupied portion of Germany had literally been swept bare and offered little in the way of food to an invading army, even if it had desired to exercise its undoubted rights. The farmers were not slow in expressing their fears that we would act in a similar manner to their own troops, and hurriedly concealed their few remaining supplies, in order to avoid starvation. Demands by the army for forage were met on every hand by deceit and subterfuges, and it required prolonged effort to make necessary quantities for animal transport available. It was soon apparent, however, that the available supplies were very short and that chief dependence in future would have to be put on our forage-dumps in France. As the army had a long line of difficult communications, it was impossible during the first few weeks of occupation to more than maintain the current army rations. Forage had to be requisitioned, and our demands, coming as they did on the heels of those of the retreating German army, worked considerable hardship on the rural population. Horses and cattle without forage for the winter, had to be sold, thus reducing the number available for draft purposes and ultimately resulting in a diminution of the meat and milk supply. On the other hand, fuel was readily forthcoming, and no Division reported any difficulty in securing the necessary quantities for their commands. The American system of receipts and the repeated assurances that requisitions would be promptly settled by cash payments, evidently made a good impression on the German population, because as the advance progressed, less and less difficulty was encountered in procuring what was needed. In one town, the Burgomaster even offered to supply more fuel than was asked for.

The policies inaugurated by the army in dealing with these problems were recognised at the time as merely temporary expedients. It was known that with the publication of General Pershing's *Anordnungen*, all orders previously issued would be rendered null and void. It was nevertheless believed necessary to attempt some solution of each problem as it arose, rather than give the German population an impression that the American army was adopting a lenient and feeble attitude.

Relations between Army and civil population.—Nearly all American Divisions in commenting on their impressions of the first few days on German soil, emphasize the apparent indifference shown by the civil population toward the occupying troops. Such comment would be surprising if we did not bear in mind that the army had just been the recipient of enthusiastic welcome in liberated France and Luxemburg. It is probable that the surprise universally expressed by officers and men was but a natural reaction to the continuous scenes of joy and welcome which had been the lot of the army in liberated France and Belgium. However friendly disposed the German people might feel toward them, it was but natural that they should remove the arches and other decorations which had been erected for their retreating troops and should attempt to appear as oblivious as possible to their conquerors. Burgomasters and officials had apparently been carefully instructed to receive us with due courtesy and to do all in their power to make us comfortable, but nevertheless warned not to make any demonstrations of friendship. Unquestionably, however, there existed a considerable fear of us among all classes, partly a result of war propaganda, but more largely, dread of the unknown. We had been consistently painted in German periodicals as a nation of barbarians, and while such stories never gained wide credence among the German people, they had attained a certain amount of success. The men, nearly all of whom had seen at some time or other the suffering of civil populations in countries occupied by the Germans, now looked forward to the possibility that their homes and families might be burned and violated, in retaliation for the acts of their armies in France and Belgium. That other nations should have a conception of military government, higher and more humanitarian than their own, seemed to many of these men incredible. The feeling, unquestionably widely prevalent, that American was really a friend of Germany, did not assert itself until all fear had disappeared that we would ravage their country and treat them as their armies had treated invaded countries.

Within a few days, however, these fears were dispelled. Stories of the correct and disciplined conduct of the American army and the kindnesses of individual soldiers spread to all corners of the American zone and paved the way for a striking change of sentiment. The attitude of indifference and dignity which had been so marked on the Sunday we entered Treves, swung rapidly to the other extreme. On all sides we were greeted with the smiles of women and the cheers of children, and if the men kept out of sight, they apparently seemed only too anxious to show no sign of hostility. One Burgomaster of a little town in Landkreis Treves, gave an American staff officer his impression of the coming of the Americans, which well illustrates this sudden change of sentiment.

> "When your troops finally reached here", he said, "we watched their every movement with mingled feelings of fear and uncertainty, but to-day, after living twenty-four hours with them, we have no longer any apprehensions. They are

wonderfully mild-mannered men and a great contrast to the domineering attitude of our own soldiers. None of your troops, not even one, has spoken a single disagreeable word to any person, and when we offered them wood for cooking and heating purposes, they accepted with what seemed a certain shyness. They sit in our rooms with us, and we smoke and try to make ourselves understood. This is a great relief when we consider how the Saxons treated us when they came through here."

No act of hostility was committed against us during the first week in Germany, and the opinion grew very general, that the civil population was anxious to create a favorable impression. While many Americans undoubtedly reacted to this friendly attitude of the people, and appreciated the efforts of the women to make them comfortable in their billets, others looked upon all expressions of friendship as mere propaganda. The First Division, in that portion of its report describing the advance to the Rhine, comments as follows on the feeling of the army toward the inhabitants:

> "In their overzealous effort to obtain the good-will of the Americans, the Germans accomplished the opposite result. Their abject servility, coupled with an utter lack of pride, and the fawning, ingratiating manner with which they strove to get off as lightly as possible from their conquerors, tended to disgust our officers and men. It is very probable that this mode of reception had been instigated by the German government."

On the other hand, German opinion was apparently unanimous that our troops were conducting themselves as well as could be desired. The *Bittburger Zeitung*, one of the largest rural papers in Bezirk Treves, in its edition of December 7th, 1918, wrote:

> "The occupation of the city and Kreis by the American troops has been accomplished so far with no restriction of liberty or circulation. The officials and business-men work in quiet and are undisturbed, and life goes on as usual. The mails and railway traffic have necessarily been disturbed. It is just to acknowledge that the American military authorities are efficient without harshness, and the behavior of the American troops is wonderfully good. Not only officers but soldiers are considerate and seem obliging. The Americans take pains not to make the occupation appear vexatious. On their part, the Americans must have observed that it is far from the intention of the local population to create any discord."

Gradually, however, as the troops approached the Rhine, a change of attitude was observed. In Kreis Mayen, there were indications that less satisfaction was felt by the inhabitants towards our occupation than in the districts nearer the Luxemburg frontier. In many places in Kreis Mayen, the Thirty-second Division found the people surly and gruff, although such an attitude was almost entirely confined to the male members of families. It was in the territory nearer the Rhine, that the attempts mentioned previously of German officials to billet our officers in poor houses, were most often made. On the whole, though, even here, the general attitude of the people was amiable, and all classes seemed anxious to please. Bolshevism had appeared to Germany as the most ruthless of enemies, and our coming, a shield to the homes and property of the people of the Rhineland. This point of view largely accounted for the surprisingly friendly attitude of the people. The Allied armies, enemies as they were, were far preferable to the anarchy of a Sparticist. The *Coblenzer Zeitung* in its editorial column of December 12th, 1918, even urged the German government to move the Reichstag to Coblence. It wrote:

> "Examining all possibilities one will find that Coblence, as a meeting-place for the Reichstag, is least accessible to outside influence, under present conditions. We have no groups of the Bolshevist-Terrorist party; we are among the quiet order-loving populace. The American occupation would assure sufficient protection against the eruption of terrorist elements from neighboring German territory. Any attempt at influence on the part of our outside opponents is hardly to be expected, as the American troops, where they have hitherto appeared, have not interfered with the political life of the occupied territory."

The most laudatory newspaper comment of all on our occupation was, however, the article in the *Trierische Zeitung* of December 12th, 1918. Its tone is so surprisingly friendly that it must either be regarded as national propaganda, or as a local attempt to lighten the burden of occupation. In part, it reads as follows:

> "The American soldiers are distinguished from one another, apart from the almost invisible insignia of rank, only by their facial expression and general appearance. All are clean-shaven and their uniforms are alike. The horses of the higher officers are no more gorgeously harnessed than those of ordinary troopers. The rations are also the same, save the officers are served a little better. Recently, as we have been informed, soup, beefsteak and trimmings and pudding were served for dinner. We should like to add that when the American soldier is not provided with rations, he draws an allowance of six marks daily. This proves that he lives fairly well. Our musketeers would have indeed been overjoyed at such an allowance. But we must take into consideration that the United States is the richest country in the world and that $1.00, or four marks, must be expended there for articles that can be obtained for one mark here. What is most conspicuous in the Americans is the assurance and the peaceful manner in which they perform their duties. We hear no strident sounds or insolence; we see no dissatisfaction, much less a haughty mien, but always a friendly smile when intercourse with citizens is necessary. There are no disagreeable relations between the

> soldier and his superior. We should draw a lesson for the citizens of our new republic from this. We cannot afford the pay of the American soldiers, but we can give our people the same position which they take in life. The great differences between the lower officers and the men do not exist with them, nevertheless, everything runs smoothly and this is a particular advantage which we would take to heart."

If more evidence is needed of the good conduct and humanity of the occupying American army during the advance, we have only to turn to the report of General C. D. Rhodes, American representative on the Permanent Armistice Commission. This report, after giving the writer's personal observation of the conduct of the American troops, concludes as follows:

> "It is most gratifying to me to be able to report that during the entire period of the existence of the International Armistice Commission, not a single complaint has reached the Commission through the German High Command or otherwise, of misbehavior or irregularities on the part of American officers or troops."

Chapter No. 3

INTER-ALLIED CONTROL

Introduction; Inter-Allied High Command; Commissions directly under Marshal Foch; Economic Control; The Luxemburg Commission; Operation of Economic Control; The Supreme Economic Council; Policy of American Military Authorities in Economic Matters; Food; The Hoover Commission; The Inter-Allied Food Commission; Duties of Inter-Allied Food Commission; Inter-Allied Railway Commission; Inter-Allied Railway Commission; Duties of Railway Commission; Inter-Allied Waterways Commission; German Inland Navigation System; Use of the Rhine by the American Army as a Supply Artery.

Introduction.—During the latter part of the great war, the armies of the various Allied and Associated Powers operated as a unit under the supreme command of Ferdinand Foch, Marshal of France. When the armistice was signed, November 11, 1918, it was not regarded as a permanent termination of hostilities, and the Marshal consequently continued to exercise his former command. He remained therefore commander in chief of the Allied armies during the occupation of Germany, and while delegating a great deal of his authority to the commanders of the individual allied armies, nevertheless retained, directly in his own hands, a large part of his former power. The power which he so retained will be referred to in this Chapter as the "Inter-Allied Control", and will be briefly described. The Marshal conducted the advance into Germany and partitioned the occupied territory in the manner described in Chapter 2 of this report.

Inter-allied high command.—On November 15th, 1918, Marshal Foch, in his capacity as Commander in Chief, issued to the Allied generals commanding the armies of occupation, a list of principles and instructions for use in the formation and exercise of the military government about to be established in the occupied area. A synopsis of these instructions follows.

According to the principles laid down by Convention of The Hague, control of German administration was to be exercised by the Commanding Generals of Allied armies, under the supreme control of the Marshal, Commander in Chief. German officials were to continue to fill the offices which they occupied at the time of the armistice and to conduct the administration of occupied territory under the direction and control of the military authorities. The generals commanding the armies throughout the territories under their control were to receive general governmental powers as far as concerned administrative decisions and the publication of regulations; these decisions and regulations to be observed by the civil population. They were also to designate the officers who, in each of the territorial divisions of occupied districts, would control the operation of civil administration through the German officials. These officers were to be subordinate to their superiors, based on the administrative decision of the territory, and also subordinate to any local military commanders to whom the general commanding an army might have delegated a part of his authority. Parenthetically, it may be stated that this plan of subordinating officers engaged in local administration to both the officer charged with the general administration of civil affairs and to the local military commander, resulted in the confusion inseparable from dual control. Decrees concerning police, public order, or the safety of the armies, were to be made by the allied military commanders, or be approved by them when emanating from the German authorities.

The Marshal realized that administration of the civil population during the period of the advance would be extremely difficult, and that no hard and fast rules could be provided for that period, but that each situation would have to be met as it arose. He therefore directed that as soon as the plans for the occupation had been drawn up by the Commanding Generals, the officers to be charged with civil administration of the various sectors should be severally attached to the tactical unit which would eventually occupy their sector, and should move forward with it. During the advance, unit commanders were merely to take necessary steps to ensure the safety of their troops, maintain order in the locality temporarily used for their cantonment, and, if necessary, lend support to the civil administration.

Marshal Foch at the same time laid down principles for the guidance of officers charged with the administration of civil affairs. A synopsis of these principles follows.

Laws and regulations in force at the time of occupation were to be respected in so far as they did not affect the occupying power or compromise its security. Requisitions were, however, to be permitted in accordance with the laws and regulations of the Allied armies. It will be noticed that the Allied and not the German system of requisition was followed. This meant, in the case of billets, that the inhabitants were not required to prepare food for Allied soldiers during their stay in a house. Requisitions were to be made only by the commanding officer of the troops in that municipality or by his representative, and then only when such requisition was served upon the chief official of the municipality. By this system, the American government became indebted to the municipality direct, and the municipality was left to settle with its own citizens. By this arrangement, also, the

Allies managed to avoid a great deal of petty bookkeeping. Furthermore, the food supplies necessary for the sustenance of the civil population were not to be requisitioned, to avoid the social unrest produced by hunger. The officers in charge of administration were directed to call on the competent German officials for all information which they might desire of an administrative or judicial nature, and to make use of the libraries and records of public offices and courts.

It was provided that public utilities should be maintained under the strict control of the Allied military authorities. Every effort was to be made to start and encourage the normal peace-time activities of communities; churches, schools, courts, and eleemosynary institutions were to be fostered, as were administrative assemblies, such as chambers of commerce, administrative councils, etc. Rules regulating rates of exchange, financial institutions, collection of taxes, payment of salaries, issues of stamps, bonds, money, etc., were not discussed in these preliminary instructions.

All German civil officials were to be temporarily confirmed in their offices, but provision was made for new appointments, removal from office or transfer, whenever such action was deemed necessary or advisable by the military authorities. Officials were not to be required to serve against their will, but all who served were to be warned that they would be held criminally responsible for misfeasance or neglect in the performance of their duties. The administrative councils and committees were to come together only upon order or under authority of the Allied authorities, and then were to deliberate solely on questions expressly mentioned in the order authorizing the convocation.

Proclamations were to be issued by the commanders of armies or groups of armies, to inform the population of the occupation, and of the duties and obligations resting upon them, and for which they would be held responsible. Police regulations in German, and either English or French, were to be published at once by the officers in charge of civil administration dealing with such matters as the police of billeting areas, travel, sale of drinks, meetings, censorship of the press, photography, carrying and sale of arms, etc. A series of model regulations were drawn up, and permission granted to modify it as circumstances might require.

The Marshal also established in these instructions rules for enforcing the regulations laid down, by means of court martial, and a limit was placed on the latters' powers of sentence. It was also provided that generals commanding groups of armies or armies, and officers in charge of civil administration might order the deportation of persons who had disobeyed these regulations and also order if it should prove necessary, the provisional or final closing of refractory commercial or industrial establishments. It was also provided that convicted Germans might appeal to the various army commanders, and finally to the Allied Commander in Chief.

Marshal Foch at the same time issued very detailed instructions regarding the declaration of residence, which was to be required of all German citizens, and the form of identity cards which they would be directed to carry. In addition, German municipal authorities were directed to keep a list of the names and domiciles of all persons within their jurisdiction. Regulations governing travel were also issued. Assemblage, except upon authorization of the military authorities was prohibited. A censorship was established of press, post, telegraph and telephone. The circulation of false information or of stories which might tend to spread disorder was prohibited. The taking of photographs without a special permit was forbidden. A large number of other rules were also published pertaining to pigeons, arms and ammunition, methods of requisition, restaurants and cafes, etc. All these matters were discussed by the High Command with great attention to detail.

Another appendix to the "Instructions for Civil Administration" dwelt on "The Measures Incumbent upon Commanding Officers of Troops Passing through Enemy Localities not yet Occupied". These measures related to posting of guards and necessary police measures, namely, protection, seizure of important places, public buildings, telephones, telegraphs, records, maps, etc., and those of "Secondary Urgency" which concerned arrest of suspicious character, obtaining of lists of names of men between 28 and 45 who had not been called to the colors, and of horses and vehicles, opening of dove-cotes, supervision of physicians and veterinarians, and publication of sanitary regulations.

Immediately on the occupation of Germany, Marshal Foch in his character of Commander in Chief of the Allied Armies became, of course, the supreme power in the Rhineland, and issued the first proclamation to the inhabitants. (This proclamation is published in Chapter 2).

This proclamation was followed by other proclamations issued by the army commanders under authority given them by the Marshal in his letter of instructions. The subsequent proclamations were based upon the general principles laid down by the Marshal, but promulgated with greater attention to detail and in conformity with the established customs of military occupation as practiced by the respective armies.

It must be admitted that in the proclamations issued by the Army Commanders, the regulations, so exactly laid down by the Marshal, were not always followed. Such action was perhaps unavoidable, as the customs of the various nations taking part in the occupation differed so widely, that a complete unity could not be expected. In some cases no attempt was ever made to enforce the regulations of the

High Command when they lay counter to the national policy of the army in question.

Commissions directly under Marshal Foch.—Paragraph XXXIV of the armistice states that "to assure the best execution of the present agreement, the principle of a Permanent International Armistice Commission is admitted. This Commission will perform its duties under the high authority of the military and naval commander in chief of the Allied armies."

(*a*) *Permanent Inter-Allied Armistice Commission.*—The Commission appointed under this paragraph of the armistice was known as the "Permanent Inter-Allied Armistice Commission" and was charged with the general supervision of armistice conditions and with formal communications to the German government. Seven Sub-Commissions operated under this permanent Commission. The first, for financial and administrative questions, was concerned with the recovery from Germany of the costs of maintaining the troops in occupied areas, under Article IX of the armistice. The second, for restitution of industrial machinery, dealt with the restoration by Germany of confiscated French and Belgian machinery. The third was for return of French and Belgian securities. The fourth, for agricultural implements received from Germany in lieu of rolling-stock. The fifth took over surrendered railway material, while the sixth and seventh dealt with the military equipment and motor transportation which Germany was required by the armistice to surrender.

It being sometimes impossible to apply the laws of war to the conditions in Germany, since, in some cases, the terms of the armistice took precedence and, in others, the views of the Allied and Associated governments differed, Marshal Foch thought it advisable to appoint additional commissions to deal with questions of policy and administration as they arose. Fifteen additional commissions were thus appointed.

(*b*) *Inter-Allied Commission for the Rhineland.*—The most important of these was the *Commission Interalliée des Territoires Rhénans*, commonly known as the C. I. T. R. This Commission was originally charged with the general supervision of occupied territories. It was authorized to appoint commissions to deal with specific phases of the occupation, and to assign to each its duties. Once it had appointed these commissions and assigned them their duties, however, it was no longer in control of their acts or decisions. The importance of the C. I. T. R. lay in the fact that General Payot, a Frenchman and a member of Marshal Foch's staff, was appointed as its Chairman and was put in charge generally of affairs in occupied territories. All the decisions of all the commissions having to be approved by Marshal Foch before they were put into execution, General Payot advised the Marshal in regard to promulgation by the High Command of any order affecting the Rhineland. As chairman of the C. I. T. R., he was expected to consult his fellow-commissioners before making recommendations to the Marshal. Consequently, this commission exercised an influence far greater than was indicated by its charter.

The C. I. T. R. appointed fourteen commissions, which it charged with supervision of the more important phases of the occupation. All matters not specifically turned over to the committees, were reserved by the C. I. T. R. to itself. There were in all, sixteen central commissions of importance, including the Permanent Inter-Allied Armistice Commission and the C. I. T. R.

(*c*) *Inter-Allied Economic Commission.*—The third commission dealing with matters pertaining to the occupation, was the Inter-Allied Economic Commission of occupied territories, commonly known as the Luxemburg Commission. This body was formed on December 13th, 1918, and was charged with supervision of the distribution of raw materials to factories in the Rhineland, with distribution of the products of factories and with regulation of interchange of goods between occupied territories and the rest of Germany. This commission had six Economic Sections operating under it, at Aix-la-Chapelle, München-Gladbach, Cologne, Treves, Mayence and Ludwigshafen,—which carried out the decrees of the main commission.

(*d*) *Inter-Allied Railway Commission.*—The fourth commission, the "Inter-Allied Railway Commission", was to administer the railways of Luxemburg and occupied territories. There were subcommissions at Coblence, Saarbrücken and Luxemburg City.

(*e*) *Inter-Allied Waterways Commission.*—The "Inter-Allied Waterways Commission" was to control navigation on the Rhine. This commission had also a branch at Coblence and control centers at Duisburg and Emmerich.

(*f*) *Road Commission.*—The sixth commission, the "Road Commission", was to provide for the maintenance of roads in occupied territories.

(*g*) *Coal Distribution Commission.*

(*h*) *Commission for Government of the Rhineland.*—The eighth commission—that for "Government of the Rhineland"—was composed of Civil Administrators and Officers in Charge of Civil Affairs.

(*i*) *Receiving Commission.*—The ninth commission—the "Receiving Commission"—was to inspect and receipt for the rolling-stock and raw material handed over by Germany under the terms of the armistice.

(*k*) *and* (*l*) *Calais Railway and Calais Waterways Commission.*—The tenth and eleventh commissions—the "Calais Railway and Calais Waterways Commission"—were to administer the Belgian waterways and railways. These two commissions only incidentally affected occupied territories.

(*m*) *Commission of Postal Control.*

(*n*) *Shipping Commission.*—The Shipping Commission was to deal with established questions pertaining to the turning over to the Allies of German shipping, including river-craft.

(*o*) *Rotterdam Food Commission.*—The Rotterdam Food Commission was established to settle commercial and other details arising from the sale of foodstuffs to Germany by the Allies.

(*p*) *Compiègne Finance Commission.*—The fifteenth commission was the "Compiègne Finance Commission". This undertook to manage financial matters between the Allies and Germany.

(*q*) *Inter-Allied Military Food Commission.*—The sixteenth commission was the "Inter-Allied Military Food Commission". This was created to distribute rations for the Rhineland supplementary to the amounts granted Germany as a whole by the Hoover Commission. The Inter-Allied Military Food Commission accomplished a great deal during the critical Spring months of 1919.

Marshal Foch had appointed these sixteen commissions to deal with such matters pertaining to the Rhineland as he desired, for reasons of national policy or the necessity for uniform control, to influence himself. All such commissions were accordingly to work under his direct supervision. They however failed to achieve the success expected of them, partly because their duties were not clearly defined and partly because, as will be seen later, other persons than the Marshal undertook, on several occasions, to issue instructions to the commissions direct, which naturally caused friction. The relative importance of the commissions also proved to be very unequal. Certain of them accomplished practically nothing, while others became dominant forces in the life of the Rhineland. The most important was without doubt the Inter-Allied Economic Commission of Occupied Territories or, as it was more commonly known, the Luxemburg Commission.

Each commission comprised American representatives, who unfortunately failed at first to maintain adequate liaison with the American Commander in Chief. In order to coordinate their work and obtain a definite control over their actions and decisions, General Pershing, on January 18, 1919, issued instructions to the Commanding General of the Third Army, directing that in future, the senior American member of every Inter-Allied Commission or Sub-Commission which dealt with economic, industrial or financial questions, or any matter affecting the government of occupied territories, should take notes of the proceedings of each meeting. Two copies of these notes were to be sent to G. H. Q., A. E. F., and one copy to Brig. Gen. H. A. Smith, O. C. C. A. in occupied American territory General Smith was directed to estab-

521572°—43——5

lish a bureau to coordinate the actions of all commissions and sub-commissions and to see that the action of the American representative was in every case in accordance with the views of the Commander in Chief. In order to avoid the danger involved in dealing with these subjects by numerous unrelated commissions, the following general principles were laid down.

When a conference was to be held which involved policies or matters tending to commit American Headquarters to a particular course of action, and it was necessary that a final decision should be reached in the conference, the American representative was directed to learn beforehand the opinion of the Commander in Chief on the matter which was to come up for decision. If such could not be obtained, the American representative was to make it clear at the conference that he must refer the matter to G. H. Q. before taking final action. All the American members of commissions or sub-commissions were furthermore directed to keep in touch with the Officer in Charge of Civil Affairs in occupied territory and to act in harmony with the general policies that had been adopted by the American Commander in Chief. A single incident will illustrate the importance of, and necessity for, these instructions. Prior to their issue, the American representative on the Waterways Commission consented in the name of his government to police of the Rhine throughout the zone of American occupation, by a joint British and French river patrol. As soon as the matter was brought to the attention of the Army Command, steps were taken to withdraw such consent and thereby much confusion and possibly friction was avoided. A river patrol under American officers and marines was established, which accomplished this work with entire satisfaction to the High Command and retained, as was entirely proper, the control in the American zone in the hands of our own forces.

Economic control.—Paragraph XXVI of the armistice stated: "The blockade of the Allied and associated powers is to be continued as at present, German merchant vessels found at sea being liable to capture." This clause apparently gave the Allies the right, as well as the power, to continue the blockade on the entire territory which was included within the boundaries of Germany as they existed in 1914. Furthermore, Article 1, Annex No. 2, of the armistice agreement, stipulated that lines of communication, up to and including the Rhine, or on the right bank within the bridgeheads, were to be under full control of the Marshal, Commander in Chief of the Allied armies, who had the right to take all measures he might consider necessary to provide for the security of the armies. This provision entitled Marshal Foch to make such regulations regarding transportation and circulation as were required by the situation. The result of a complete exercise of the right of blockade would have been to isolate Germany from the rest of the world and, further, to isolate occupied territory from the rest of Germany, in fact, this was accomplished only too well during the early days of the armistice. Circulation within occupied territories was greatly impeded by an elaborate and cumbersome system of passes which were required not only between occupied and unoccupied territories, but even between the zones of the different armies.

The last paragraph of section VI of the armistice stated: "There shall be no general measure or official order having as its consequence an impairment of the industrial establishments or a reduction of their personnel." This clause was in reality a contradiction of Section XXVI; the complete commercial isolation of occupied territories and unoccupied Germany would have as its direct "consequence, an impairment of the industrial establishments and a reduction of their personnel", since the industrial establishments could not hope to continue operations with only local markets for their source of raw materials and for the sale of their products. Furthermore, the closing of factories in occupied territories by means of the blockade presented a very serious military problem, for the employes thus thrown out of work would have shortly been reduced to a condition of starvation. Rioting and disorder would have been the inevitable consequence of such action.

The Luxemburg Commission.—This was established to supervise the distribution of raw materials to factories in the occupied territories, to control their output of manufactured goods, and so to regulate the economic relations of occupied territories with the rest of Germany that industrial plants could be kept operating and yet prevented from competing unfairly with similar factories in France and Belgium. It must be remembered that Germany, having almost entirely demobilized her army after the armistice, had thereby obtained a considerable potential economic advantage over her enemies, who were still obliged to maintain great armies in the field.

On December 13th, 1918, Marshal Foch made the first tentative plan for organization of the Luxemburg Commission. Pending its formation, he delegated power to authorize import and export of certain commodities to the president of the Inter-Allied Railway Commission. This authorization was at first limited to import of raw material, fuel, food supplies and spare parts necessary for operation of factories. All exports were forbidden.

The Luxemburg Commission came into being on January 6th, 1919. It was composed of delegates from the American, Belgian, British and French armies and of a French civil official. General Maugas, representing France, was appointed president. This commission was expected to be the supreme authority on economic matters in the Rhineland. It actually acted only from January 6th until May 15th, when it was superseded by the Inter-Allied Rhineland Commission, a civilian body.

Operation of economic control.—The Luxemburg Commission directed the organization of a number of Economic Sections scattered throughout occupied territories. These sections were charged with the duty of carrying out the instructions issued by the Commission, and with the collecting of certain statistical data. Economic Sections were established at:

Aix-la-Chapelle (Belgian)
München-Gladbach (")
Cologne (English)
Treves (American)
Mayence (French)
Ludwigshafen (")

Each Economic Section was given authority to determine upon its own method of work, and consequently there was considerable divergence in the methods employed. The duties of the sections were however very similar, consisting of granting authorization for export and import of certain commodities, and refusing it for others. In the case of a third class of commodities, Economic Sections would not act themselves, but forward requests to the Central Commission at Luxemburg. During the first months of the occupation, separate authorizations were required for each individual shipment, but after April 15, 1919, large industrial establishments were given authorizations covering their monthly requirements. This change of policy, of course, considerably reduced the work of Economic Sections.

The American Economic Section at Treves by informing the Chambers of Commerce of Coblence and Treves of all rules and regulations issued by the Luxemburg Commission and of alterations of policy as they became effective, placed responsibility for proper working out of requests on these chambers of commerce. This method also greatly lessened the work of the Economic Sections and permitted them to keep up a maximum efficiency with a minimum personnel. The Economic Sections of the various armies constantly exchanged statistical data on economic matters.

In spite of the fact that the six Economic Sections were theoretically identical and worked under the directions of the Luxemburg Commission, they soon began to develop certain differences of policy. This was due partly to the different points of view of the Allies, and partly to the fact that a plural control began to assert itself strongly in economic matters.

The Supreme Economic Council.—This was a Sub-Commission of the Peace Commission at Paris. It was originally charged with studying and drawing up the economic terms of the peace treaty.

However, it soon commenced to lay down rules for the economic management of the Rhineland and to give orders to the Luxemburg Commission which, in theory, that organization should have received only from Marshal Foch. Sometimes the Luxemburg Commission carried out these orders and sometimes it ignored them. In still other cases, the Luxemburg Commission disapproved them, while one or more of the Economic Sections formulated their policy in accordance with the Supreme Economic Council. It was therefore natural that slight differences of economic policy arose.

Policy of American military authorities in economic matters.—The tendency of the early decisions of the Luxemburg Commission was to throttle German trade, a tendency which the American military authorities foresaw would result in the complete closing of all factories, if it was pursued to its logical end, since large numbers of workers would be thrown out of employment. Unemployment ever breeds social unrest, which is a condition that a military government always strives to avoid. The American military authorities therefore decided that they should have the final decision in all matters vitally affecting industrial life and commercial activity in the American zone, and not be subject entirely to the instructions of the Luxemburg Commission. This decision was taken because it was believed that the commission was not in so favored a position to judge economic conditions as were the American officers who were assigned specifically to these duties. However, the principles and policies announced by the Luxemburg Commission were carefully followed, whenever they seemed consistent with the responsibility of the Commanding General, and with his desire to keep the population engaged in productive efforts. The results of this decision may be judged by the fact that the number of unemployed in American territory, instead of increasing, shrank from 4073 on January 1, 1919, to a negligible figure on July 1, 1919.

Food.—The second paragraph of the XXVI section of the armistice terms states: "The Allies and the United States are to consider the provisioning of Germany during the armistice to the extent recognized as necessary". Such action proved to be urgently necessary, as the German food situation showed signs each day of becoming worse instead of better. This was due partly to the continuation of the blockade and partly to the fact that the greater part of the agricultural laborers had been in the army for more than four years and that the productivity of the soil had decreased owing to the lack of fertilizers.

The Hoover Commission.—This was a commission originally appointed and presided over by Mr. Herbert Hoover. Its work during the war had been the study, development, distribution and control of the food supply of the world. After the armistice, when it became necessary for humanitarian reasons to ration Germany, this task was assigned to the Hoover Commission, it being the only agency in existence which was really competent to carry it out. The Hoover Commission had in theory nothing whatever to do with any of Marshal Foch's commissions for occupied territories of Germany, since it was a purely civil organization and in no sense connected with the military. In practice, however, since it supplied food to both occupied and unoccupied Germany, it came in close connection with the Allied armies in the Rhineland. The Inter-Allied Food Commission finally came to use bases of the Hoover Commission at Rotterdam and Antwerp as a source from which to draw food.

The Inter-Allied Food Commission.—The food situation in Germany grew more and more serious in the months following the armistice, until it approached a point at which hunger-riots and disorders were to be feared in both occupied and unoccupied territories. The Supreme Economic Council, to avoid this, decided on February 24th, that the food supply of occupied territories of Germany should be considered as distinct from that of unoccupied ones and should be administered under auspices of an Inter-Allied Military Food Commission to be organized by Marshal Foch. This organization was the sixteenth of the Commissions, through which Marshal Foch endeavoured to exercise a personal control in the Rhineland.

While the question of food-supply is treated at greater length in another chapter of this report, it is advisable to speak here of the functions of the Inter-Allied Food Commission, since it was of the greatest importance as an agent of the Inter-Allied Control. An inequality or partiality in the distribution of food by any of the Allied or Associated powers would at this time have created a lasting prejudice in Germany against the nation responsible.

Duties of Inter-Allied Food Commission.—These were: to ascertain the actual food-situation in the various occupied territories; to calculate on a ration-basis, the amount of food thus found; to determine a reasonable ration for the population; to supplement the actual ration by importations, until a reasonable ration was obtained. This last duty required the armies to make arrangements for supply, importation and sale. The distribution of the food, after its arrival, was directed almost entirely by the German authorities, though the Allies reserved the right to supervise the distribution, if they so desired. The source of supply was originally intended to be solely the Hoover stocks which were held by the Permanent Supply Commission at Rotterdam and Antwerp. For several months, however, nothing was received from this source. This was due partly to the fact that the Belgians could not guarantee the safety of German crews in Belgian waters, and partly because the Germans and Belgians could come to no terms for its shipment in Belgian bottoms. Finally, in order to avoid a very grave crisis, the British army was obliged to sell a portion of its reserve stocks to save the population of Cologne from starvation. The French soon did likewise and the United States followed these examples. All such sales were of course from stores belonging to reserve supplies of the armies, and were independent of the Hoover supplies. The various food commissions made their own arrangements with the local German authorities and financiers as to payment, but it was arranged by inter-allied agreement that each army should supplement the German per capita ration in equal amounts.

Inter-Allied Railway Commission.—Section VII of the armistice terms states:

> "Roads and means of communication of every kind, railways, waterways, highways, bridges, telegraphs and telephones shall be in no way impaired.
>
> "All civil and military personnel at present employed on them shall remain.
>
> * * * * *
>
> "Further, the material necessary for the exploitation of the territory on the left bank of the Rhine shall be left *in situ*.
>
> "All supplies of coal and fuel, of material for roadways and for signals shall be left *in situ* and maintained by Germany for the operation of the means of communication in the countries on the left bank of the Rhine."

The Allies, under authority granted them in this provision, took over the waterways and railways with their complete personnel and material as soon as they entered the Rhineland. Means of transportation had necessarily to be kept under Inter-Allied control, because not only their use, but also their physical condition made them of common value to all the occupying armies. The divisions of railways into zones following military delimitations would have been utterly impractical. Furthermore, as transportation is one of the most vital factors in modern war, the Commander in Chief believed it advisable to keep it under his control. He therefore appointed an Inter-Allied Railway Commission soon after the armistice to supervise railway transportation in Luxemburg and the occupied portion of Germany. He also appointed an Inter-Allied Waterways Commission to supervise navigation on the Saar, Rhine and Moselle rivers.

Inter-Allied Railway Commission.—This commission, formed immediately after the armistice, consisted of an American, a Belgian, an English and a French delegate. Lt. Col. Guitry of the French army was appointed president. This commission, which sat at

Treves, promulgated a large number of carefully drawn and highly technical orders, minutely covering every phase of railway activity. Five Sub-Commissions—in Luxemburg, Saarbrücken, Mayence, Cologne and Coblence—were under it. These sub-commissions were charged principally with administrative duties, although they exercised a certain amount of executive power.

In order to carry out orders of the Railway Commission, each Sub-Commission was provided with a more or less technically trained personnel, which was detailed from the army in whose area it was operating, and distributed at the more important railway centers and at points along the boundaries of occupied territories.

German railways, unlike American ones, are publicly owned, the main lines in the Rhine Province being the property of Prussia. For purposes of administration, they are divided geographically into units known as "Direktionen"; for example, the Direktion of Cologne, Direktion of Saarbrücken, etc. Each Direktion has its president and board of directors, who manage it, much as an American railway is operated, although all railway employes are government employes.

The first order of the Inter-Allied Railway Commission requisitioned all labor on all railways of occupied territories, with assurance of proper pay and treatment as long as the workers' duties were properly performed, but threatening severe punishment in case of strikes or other misfeasance.

Duties of Railway Commission.—These fell naturally into two classifications: first, that of attending to operation and maintenance of railways; the second, that of enforcing purely military orders, such as those pertaining to the blockade and to circulation.

Duties under the first classification consisted in supervising German railway officials, planning time-tables, maintaining train schedules, taking necessary measures for prompt execution of repairs, keeping the amount of rolling-stock in occupied territories at the required level and, finally, insisting that a supply of fuel and oil sufficient to last thirty days was kept on hand by the railways. In addition to ordinary administrative duties, careful study was made of the entire railway system in occupied territories, maps were prepared of railways and yards, and plans were drawn up so as to permit the Allies to take over the entire operation of railways in the event of an outbreak of hostilities or of a strike of the German personnel.

Duties under the second classification pertained to the enforcement of purely military measures. At first, before the formation of the Luxemburg Commission, the Inter-Allied Railways Commission promulgated and enforced regulations concerning such commodities as were permitted to pass in and out of occupied territories. It contented itself later with enforcing the rules laid down by the Luxemburg Commission, placing small detachments for that purpose at stations along the borders of occupied territories. These detachments examined all incoming and outgoing consignments, and confiscated those not properly authorized. The Commission issued orders to the German personnel, forbidding them to sell tickets to civilians not provided with proper passes, thus enforcing circulation regulations and, at the same time, making the blockade effective.

Inter-Allied Waterways Commission.—This worked hand in hand with the Railway Commission, to maintain the blockade. It was appointed under the provisions of Article I, Annex No. 2, of the armistice, which stipulates that lines of communication up to and including the Rhine or on the right bank within the bridgeheads, should be under the full control of the Marshal, Commander in Chief of the Allied armies, who was to be given authority to take all measures he believed necessary for the safety of the armies of occupation. This commission was composed, as were all the others, of delegates from each of the Allied and Associated powers. As with the other commissions, its president was a Frenchman. It was charged with the responsibility of regulating navigation on the Saar, Rhine, and Moselle rivers, in the same way that the Railway Commission was to regulate land traffic of railways. As a matter of fact, in practice it concerned itself only with navigation on the Rhine, because the other rivers of the Rhineland are navigable only to a small extent.

German inland navigation system.—Inland navigation prior to the war was practically as free in Germany as it is in the United States. The pressure of war forced Germany to organize intensively her means of transportation, and a department, the Ministry of Public Works was instituted at Berlin for the management of navigable waterways. This department was known as the "Schiffahrtsabteilung". It was intended that it should so manage inland water transportation as to produce maximum results with a minimum outlay of material, always insuring priority to military supplies. This department, which had developed a strong and efficient organization by the beginning of the Allied occupation, had its headquarters for the western group of rivers at Duisburg, at the mouth of the Ruhr.

The Allies, having decided to avail themselves of this organization, moved the headquarters of the commission to Cologne, and thereafter issued all orders to German navigators through its agency.

Control posts for regulating transportation and circulation were established by the Inter-Allied Waterways Commission as follows:

Belgian Control Post at Emmerich
English " " " Ruhrort-Duisburg
English " " " Cologne
American " " " Coblence
French " " " Mayence
French " " " Ludwigshafen Mannheim.

These control posts were charged with the duty of carrying out orders of the Inter-Allied Waterways Commission, which was accomplished by giving clearance papers to vessel-commanders and by organizing and supervising the operations of armed patrol boats. Police, transportation, and circulation regulations along the river were enforced in this manner.

The French and British navies furnished suitable craft for these purposes, but the American control post, on authority of the Inter-Allied Waterways Commission, requisitioned German boats, manned them with marines and armed them with machine-guns. Each nationality patrolled that portion of the river included within its boundaries, but the American control post at Coblence controlled the French patrol-boats between Horchheim and Bingen, in the area which originally intended to form a part of the American zone.

The original duties of the Inter-Allied Waterways Commission were carried out as indicated above. Its duties were later considerably extended by the inclusion of matters pertaining to the transportation and supplies of the United States army.

Use of the Rhine by the American Army as a supply artery.—Soon after the Allied armies had settled in the Rhineland, an effort was made by France to reduce her army to a partial peace-basis. Demobilization was begun and every effort made to assist her factories to resume normal production. One of the chief difficulties encountered was transportation, for the French railways had greatly deteriorated during the war and a large part of the available rolling stock was still retained by the army in order to maintain its supply. The American army was therefore approached by France, and requested to divert as much of its traffic as possible from French railways.

The Rhine, as the time-honored commercial river of central Europe, presented obvious advantages as an artery of supply for the American army. Its use would not only relieve the railways of France, but also reduce the cost of transportation for the American army. It was therefore decided to ship supplies from the United States to Antwerp or Rotterdam, according to the port facilities available, to transfer them from ships to river barges, and tow them up the Rhine to suitable ports within American occupied territory. The ports chosen were Bendorf on the right bank, and Andernach and Coblence-Lützel on the left bank. Large ration dumps were established at Andernach and Bendorf and a quartermaster depot at Coblence-Lützel.

It having been resolved to make use of the Schiffahrtsabteilung, Gruppe West, the Inter-Allied Waterways Commission gave orders to that organization that it should unload all supplies which were to be sent up the river. It was further to provide coal, derricks and stevadores in sufficient quantity to insure proper performance of the work.

The Inter-Allied Waterways Commission merely notified the Schiffahrtsabteilung of the number of barges needed, and the amount and destination of the expected supplies. The Schiffahrtsabteilung, Gruppe West, performed all the work and paid all the costs, including those of labor, merely requesting receipts to show the amount of work done. This system greatly eased the task of the American army, and at the same time was entirely agreeable to the German government, inasmuch as it was thereby permitted to control expenditures which it was in any case required to pay under the terms of the armistice. The first supplies for the American army came up the river on March 23rd, 1919, and continued to arrive thereafter at an average rate of 1216 tons per week at Andernach, and of 1912 tons per week at Bendorf. One shipment of 977 tons of quartermaster stores was sent to Coblence-Lützel. Each port-commander was required to render a daily report of goods unloaded. These reports enabled the Coblence delegate of the Inter-Allied Waterways Commission to verify the amount of work credited to the Schiffahrtsabteilung.

Labor difficulties were experienced early in April, 1919. A large strike occurred at Andernach, possibly encouraged by the authorities of that town in the hope of getting the contract for unloading of barges transferred from the Schiffahrtsabteilung to a local contractor. The only result of the strike, however, was that the strikers were tried before American provost courts and sentenced to ninety days hard labor for conspiring to interfere with the supply of the army. There was also a limited sympathetic strike at Bendorf, which resulted only in the loss of a half day's work. No arrest was made at Bendorf.

It is interesting to note in connection with the work of the Commission, that it required an average of five and a half river barges a week to transport the necessary supplies for the American army, and that the gross river tonnage passing through the American sector of the river averaged about 300,000 tons a week.

When the American combat divisions began to leave Germany for the United States, all heavy ordnance property was shipped down the Rhine in barges, instead of sent by rail through France. Holland permitted this material to pass through her borders, Germany having informed Holland that she would raise no objection thereto.

All river shipping was at first managed from Rotterdam, by the Schiffahrtsabteilung, Gruppe West. Later the Belgians made a strong plea to have it managed from Antwerp. Although Antwerp was almost three days further from Bendorf and Andernach than was Rotterdam, yet its storage facilities were better than those at Rotterdam, and the Belgian government guaranteed that the cost of handling goods from Antwerp would be no greater than from Rotterdam. A change of base was therefore made to Antwerp. This involved using Dutch boats and crews for transportation of supplies, inasmuch as the Belgian government would not guarantee safe conduct to German crews operating in Belgian waters. The Schiffahrtsabteilung, Gruppe West, continued to unload supplies whenever they were carried on German or Dutch boats.

CHAPTER No. 4

ADMINISTRATION OF MILITARY GOVERNMENT IN AMERICAN OCCUPIED TERRITORY

Formative Period of Military Government December 1st–13th, 1918; Outline of Organization Adopted to Permit Exercise of Civil Powers by Military Authorities; Office of Civil Affairs at Advanced G. H. Q.; Third Army as a Unit of Civil Control; The Corps as an Administrative Agent of Military Government; Administration of Civil Affairs by the Division; Military Government within the Division; analysis of Powers and Limitations of Civil Affairs Organization thus Established; Reorganization of Military Government.

Having undertaken by the terms of the armistice to occupy a portion of enemy soil, America could not free herself from the responsibility and obligation of asserting governmental powers thereon. Although there may well have been illusions in certain quarters, and a belief that the occupation could proceed without undue interference with the civil body politic, such convictions, if ever seriously entertained, must have been shortly dispelled. Military government under even the most favorable circumstances, is a formidable task, and although the provisions of the armistice assured the retention at their posts of the German civil officials, other factors, such as the critical food and industrial situation, portended no rosy path for the armies of occupation to tread.

The necessity for the American army to exercise governmental suzerainty in the Rhineland could not but be regarded as an accepted fact, once the higher Command had decided to include a contingent of American troops among the occupying forces. The exercise of sovereign powers, however, no matter how light the touch may be, and how closely confined to a mere supervision of local functionaries, requires the most careful study. Efficient administration has ever required a smooth-running organization especially a chosen and trained personnel, and usually a thorough knowledge of the psychological, economic and political factors which may bear on the task in hand. Thorough organization is as vital to the successful discharge of civil government as it is for the conduct of operations. Training of personnel for bureaucratic duties and education in all phases of local problems will, to a degree at least, be learned as time goes on. An imperfect organization of the form of government, on the other hand, becomes difficult to change, as the weight of custom more and more asserts its influence. To be sure, any outline of a system of administration for an occupied territory, presumes knowledge of its political and economic structure. Scholastic knowledge relating to any nation of the globe, must certainly have been at hand in an army of the size which America had in Europe, and if placed at the disposal of the proper persons, need not have been widely distributed to constitute a basis on which civil machinery could be erected.

All of the energy of the American army had been centered on an early decision in the field and there had been no opportunity to study the civil problems involved in an occupation of German territory. The American army therefore began its duties in occupied territory with only the scantiest information both of the particular situation confronting it and even of a broader nature, such as would permit it intelligently to frame an organization commensurate with its wide governmental powers. From the beginning therefore there was a crying need for personnel trained in civil administration and possessing knowledge of the German nation.

It is extremely unfortunate that the qualifications necessary for a civil administration are not developed among officers in times of peace. The history of the United States offers an uninterrupted series of wars, which demanded as their aftermath, the exercise by its officers of civil governmental functions. Despite the precedents of military governments in Mexico, California, the Southern States, Cuba, Porto Rico, Panama, China, the Philippines and elsewhere, the lesson has seemingly not been learned. In none of the service-schools devoted to the higher training of officers, has a single course on the nature and scope of military government been established. The majority of the regular officers were, as a consequence, ill-equipped to perform tasks differing so widely from their accustomed duties. There were a considerable number of officers, largely of high rank, who had received a valuable, practical training in Cuba and the Philippines. There were probably also many reserve officers with administrative experience along similar lines, either in the government service, or in large industrial enterprises. However, a very limited number of trained regular officers was assigned from the former group to the occupying forces, and none at all from the latter group.

With the signing of the armistice, the prospective occupation became a real factor in the situation, and the problems to which it gave rise could no longer be evaded. On November 11th, 1918, a month still ensued before our armies would reach German soil,—a period sufficient to at least lay a solid foundation for the future military government. The magnitude of the responsibilities assumed, appear to have been greatly underestimated. There can be no doubt that the belief, felt in many quarters, and to which reference has been made, that the armies could occupy enemy soil and yet divest themselves of the responsibilities of government, was both prevalent and powerful. It is also not outside the range of possibility, that the Allied negotiators at Compiégne, were themselves a party to this illusion when they formulated the terms on which hostilities were to cease. The military situation, serious enough when we consider that G. H. Q. was at this time faced with the problem of moving 300,000 men to the Rhine, and the uncertainty in regard to the nature of the occupation, were no doubt responsible for the American failure to prepare for the task at hand. The failure, however, laid a heavy burden on the shoulders of the Third Army Commander and his subordinate officers, charged with the security of their several commands. In all this force, with the exception of perhaps a half-dozen men, there was probably no one who had the faintest conception of the German governmental system, of its functions, limitations or channels of communication. The Second Section of the General Staff at G. H. Q. had, it is true, in November, prepared a pamphlet dealing with this subject, but its material was antiquated and its treatment inaccurate. More from the lack of other textbooks than because of inherent value, this pamphlet did prove of some use, and during the march to the Rhine constituted for both staff and unit commanders their sole source of information in regard to the government of the country which they were occupying.

The conclusion from these facts is incontestable; the American army of occupation lacked both training and organization to guide the destinies of the nearly 1,000,000 civilians whom the fortunes of war had placed under its temporary sovereignty.

Formative period of military Government: December 1st–13th, 1918.—The preparation for military government undertaken by General Headquarters, prior to the crossing of the Luxemburg-German frontier by the Third Army, was limited to the following:

(a) The publication and distribution among the Third Army of the pamphlet referred to above, on German governmental organization;

(b) The appointment of Brig. Gen. H. A. Smith, as Deputy Chief of Staff for Civil Affairs in occupied territory. This officer was later given the title of Officer in Charge of Civil Affairs in Occupied Territory and placed at the head of a separate office of the Advance General Headquarters when they were set up at Treves;

(c) The appointment of Colonel I. L. Hunt, Infantry, as General Smith's assistant. When Third Army Headquarters was moved forward to the Rhine, it was deemed advisable by General Smith to have an officer to represent the civil administration with that body. Colonel Hunt was therefore attached to the staff of the Commanding General, Third Army, as an advisor in civil matters, and was shortly thereafter appointed its Officer in Charge of Civil Affairs.

Work had also been begun by G. H. Q. on a proclamation called "Anordnungen" (ordinances), containing regulations for the guidance of the civil population of the occupied territory. This, however, had not been completed on December 1st, 1918, the day on which the Army began its forward movement from Luxemburg onto enemy soil.

The officers appointed by G. H. Q. to organize military government in occupied territory, were therefore placed in a difficult position. Without information, either in regard to the local situation or the future code of laws being drawn up at Headquarters, they yet had to devise temporary expedients, which might continue to tide over the situation until the army eventually settled down in what were to be its permanent billeting areas. It is not proposed to repeat the story of military government during the march to the Rhine in this chapter. The provisional policy adopted by the army towards the civil population was outlined in Memorandum No. 4, Third Army, November 30th, 1918, and was issued by the Army Commander after consultation with General Smith. The method of enforcing the rules therein laid down was purposely left vague, as it was desired to leave as much authority as possible to Corps and Division Commanders and it was believed unwise to anticipate action by G. H. Q. This memorandum further made no attempt to organize or outline any proceedure of military government. This task was left to the future, when the local situation could be studied, and a system outlined on the basis of local investigations and the wishes of the Commander in Chief.

Outline of organization adopted to permit exercise of civil powers by military authorities.—Orders No. 1, Advance General Headquarters, Treves, December 13th, 1918, became, together with General Pershing's Anordnungen, the two fundamental charters of the military government. The Anordnungen announced the regulations which were to govern the relations of the civil population to the occupying army. Orders No. 1, on the other hand, announced the organization of the military government which was to make these regulations effective. The system of military government therein established, together with its later modifications, are the subject of this chapter.

Two fundamental principles which profoundly affected the whole course of military government were announced by G. H. Q. in Orders No. 1.

1. *Establishment of an authority in occupied territory, directly representing the Commander in Chief, and with full power, subject to his wishes, to promulgate the orders of the military government and to supervise their execution.* This office was entirely independent of Third Army Headquarters at Coblence, and therefore in a very real sense, limited the authority of the Commanding General of the latter force. That discord did not arise from such an awkward division of authority,

must be ascribed largely to the spirit in which the several headquarters subordinated personal matters to the public welfare.

2. *Use of tactical formations of the Third Army as units of military government.*—In this connection, looking at it purely from an historical standpoint, the American army committed its first grave mistake, and one which it found impossible to rectify until June of the following year. The mistake was due entirely to inability, in the time at our disposal, thoroughly to grasp the importance of the German governmental system and the facility with which our military government could be adopted to it. German government is more minutely and carefully organized than in any other country in the world. The Provinces, Regierungsbezirke and Kreise all offered a civil hierarchy to which a foreign military government could easily be adapted. Each German political sub-division was therefore provided with its civil commander and its quota of officials. The advantages of organizing the military government along already established civil lines was, however, recognized too late. The decision to use tactical formations as units of control had already then been adopted, and the system so clearly established, that the resulting confusion would not have justified the change.

Both the French and British armies, possibly due to their better knowledge of German local government, never committed our mistake, and modelled their military government from the outset along lines paralleling the civil system.

Office of civil affairs at advanced G. H. Q.—The Office of Civil Affairs at Advance G. H. Q. was designated by the Commander in Chief as the fountain-head of civil authority within the American zone. Its chief, the Officer in Charge of Civil Affairs in Occupied Territory, stood within American boundaries, in relation to the temporary sovereign powers of America, as the direct representative of the Commander in Chief. His authority and duty was of a threefold nature. In a legislative capacity, and after approval of his decisions by the Commander in Chief, he drew up the ordinances, which were to govern the civil population during the occupation. These were issued, in first instance, as a part of the Anordnungen, later, as "Orders, Advance G. H. Q." These ordinances outline the methods by which the American army was to make the military government effective.

Another primary duty of the Officer in Charge of Civil Affairs was that of supervising the execution of orders already issued. It should be borne in mind that the officers charged with the execution of these orders were the unit commanders of the Third Army, who in all other matters were directly responsible to the Commanding General. Such supervisory powers as G. H. Q. cared to assert, has to be exer-

521572°—43——6

cised through the medium of the Army Commander. The alternative was considered, but discarded, of having G. H. Q. directly supervise subordinate commanders of the Third Army in their conduct of civil affairs. It was, however, not believed feasible to place officers in the embarrassing situation of serving two masters. Advance G. H. Q. therefore decided that it was impracticable for it to assert its supervisory power below the Army Commander. As that officer was directly charged with the security of troops, it ought to be exercised in his case only with the utmost discretion. The opportunity to have the same authority issue and execute the ordinances was lost, when it was decided to utilise the tactical formations of the army as units of civil control. Thereafter, supervision of the control of civil affairs in lesser units was exercised by the army commander, through his Officer in Charge of Civil Affairs.

A third field of activity of the Officer in Charge of Civil Affairs, though of less importance lay in the actual administration of the city of Treves, which was placed directly under Advanced G. H. Q.

The powers of legislation and promulgation still residing in G. H. Q. were sufficiently important and numerous to require an extensive organization. Its office organization and, to a certain extent, the whole machinery of civil administration, was outlined in Orders No. 1, Advance G. H. Q., December 13th, 1918.

Five main departments were thereby established:

1. Public works and utilities;
2. Fiscal affairs;
3. Sanitation and public health;
4. Schools and charitable institutions;
5. Legal department.

This organization was essentially theoretical, destined to meet any situation which might in future confront the American army in its government of occupied territory. Its comprehensiveness hinted at a definite reversal from the theory of non-interference which had existed prior to actual occupation. In fact, the veer was more pronounced than actual circumstances warranted, and several of the departments never came to function in a more than nominal sense. Nor was it found possible to lay down hard and fast limits of competence for the other departments. Problems arose of which the army had never dreamed, or, if it had, believed could be shunted on the shoulders of the civil officials. The above outline of office organization can therefore be better understood if considered as a temporary expedient, to be tested and altered in the light of practical experience.

The Department of Public Works and Utilities was expected to supervise the municipal light and gas plants, street-railways and telephone lines with the American zone. American jurisdiction over the railways was limited by the superior authority of the Inter-Allied

Railway Commission, of which, representatives of all the Allied nations taking part in the occupation were members. American control of other public utilities was also limited by the lack of coal, and by an Inter-Allied Commission, with sole authority over the production and distribution of coal.

When the Department of Fiscal Affairs was created, it was believed that supervision would be necessary over banks, financial institutions and German taxation. Such supervision was found to be advisable only to a very limited degree in the case of banks, and never in that of civil taxation. This department, however, was charged at a somewhat later date, with the receipt of the fines imposed by our provost courts, and with the accounting for them to the Treasury of the United States. This proved to be a considerable task, as the amount of fines often ran over 100,000 marks weekly.

The officers of the Department of Sanitation and Public Health were selected from the Medical Corps. This department of the Office of Civil Affairs was charged almost immediately with an extensive and vital work, particularly necessary in view of the conditions of German villages, crowded often to double their normal capacity, with an uninvited occupying army, thus facilitating the spread of disease. The results of the work of the department, which was duplicated in the civil sanitary sections of Army and Division Officers of Civil Affairs, has been made the subject of a special chapter. To the other duties of the department, were later added the many problems concerning civilian food supply. The very critical situation in Germany during the winter of 1918–1919 rendered American relief action advisable, especially as the army was charged with the maintenance of public order and the security of troops.

The functions of the Legal Department covered two separate fields. In the first instance, it exercised general supervision over all provost courts, and served as a repository for the very numerous court-records. In the second instance, the Legal Officer served as a technical advisor to the Officer in Charge of Civil Affairs, particularly in drafting the rules governing the operation of provost courts. Later in the occupation when a certain amount of supervision was found to be necessary over the German civil courts, this, too, became a duty of the Legal Department. The department was duplicated in the Office of Civil Affairs of the Army.

An additional department was shortly afterwards added, termed the Economic Section. This grew out of the decision of the Luxemburg Commission, mentioned elsewhere, which required licenses for the export of all merchandise from occupied territory. This necessitated considerable personnel and regulating machinery, and a separate department was deemed necessary to deal with the volume of business. It never, however, attained the importance of the corresponding offices in the armies of our allies. The Luxemburg Commission imposed many restrictions on the German dye-industry, but as the American zone contained no dye-factory, the Economic Section was relieved of its heaviest responsibility.

It would be expected that the Officer in Charge of Civil Affairs in Occupied Territory, considering the far-reaching powers accorded him, would have retained the exclusive right to give directions to the chief civil official of occupied territories, viz., the Oberpräsident of the Rhine Province. Such procedure was, however, prevented by geographical reasons. Treves, which was the seat of Advanced G. H. Q., was a relatively important governmental town, but in no sense comparable to Coblence, which was the capital not only of the Rhine Province, but also of Regierungsbezirk Coblence. Coblence was the Headquarters of the Third Army, and it seemed entirely proper for the Officer in Charge of Civil Affairs, Third Army, to transmit to the civil hierarchy the orders and decisions promulgated by the American authorities. It therefore was inevitable that this office, more closely in contact with the high civil bureaucracy than any other, would, in German minds at least, come to occupy a predominant position.

Third Army as a unit of civil control.—It would be well to emphasize here that in speaking of the Third Army as a unit of civil control, reference is alone made to its position during the period December 9th, 1918, to June 1st, 1919, when it served in a subsidiary capacity to Advance G. H. Q. During the two weeks prior to December 9th, 1918, on which day both the Anordnungen and Orders No. 1, were published, the army maintained an entirely independent position, with full powers of military government. After June 1st, 1919, it resumed complete civil authority, Advance G.H.Q. having been on that date dissolved.

Inasmuch as Orders No. 1 failed to outline duties for the Army Officer in Charge of Civil Affairs, contenting itself with his appointment, the growth of the office became in large measure a question of expediency. That it perforce assumed the important position which it did in the American military government of Germany, was due to circumstances outside the scope of knowledge of the authors of Orders No. 1.

There were, at the outset, four fields of activity to which the Army Office of Civil Affairs was compelled to devote its energies:

1. In an advisory capacity to the Commanding General, in relation to his powers of military government. These were at first presumed to be of small importance, since with the exception of the Commanding General's power to appoint military commissions, the initiation of measures was to be reserved for Advance G. H. Q.;

2. In a capacity similar to that of the Officer in Charge of Civil

Affairs of a Division, in regard to territory occupied by Third Army troops. With the appointment of a Garrison Commander for the area, the control of Civil Affairs therein was removed from the direct jurisdiction of the Army and from its office of Civil Affairs;

3. In an advisory capacity to the Officer in Charge of Civil Affairs in Occupied Territory, presenting to him the practical problems confronting the army and suggesting solutions;

4. As a pure and simple office of transmission between smaller tactical units and the Officer in Charge of Civil Affairs in Occupied Territory.

These were in no sense activities which foreshadowed its evolution into the important office which it later became. New responsibilities, however, defined themselves almost as soon as Headquarters were fairly established in Coblence.

Actual supervision over the execution of American orders was performed, in the first instance by the commanders of towns and villages. These officers were in no sense members of any separate hierarchy of military government. They were primarily unit commanders, to whom their civil duties were incidental. For the security of their troops, and in all but civil matters, their channels of responsibilities converged in the person of the Army Commander. In practice, the Office of Civil Affairs at Advanced G. H. Q. found that it could supervise the execution of its orders only through the Army Commander. This officer, therefore, having been made responsible to the representative of the Commander in Chief for the execution of the orders, had perforce to supervise the control of Civil Affairs by his own subordinates. It seemed therefore only natural for the Army Commander to utilize his Officer in Charge of Civil Affairs as his representative in such matters.

It is to be noted, however, that in apportioning the supervisory powers of the military government, the Officer in Charge of Civil Affairs did not exercise control over civilian circulation, which was turned over to the Second Section of the General Staff, as was censorship of post, telegraph and telephone. Other matters were assigned to G–1 and G–4. While the Office of Civil Affairs at Advanced G. H. Q. thus had legislative powers on all subjects, the supervisory powers of the Office of Civil Affairs in the Army were by no means analagous thereto.

The distance and lack of contact between Advanced G. H. Q. and town and village commanders also tended to place responsibilities on the shoulders of the Army Commander. Orders are seldom free from misinterpretation, and in view of the general ignorance at first displayed by the majority of officers of military government, it was not unnatural that enquiries and requests for explanations constantly reached the Army Commander. Assumption by the Army of the responsibility of interpreting the orders of Advanced G. H. Q. was the logical consequence of this situation. These interpretations appeared as Letters of Instructions, Third Army, and although prepared by the Office of Civil Affairs, were issued over the signature of the Chief of Staff. They embraced a wide variety of subjects—rate of exchange, public health, provost court procedure, traffic regulations, etc. While principally confined to the interpretation of regulations promulgated by Advanced G. H. Q., they occasionally made use of the authority, vested in the General to provide for the security of his troops, to promulgate new regulations. Thus, in Letters of Instructions No. 19, April 12th, 1919, the Army published a series of traffic rules, which was a matter covered by none of the orders of Advanced G. H. Q. In Letters of Instructions No. 7, it adopted a similar policy, by forbidding the sale by civilians to soldiers of habit-forming drugs. Both these matters were of such a nature that orders were necessary to protect the troops.

Had a strict interpretation of Orders No. 1 been adopted, the Army Office of Civil Affairs would have confined its actions in provost court procedure to the forwarding of records to Advanced G. H. Q. The power to approve, disapprove or mitigate sentences imposed by such courts was exercised in first instance by the officer appointing the courts,—Division commanders in the case of all superior provost courts, and town commanders in the case of inferior courts. It was contemplated that ultimate approval or disapproval of the findings would be expressed at Advanced G. H. Q. Because, however, of the time which would elapse before cases were finally settled, the O. C. C. A. in Occupied Territory delegated this power of ultimate review (Orders No. 14, Advanced G. H. Q., March 5th, 1919) to the Third Army. It should be mentioned in this connection that the records of military commissions were handled in a very different matter, and did not come under this ruling.

Possibly the most influential reason for the increase of the authority of the Army in military government, was the decision of Advanced G. H. Q. to leave all matters pertaining to the Oberpräsident of the Rhine Province to the Army Commander. All correspondence with this high official was thenceforth carried on by the Officer in Charge of Civil Affairs, Third Army. This proved to be a factor of vital importance. The Oberpräsident was by far the most important civil official in any part of occupied territory. His authority extended over the entire area occupied by the American army, with the exception of Kreis Montabaur, while not long after American troops had occupied the bridgehead, this district too, was placed within his jurisdiction. The German government had been quick to realize that Kreise detached by artificial boundaries from their normal superiors, could not be left entirely on their own responsibilities. The authority of the

Oberpräsident among civil officials was also supreme throughout the entire British and Belgian zones and a large part of the French zone. An order transmitted to him would reach every Bürgermeister and Gemeindevorsteher in the American area, and offered the surest and quickest method of disseminating information. This method of transmitting orders was therefore adopted, both in regard to the "Orders, Advanced G. H. Q." and the Letters of Instructions". On his part, the Oberpräsident always had a large number of queries, petitions and suggestions to submit to the Army Commander. He was required to submit these direct to the Officer in Charge of Civil Affairs, as the channel of communication for reaching the Army Commander. As a result of this procedure, the Office of Civil Affairs of the Army came to be looked upon by the civil population as the real source of American authority. The duties and responsibilities of the office, by a natural process of evolution, had become both numerous and important, and the Chief of Staff kept it under his own personal direction.

Organization of the Office of Civil Affairs developed as its duties expanded. An organization for the Office of Civil Affairs in Occupied Territory, had been outlined in Orders No. 1, which, when modified in accordance with the less numerous duties of the Army office, might serve as a model as these duties expanded. Experience had already taught, however, that there was no need, even at Advanced G. H. Q., for Department 4 (Schools and Charitable Institutions). With a few exceptions, all questions pertaining to the Department of Fiscal Affairs (Accountability for fines, bank inspections, etc.) were entirely handled by G. H. Q., and the few matters on this subject which came within the jurisdiction of the Army did not warrant the creation of a special department. The many problems affecting Public Utilities, which at Advanced G. H. Q. were handled in the Office of Civil Affairs were in the Army turned over to the Office of the Chief Engineer, functioning under G-4.

The internal machinery of the Office of Civil Affairs, Third Army, was therefore neither so large nor so complicated as that of its prototype at Treves. There were, under the section chief, an adjutant (at first designated as "Secretary," later as "Executive Officer"), and a number of departments.

These departments were, but two in number in December, 1918, but were eventually increased to four. The original departments,—Legal, and Sanitation and Public Health, were created to perform duties within their spheres along the same lines as those performed in the analogous departments at Advanced G. H. Q. Lt. Col. Hugh L. Ogden was transferred from the Judge Advocate's Office of the 42nd Division, to head the newly created Legal Department of the Office of Civil Affairs. Lt. Col. Walter Bensel, a medical reserve officer with long experience in the Department of Health of New York City, was placed in charge of the Department of Sanitation and Public Health. The third department created corresponded in no way with any section of Advanced G. H. Q. Correspondence between the Office of Civil Affairs and the higher civil functionaries increasing in volume week by week, it became absolutely necessary, not only for the American army to keep in close contact with the work of the Oberpräsident and the Regierung, but to investigate and supervise the execution of American orders by those bodies. The officer placed at the head of the department charged with this, was in reality a liaison-agent, but in view of the authority vested in him over civil officials, this term was not used, and he was given the title of "American Representative with the Oberpräsident of the Rhine Province." The fourth department added at the end of March was termed the Historical Section.

Although the several departments are treated at length in other parts of the report, a brief outline of the duties performed by each during the period of December, 1918, to June, 1919, is not out of place at this point.

Legal Department.—Review of the records of all cases tried by the provost courts, had, as has previously been mentioned, been made the duty of the Army by Advanced G. H. Q. These duties were now delegated to the Legal Department of the Office of Civil Affairs, its decision, in case of reversal of findings, being promulgated by the Commanding General.

Even more important were the functions of the Department in advising the Commanding General and the O. C. C. A. in regard to legal aspects of the occupation. There were always a large number of questions referred to the Office of Civil Affairs for solution, which demanded a legal study of the provisions of international law and of the Armistice agreement. Mistakes could not be made, since Germany after the ratification of peace would have legal means of seeking redress. Among the more knotty of these questions were those of abandoned enemy war-material and the method of paying requisitions and claims. In view of the constant allegations on both sides during the war of violations of international law, (thus showing a great advance in international public opinion), great care was exercised to see that all our acts were justified either in international law, or by the terms of the armistice, and it is not believed that a single official act can be alleged which can be construed as in contravention of either.

Department of Sanitation and Public Health.—As the health of American troops would be affected by any epidemic or disease among the civil population, the army could not divest itself of responsibility in what was primarily a civil matter. Supervision of the civil sanitary affairs was found to be essential. The problem was too large and too important to be dealt with by a single office, such as that of the Chief Surgeon, and all matters pertaining to public health were therefore referred to this Civil Affairs department. This brought the subject of prostitution under the jurisdiction of the Office of Civil Affairs, which was a matter requiring serious study before even an attempt could be made to lessen its evil effects.

A hardly less important duty of this department was the investigation of civilian food conditions. This finally brought to light such overwhelming evidence of the crisis confronting the population, that the Army undertook to supply a certain percentage of their needs. Later, when deliveries actually took place, we undertook both to assure a fair system of distribution and to prevent profiteering. Necessary supervision over both these matters was exercised by this department, acting in cooperation with the supply departments of the army.

Liaison Department.—The duties of this department were so inextricably confused with the supervision of the entire German civil administration, that they will be treated in the chapter dealing with that phase of the occupation.

It must be noted that the organization here outlined continued only until June, 1919. With the dissolution of the Office of Civil Affairs, Advanced G. H. Q. about June 15th, a complete reorganization of the conduct of Civil Affairs within the Army took place. This reorganization will be discussed in the latter part of this chapter.

The Corps as an administrative agent of military Government.—In an administrative capacity, and in a strictly military sense, the Corps does not play so important a role as either the Army or the Division. Its field of activity lies largely in strategy and tactics, for in administrative matters pertaining to its Divisions, it merely transmits the correspondence of higher and lower units. Orders No. 1, Advanced, G. H. Q., directed the appointment by each Corps Commander of an Officer in Charge of Civil Affairs, but outlined no duties for him to perform. The ability of the Corps to supervise the administration of military government among the Divisions was limited usually by the distance separating Corps headquarters from Division headquarters. Corps Commanders therefore contented themselves with appointing an Officer in Charge of Civil Affairs and permitting the Army to supervise the work of their Divisions, either directly or through them. The officers appointed were in two instances Corps Judge Advocates, who were expected to conduct matters pertaining to civil affairs in addition to their other duties.

There were within the Corps a certain number of units unattached to Divisions and serving directly under the Corps Commander. These units, in all cases, were billeted in compact areas,—the Third Corps troops, in that part of Landkreis Coblence lying on the right bank of the Rhine, the Fourth Corps troops around Cochem, and the Seventh Corps troops in Kreis Wittlich. It was in the administration of these areas that the activities of Corps Officers in Charge of Civil Affairs principally centered, and their authority and duties in connection therewith were similar to the authority and duties of the Division O. C. C. A. in respect to the Division area.

Aside from the administration of the Corps troops area, the O. C. C. A. of the Corps played no other important *role* in our administration of military government, merely serving to delay matters by forcing both Army and Division to transmit communications through his office.

Administration of civil affairs by the division.—Although the influence of the Corps on American military government of occupied Germany never assumed important proportions, the Division Commander, through his O. C. C. A., became the real administrator of local affairs under the armistice. The provisions of Orders No. 1 not only charged the Division Commander with responsibility for the enforcement of military orders in his area, but endowed him with very extensive control of the judicial organization set up to enforce them.

In respect to Military Commissions, his authority was paramount, subject only to that of the Commander in Chief. If he deemed an offense by a civilian to be sufficiently grave, he could convene a Military Commission, approve its sentence and forward it to G. H. Q., without subjecting it to review in either Corps or Army Headquarters. This, however, was a mistake, as the trial of civil officials frequently involved questions of policy regarding which the Army Commander should have been consulted. He also was given the power to appoint Superior Provost Courts, and to approve, disapprove or mitigate the findings of such court, though the records in all Provost Court cases were forwarded to the Army Commander, who has the right to express a final opinion.

The Officer in Charge of Civil Affairs for the Division became a very important member of the Division Staff, occupying a position equivalent to that of the Chief of a General Staff Section. Unfortunately, however, Orders No. 1 did not prescribe that the O. C. C. A. should be an additional member of the Division Staff, and when he was not so assigned by the Division Commander, the results were far less satisfactory than when these important duties were assigned to an officer who could give his exclusive attention to them. His duties required him in first instance to supervise the work of the various Civil Affairs officers of the towns and villages, and to act as an advisor to the General in all Civil matters. Advance G. H. Q. had never contemplated Division Commander deciding questions of policy on their own initiative, or even asserting an "interpretive" power analogous to that of the Army. It was perhaps to be expected, that a Division Commander, being charged with the security of the officers and soldiers of his Division, should at times attempt to exercise such authority. In one instance, a Division Commander, being confronted with a particularly unruly civilian element within his district, instituted an earlier closing hour for public houses than that announced by the Army. Although, from a local viewpoint, this action may have appeared justifiable, it could not in a broad sense but have wrought havoc on the Army's attempt to maintain that unity of policy so desirable in a military government. The order of the Division Commander was subsequently revoked at the wish of the Army Commander. In all cases thereafter, Division Commanders were expected to obtain the consent of the Army before issuing orders to the civil population.

By his almost complete control over the Provost Courts, the Division Commander wielded immense authority within the area occupied by the Division. His *régime* could be either strict or lenient, as his personal feelings dictated. He could have either enforced the regulations of higher authority, in a manner harsher than intended, or erred on the side of leniency and thereby weakened the prestige of the American army. That in every case, Division Commanders grasped the spirit which the Commander in Chief intended to infuse into the occupation, is greatly to their credit, and accounts in no small degree for the feeling which grew up among the civil population, that the American army was just, humane and disinterested.

Military Government within the Division.—Orders No. 1, Advance G. H. Q., prescribed that, for the conduct of the administration of civil affairs within Divisions, the Commanding General should detail officers, preferably the commanding officer of the troops therein billeted, to be in charge of each occupied town. Towns not garrisoned were to be inspected and regulated by officers appointed for that special duty.

These provisions of Orders No. 1 were of such a broad nature, that they gave opportunity to the Division Commander to constitute either an elaborate or simple system of military government. Advanced G. H. Q. made no provision for any intermediate Office of Civil Affairs between the Division and the town garrison. In a number of Divisions, such an intermediate organ was, however, organized. The 89th Division, garrisoning the Treves region, which included Kreise Prüm, Bitburg, Saarburg and Treves (Land), occupied a very extensive area. It proved impracticable for the Division Office of Civil Affairs to make its influence felt directly in each town and village. The three brigades of the Division were billeted in areas roughly corresponding to German Kreise. It therefore was only natural to appoint an Officer in Charge of Civil Affairs in each Brigade, who, by reason of his location, could more closely observe the work of the town commanders. These Brigade Headquarters being at Kreis capitals, the Brigade O. C. C. A. could also supervise the work of the Landräte and their officials, whose jurisdiction was approximately the same as the area under control of the Brigade. In the Third Division area, which corresponded with the borders of Kreis Mayen, the Division Commander divided his area into districts which corresponded with the boundaries of the rural townships (Landbürgermeistereien). These district O. C. C. A.'s served on the staff of the Regimental or Brigade Commander, whose troops were garrisoned in the vicinity. The Officer in Charge of Civil Affairs was in no sense independent of the tactical units, and his relations with the Division O. C. C. A. were carried on only through military channels.

In no Division was an attempt made to set up a special Civil Affairs organization, independent of military channels of responsibility. To the officers of the army as a whole, the occupation came as a relief from the activities of the western front, and as an opportunity to train the thousands of replacements which had been put into their battle-scarred units. There was little or no interest in military government in its strict sense. Very few anticipated a lengthy stay on the Rhine, and all seemed willing to allow Civil Affairs to get along as best they could, on condition billets were provided and requisitions complied with.

The Office of Civil Affairs in the towns and villages was usually made an appendage of some staff officer of the Battalion or Regiment. In case a special officer was selected for these duties, he could seldom devote the greater part of his time to them. A knowledge of German was usually considered the only requirement for the office, and except for occasional conferences, the Division O. C. C. A., overloaded with work himself, had no opportunity to give such an officer the special training so necessary for his civil duties. By far the larger number of injustices committed by the American army can be traced to the ignorance of inexperienced O. C. C. A.'s. and Provost Court officers, who suddenly found themselves in responsible and unaccustomed positions. Punishments, totally out of proportion to the crimes involved, were meted out by the courts, and could be reversed only at a later date, after an impression of undue harshness had already been created among the civil population.

Another serious handicap to efficient administration arose from the constant transfer of troops from town to town within the Division area. As the O. C. C. A. and the Provost Court were primarily troop officers, even transfer of a unit resulted in the organization of an entirely new *regime*. This method proved as inefficient in the long run for the military government as it was distressing to the civil population. O. C. C. A's had no sooner left their community than they found themselves set down in entirely new surroundings, with all the knowledge of the local situation, gained by weeks of investigations, to be learned again. It must be admitted that due to lack of previous preparation and appreciation of the problems facing a Military Government, the structure reared by the American Army at this state was far from homogeneous, and violated a number of fundamental principles which were later realized to be sound.

Analysis of powers and limitations of civil affairs organization thus established.—It is at best difficult to define and bring into harmony the functions of the various Officers in Charge of Civil Affairs. So intermingled did the military government become with the ordinary administration of the army, that the duties and powers of an O. C. C. A. in any unit were seldom analogous to the duties and powers of the O. C. C. A. in the unit immediately superior or inferior. Some of the discrepancies, however, ought to be carefully pointed out, if a clear understanding of the American military government in Germany is to be obtained. This is particularly necessary if America is to avoid the same mistakes in a future military government.

The legislative powers of military government withheld from the Army Commander who was charged with maintaining public order and with the security of troops.—During an actual state of war, or in the expectation of a renewed state of war, the first duties of any military government must be to protect the occupying army. Peace was not so certain in the winter of 1918–1919 that we could afford to neglect precautions against either renewed attacks or an uprising of the population in our midst. It had been contemplated that legislative power should be lodged in the Office of Civil Affairs at Advanced G. H. Q. The Commanding General was on the other hand to be directly responsible to General Pershing for the maintenance of public order within the American zone. Separation of executive and legislative functions, while a dogma of our government, is wholly unsuited to a military occupation. One outbreak of violence by the civil population, and the structure so carefully erected will collapse, legislative power passing automatically to the commander responsible for public order.

The difficulties of securing adequate liaison between the Commanding General, Third Army, at Coblence and the O. C. C. A. in Occupied Territory, at Treves, separated by two hundred kilometers, was also found to be a handicap. These factors are sufficient to account for the decision of G. H. Q. to permit a limited legislative power to Third Army Headquarters. This has been referred to in these pages as "interpretive power".

Advanced G. H. Q. without authority over rail and water communications.—It was unquestionably necessary to make the control of communications of occupied territory an Inter-Allied matter. The Rhineland was economically a unit, and the creation of independent railway-systems in each Allied zone would have been disastrous to the revival of its industry, as well as injurious from the view-point of Allied strategy. The supervision of railways and waterways was exercised by Inter-Allied Commissions, serving directly under Marshal Foch, on which, each of the occupying armies was represented. These Commissions served as a personal link between the Inter-Allied Commander in Chief and occupied territory, permitting the Marshal to make his influence felt in any of the occupied zones. Advanced G. H. Q. maintained liaison with the American members of these Commissions, informing them of the views held by General Pershing, which they were expected to present to the Commissions.

While these Commissions limited the powers of military government ordinarily possessed by an invading army, their creation was yet an unavoidable necessity. Their work throughout the occupation, despite the difficulties incident to their international character, proved entirely satisfactory from the viewpoint of the American military government. In any future military government, separation of control of railways from the authorities charged with maintaining order, might well prove disastrous.

All other O. C. C. A.'s merely advisors of Unit Commanders, without direct channels of communication with the O. C. C. A. in Occupied Territory.—Precaution should be taken to avoid the inference, that because of similarity of titles, the various O. C. C. A.'s were all part and parcel of an administrative system apart from the Army. The O. C. C. A. in Occupied Territory alone maintained a position independent of the Army, without the means, however, of making his regulations effective. All other O. C. C. A.'s served merely as advisors in civil affairs to their respective commanders, who possessed in their own person such powers of military government as were vested in them by Advanced G. H. Q.

The commander of a tactical unit might, as a matter of convenience, lodge as much or as little responsibility in his O. C. C. A. as he felt advisable. In the Army, the position of the Officer in Charge of Civil Affairs was coordinate with the General Staff Sections. In several Divisions, though not in a majority, the Officer in Charge of Civil Affairs was regarded as of minor importance and placed under a Staff Section, thus interposing two links in the chain of responsibility between the Commander and the officer upon whom depended the reputation which the unit would acquire in matters of military government.

Next to the principle of undivided and centralized control of the military commander over both military and civil matters within his jurisdiction, there is no principle of military government so clear as this: (a) that the military staff should not have its efficiency impaired by the assignment of civil duties; (b) that the civil administration must be conducted by a separate staff; (c) that this staff should be coordinate with the general staff sections on the staff of each commander. So important is this principle that, in the French and Belgian armies, this staff organization was made an additional section of the General Staff, and in the British, it was made coordinate with their existing three staff systems, viz.: General Staff (G), Supply (Q) and Administration (A).

Reorganization of military Government.—It must not be supposed that the difficulties inherent in the organization of military government, described above, were not perceived by those in authority. As early as February, 1919, it existed, and steps were taken gradually to remedy it. The organization of the military government in its eventual shape as of December, 1919, was the result of long investigation, changes being adopted as the military situation cleared. These changes in organization were introduced very gradually, and took their final form only when the great Army of 1918 had evacuated the territory, leaving the normal garrison provided for in the terms of peace. The gradual evolution of the military government will be treated chronologically in the following pages.

March, 1919.—The first Divisional unit of the Third Army left occupied territory for the United States in March. This was the Forty-second Division, which had been billeted in Kreis Ahrweiler. It was not the intention at this time to leave any district in the American zone without a garrison, and the Fourth Division was moved from the Adenau-Cochem area to Ahrweiler. It was intended to replace the Fourth Division by the Sixth, one of the units of the Second Army. Certain regiments of this Division (about a Brigade) actually did reach the Adenau-Cochem area and Division Headquarters were established at Bad Bertrich. However, orders for the Division to return to the United States were received before its Headquarters had had a chance to become familiar with the requirements of military government. The disadvantages involved in having the personnel for military government attached to tactical formations was plainly shown during these changes. Not only had the Office of Civil Affairs of the Fourth Division to begin work in an entirely new field of activity, but a complete Civil Affairs organization had to be formed within the Sixth Division, among officers without previous experience in military government. No actual change in the organization of Civil Affairs took place during the period, but the advisability of such a reorganization became recognized as a necessity of the near future.

April 1st to July 2nd, 1919.—During these three months, the return of the Eighty-ninth, Ninetieth, Thirty-second and Sixth Divisions to the United States, left a large portion of the American zone ungarrisoned, including nine Kreise in the Treves region. The departure of the Thirty-second Division from the bridgehead was less important from the standpoint of military government, as the Second and First Divisions merely spread out and placed garrisons of their own troops in the evacuated territory. In the western Kreise, however, it was an entirely different problem, since the area evacuated was so large, and the Divisions remaining in the Third Army so few, that further spreading out was not feasible. The Third Army, with the approval of Advanced G. H. Q., therefore undertook the task of maintaining American military government in these unoccupied districts. It was decided to base the reorganization on two major principles, which the experience of five months had shown to be essential. These were:

(a) Removal of the personnel for military government, from the control of the commanders of tactical formations and their subordination to the Office of Civil Affairs of the Third Army;

(b) Co-ordination of military government with German governmental bodies, particularly the Kreise.

The first Civil Affairs Bulletin, May 7th, 1919, published the outline of the new reorganization. In Kreise Adenau, Ahrweiler, Cochem, Daun, Prüm, Wittlich, Berncastel and Saarburg, Offices of Civil Affairs were directed to be installed and to take over military governmental powers at once from the Division returning to the United States. It was contemplated that eventually all territory within the American zone would receive this new form of government. At the head of these territorial units was placed a field officer with the title of Kreis Officer in Charge of Civil Affairs. He was given several assistants to assist him in conducting provost courts, sanitation, etc., among them being a medical officer. O. C. C. A.'s were charged primarily with supervising the execution of regulations promulgated by higher authority. In exceptional cases only were they authorized to publish regulations of their own to the civil population and, in such cases, a copy of the order was to be forwarded to the Army. Although in most matters the Kreis Officer in Charge of Civil Affairs received his orders directly from the O. C. C. A. of the Army, he was also charged with carrying on the business of other sections of the Army Staff in his district. For instance, matters pertaining to censorship of the press and to circulation, which in the Army were handled by G–2, were in the Kreise, where territorial governments had been set up, performed by the Officer in Charge of Civil Affairs. The large volume of business in every Kreis, pertaining to the recording and investigation of claims, was in some instances handled by zone majors appointed by G–1, but in far the larger number of cases was also performed by an assistant to the Officer in Charge of Civil Affairs.

The problem of garrisoning the Kreise was solved by the assignment to the Third Army of a number of military police companies, which had been set free in France by the departure of Divisions for the United States. Paragraph 290 of the Rules of Land Warfare states that occupation of an occupied territory must be actual and effective if the enemy citizens are to be expected to recognize the temporary sovereignty of the invading army. It was not deemed wise in May, 1919, to leave such a large proportion of the American zone bare of troops, as it was not known even at this time whether Germany would assent to the peace-terms then being prepared in Paris. The arrival of the military police companies from France was therefore very opportune, and they were assigned to the vacated Kreise as follows:

		Date of termination of service in the Rhineland.
Saarburg	286th M. P. Company	August 4th, 1919.
Bitburg	278th M. P. Company	September 12th, 1919.
Prüm	277th M. P. Company	August 25th, 1919.
Daun	282nd M. P. Company	August 25th, 1919.
Wittlich	VIIth Corps M. P. Company	July 22nd, 1919.
Berncastel	VIth Corps M. P. Company	July 22nd, 1919.
Cochem	IVth Corps M. P. Company	July 23rd, 1919.
Adenau	281st M. P. Company	September 3rd, 1919.

The signing of peace at Versailles on June 28th, 1919, rendering the presence of these units in the Kreise no longer necessary, the Army permitted them to return to the United States. During the period, however, in which they served under the direction of the O. C. C. A., they performed the duties assigned them in a very efficient manner. The calibre of their personnel and their previous experience in police work, made them particularly valuable for the duties which they were called upon to perform. Garrisoned in the capitals of the Kreise, they patrolled the surrounding countryside daily in motor-cars, thereby bringing home to the civilians the fact that despite the departure of the Divisions, American military orders could not be violated with impunity. Instructions as to the duties of these M. P. Companies were given by the local Officer in Charge of Civil Affairs. Their internal administration was entirely handled by their own officers.

A separate Kreis administration had been set up on May 7th in Kreis Ahrweiler, in the expectation that the Fourth Division would follow the 89th and 90th to the United States. When its departure, due to the uncertainty of the advent of peace, was put off from week to week, the Kreis Office of Civil Affairs was again placed under the authority of the Division Commander.

On June 1st, 1919, the Office of Civil Affairs, Advanced G. H. Q., was abolished, and all powers incident to military government until then resident in the Office, were transferred to the Commanding General, Third Army. The change of authority, however, brought only

521572—43——7

minor changes of organization in its train. The Stadt and Landkreis Treves, formerly administered by the garrison commander of Treves, who functioned directly under Advanced G. H. Q., was transferred to the Third Army's jurisdiction. The Treves Office of Civil Affairs was then organized along lines similar to the Offices in other Kreise.

The Civil Affairs Bulletins which had begun to appear on May 7th, and which had replaced the Letters of Instructions, now took over, in addition, the functions of "Orders, Advanced G. H. Q.," serving as the mouthpiece of the Commanding General in all matters pertaining to military government.

The only departments of Advanced G. H. Q. which did not already exist in the Army, and which it was believed necessary to retain, were the Finance and Economic Sections. The first of these sections had been accountable for all fines turned in by the Provost Courts. The latter department granted export and import licenses for merchandise to and from the American area. These departments, now that Advanced G. H. Q. had been abolished, were transferred to the Office of Civil Affairs, Third Army, where they continued to perform their former functions.

No attempt was made at this time to institute territorial officers in any part of the area still occupied by combat Divisions. Their departure from Germany was expected in the near future, and it was not believed wise to complicate their last days by setting up new authority in their areas, independent of the Division Commander. On July 2nd, 1919, the Third Army ended its existence, the troops remaining in Germany taking the names of "American Forces in Germany".

July 2nd, 1919 to January 10th, 1920.—The new title of the Army, "American Forces in Germany", brought with it, for a short time at least, merely minor changes in the organization of military government. Pursuant to the policy of adapting the American civil organization to the German administrative system, the Commanding General in Paragraph 5, Civil Affairs Bulletin, July 16th, 1919, appointed the Officer in Charge of Civil Affairs, Stadtkreis Treves, as the superior American representative within that Regierungsbezirk. All Kreis representatives within the district thereafter received their instructions through the Treves office The movement of the remaining combat Divisions to the United States continued unabated.

The Fourth Division left Germany on July 12th, its departure being followed on the 21st by the Second Division. The Third and First Divisions left in August, on the 11th and 21st respectively. This cleared the American area of all combat Divisions, leaving only a few minor units of the old Army behind, together with the permanent garrison.

The time had now come for the extension of territorial organization over the remaining area. Simultaneously with the departure of the

Third Division from the Mayen area, a territorial civil affairs unit was organized from the Divisional personnel, which had been on identical duty. It was also a very simple matter for the Officer in Charge of Civil Affairs of the Fourth Division,—who for a short time had been independent of the Fourth Division Commander,—to resume his former status.

The reorganization of military government in the remaining area,—the bridgehead and the Land-and Stadtkreis Coblence,—took place on September 12th. It was not feasible to adopt a strict Kreis organization in the territory on the right bank of the line. The adoption of a bridgehead perimeter, 30 kilometers distant from Coblence, had placed a number of townships and Gemeinden in occupied territory which belonged to Kreise whose capitals were in the neutral zone. For purposes of American administration, the bridgehead was divided into two zones, called by the names of the Kreise which formed the larger part of each area, Neuwied and Unterwesterwald. That part of Kreis Altenkirchen which lay within the perimeter was added to Neuwied. Other territory belonging to Kreise Marienburg and Westerburg was attached to Unterwesterwald.

By the provisions of the same order (Par. 22, Civil Affairs Bulletin), the Stadt-and Landkreis Coblence was also given a territorial organization. Civil affairs in this district had previously been administered by an Officer in Charge of Civil Affairs, on the staff of the City Commandant. His relation to the City Commandant had been similar to the relation of the Division O. C. C. A. to its commander. The new Office of Civil Affairs of Coblence, however, differed in a number of respects from the organizations previously set up in the Kreise. Due to the fact that nearly the whole garrison as well as Army Headquarters was situated in the city, there was nowhere nearly the same opportunity for independence and concentration of civil matters in one office. The police of the city, both civil as well as military, functioned under the Provost Marshal and therefore, eventually, under G–1. Even the Provost Court, who in the outlying Kreise, was a member of the Civil Affairs organizations, in Coblence, also served under the Provost Marshal. The wisdom of placing the courts under the control of the police is in any case a dubious experiment, and Provost Court matters were eventually turned over to the Office of Civil Affairs. The control of circulation, and the censorship of the post, which in other Kreise were performed by the O. C. C. A.'s in their capacity as agents of G–2, were actually performed in Coblence by agents of that section.

Among the duties to which Officers in Charge of Civil Affairs in the Kreise were called upon to perform were the following:

1. *Administration.*
   Supervision of prisons;
   Investigation of appeals for clemency;
   Issuing orders for deportation of vagrants;
   Supervision of civil police;
   Investigation of complaints of civilians against the military;
   Supervision of local circulation;
   Supervision of civil administration;
   Supervision of sanitation.
2. *Judicial.*
   Provost Court authorities, actually performed by an officer designated as Superior Provost Court and attached to the local Office of Civil Affairs;
   Supervision of Civil Courts.
3. *Economic.*
   Surveillance of local food and fuel situation;
   Supervision of public utilities;
   Supervision of industrial meetings;
   Mediation between industries and the military forces;
   Mediation (in emergencies) between employers and employees;
   Action on requests for authorization to import and export;
   Surveillance of labor conditions and unemployment.
4. *Political.*
   Supervision of political meetings;
   Surveillance of local political tendencies;
   Liaison between military forces and civil officials.

The Kreis organization was a vast improvement on the old method of exercising control through tactical units. Injustice was rarer, the application of our rules more uniform, and officers connected with the military government were better informed and better trained for the duties incident thereto. Under the old system, the Officer in Charge of Civil Affairs, Third Army, had since July made it a practice to hold weekly conferences with his subordinates, in which the problems confronting the army were discussed as they arose. These conferences undoubtedly contributed to the growth of efficiency, so marked during the final six months of the military government. There were still, however, a factor which prevented the American military government from achieving a complete success. The Army had failed to realize that efficient government can be conducted only from a single source. The failure to centralize in one office all matters pertaining to military government, resulted in untold friction, and often also caused divergent policies to be put into force. Despite these disadvantages, the

Kreis organizations were a vast improvement over former unwieldly methods.

The changes in the organization of the Office of Civil Affairs, brought about by the reduction of the American area, should be briefly mentioned here. In view of the decision of the War Department to retain a merely nominal garrison on the Rhine, it was inevitable that America would have to give up to the French a considerable part of the area which it had formerly occupied. The agreement finally reached by the respective armies was that the French should take over the American area of Regierungsbezirk Treves, including Kreise Treves-Stadt, Treves-Land, Saarburg, Bitburg, Prüm, Berncastel, Wittlich and Daun. The evacuation began during the last week of August, each Kreis Officer in Charge of Civil Affairs turning over all business to the succeeding French administration. Treves was occupied by the 38th French Division on September 3rd, and by the 10th of that month, all Officers in Charge of Civil Affairs throughout the Treves region had been withdrawn.

Military Government in the strictest sense of the term passed out of existence on January 10th, 1920. Although the Office of Civil Affairs still continued to function, the scope of its activity was from that time on restricted by the assumption of a number of its former duties by the High Commission, which as a result of the Commanding General's decision, was permitted to exercise its authority in American occupied territory.

Chapter 5

MILITARY TRIBUNALS

Basis of Civil Administration; Jurisdiction of Military Tribunals—1. Provost Courts, 2. Military Commissions; Review of Judgments of Provost Courts; Review of Judgments of Military Commissions; Procedure in Provost Courts and Military Commissions; Proposed Draft of Order Establishing Military Courts and Defining their Procedure; Method by which Uniformity of Court Action was Established; Vagrancy and Juvenile Courts; Statistics of Trials; Disposition of Funds; Abandoned Enemy War Material; Superiority of Territorial over Tactical Organization of Courts; In what Authority Final Control over Judgments should be Reposed; Acts which Constitute Punishable Offenses should be Clearly Specified; Proposed Draft of Orders to the Inhabitants of an Occupied Territory; Orders Defining Offenses should Emanate from One Authority Only.

On December 9th, 1918, General Pershing issued his proclamation to the inhabitants of the district of Germany occupied by the American forces. This proclamation advised the inhabitants as to the conduct and mode of life which would be required of them during the occupation, and warned them that punishment by military tribunals would result from their failure to meet the requirements set forth. As compared with the ultimate scope of the rules and regulations for the government of the inhabitants, this original proclamation is rather limited. It covered matters of identification, circulation, arms and ammunition, alcoholic drink, assemblies and meetings, censorship, carrier pigeons and photographs, and, in addition, declared as an offence, punishable by a military court, any interference with the American troops; any destruction or damage to railway, telegraph or other public service property; the purchase of, or having in possession, articles of clothing, equipment or rations furnished American soldiers or belonging to the American Army; the destruction, damage, secretion or polution of fodder or water or other things of use to the American Army or in obstruction of the military government; the commission of war treason, or the attempt to do any of the things specifically enumerated. To the inhabitants of such territory as was then in fact occupied by the American forces, this proclamation became a rule of conduct. Concurrently with the daily advance of the troops, the proclamation was promulgated to the inhabitants thus coming within the actual domain of occupation, until the whole area ultimately occupied became subject to its provisions.

THE BASIS OF THE CIVIL ADMINISTRATION

On the 10th day of December, General Pershing issued at General Headquarters of the American Expeditionary Forces, General Orders No. 225, setting forth and defining the organization and the means and the manner by which the mandates of his proclamation issued on the previous day, and all subsequent orders properly issued to the inhabitants, should be executed. Under this order there was appointed an Officer in Charge of Civil Affairs for the entire occupied area, with headquarters at Advance G. H. Q., A. E. F., Treves, Germany. This officer became the personal representative of the Commander in Chief with respect to all matters affecting the government of the civil population. He was authorized to issue, and did in fact issue, special orders and instructions for the control and management of the civil population and, when experience had shown its necessity, modified to some extent the mandates of the original proclamation and of the organizational order of the 10th of December.

The latter order provided that the Division Commander should be responsible for the administration of Civil Affairs in his district. By this provision, the principle of decentralization was announced. As originally occupied, the American zone in Germany was divided into nine areas. Seven of these were occupied by Divisions, one was occupied by Third Army troops and the other by Third Corps troops. In addition, the Fourth Corps controlled the city of Cochem, and the Seventh Corps, the city of Wittlich. The Commanding General of the Army and the Commanding General of each Corps were granted the powers and jurisdiction as to civil affairs in their respective areas that were specifically granted to Division Commanders as aforesaid.

Jurisdiction of Military Tribunals.—1. Provost Courts.— Superior and Inferior Provost Courts were established in each area under the provisions of G. O. 225. The jurisdiction of the Superior Provost Courts, under the terms of the order, was limited to the imposition of imprisonment for six months or a fine of 5000 marks, or both such fine and imprisonment. The jurisdiction of the Inferior Provost Courts was limited by the order to the imposition of a sentence of imprisonment for three months or a fine of 1000 marks, or both such fine and imprisonment. The Superior Provost Courts were appointed by the Commanding Generals of the various Divisions, and in the case of the areas occupied by Army and Corps troops, by the Commanding General of the Army and the Commanding General of the Corps, respectively. The Superior Provost Court sat only at the Headquarters of the Commanding General. The respective Commanding Officers of each garrisoned city, town or other occupied place, appointed Inferior Provost Courts for the trial of offenses within their respective jurisdictions. The action of the Superior Provost

Courts went to the Commanding General for approval or disapproval. The action of the Inferior Provost Courts went to the appointing officer for approval or disapproval. By the terms of the order, the various courts were to make weekly reports and to forward them through channels to the Officer in Charge of Civil Affairs at Advance G. H. Q.

It is apparent from the foregoing that the Commanding Officer in each area was the final authority in all matters connected with Provost Court prosecutions, except, of course, that the Commander in Chief, either directly or through his personal representative at Treves, might approve or disapprove the judgments rendered. Strictly speaking, under the organization originally set up, the Commanding General of the Army was without jurisdiction to act in review upon cases tried within a Division or Corps area. By a subsequent order (March 5th, 1919), issued at Advance G. H. Q., the Commanding General of the Army was given the power to revoke, mitigate or commute any sentence adjudged by a Provost Court which, in his opinion, was irregular or excessive. In the practical working of the system, the power thus lodged in the Commanding General of the Army has been, with slight exception, the only revisory power exercised (after that of the authorities appointing the courts), with respect to judgments of the provost courts; for the Officer in Charge of Civil Affairs at Advance G. H. Q. limited his supervision over Provost Courts, after the first few weeks of the occupation, to verifying the fines imposed and collected, and making certain that these fines were turned over to the Officer in Charge of Fiscal Affairs.

Under the decentralized plan of G. O. 225, orders for the guidance of the civil population were issued by the Commanding Officers in each of the various areas. The Officer in Charge of Civil Affairs at Advance G. H. Q. also supplemented the original proclamation by various orders issued in the name of the Commander in Chief. These latter, of course, constituted the fundamental law for the occupied territory, and no Commanding General of an area might issue orders contrary to these or to the proclamation. With respect to all matters however, which were not covered either by orders at Advance G. H. Q. or by the original proclamation, the Commanding Generals of the various areas were free, under G. O. 225, (making them responsible for the administration of civil affairs), to issue such orders as to them seemed advisable or necessary to meet existing local conditions. Thus the law for the guidance of the civil population was to be found in the original proclamation, in subsequent orders issued at Advance G. H. Q. and in the Special Orders issued in the eleven separate areas of occupied territory, with the result of more or less confusion to the final reviewing authority and large confusion to the civil inhabitants, who were never in a position to know what was and what was not prohibited.

Jurisdiction of Military Tribunals.—2. Military Commissions.—G. O. 225 provided that Army, Corps and Division Commanders should convene Military Commissions for the trial of inhabitants for offenses against the laws of war or the military government. This provision was interpreted to mean that for an offense committed within a particular area, the Commanding General of that area should appoint the Military Commission. No limitation was placed upon the Military Commission as to the penalty which might be imposed. It could properly act in accordance with the laws of war, and was thus empowered to impose any penalty that to it should seem proper. The order did provide, however, that no death-sentence should be carried into execution until approved by the Commander in Chief. It was readily apparent from the whole order that for any offense for which the maximum penalty imposable by a Superior Provost Court was deemed insufficient, the procedure to be adopted was to appoint a Military Commission for the trial of such offense.

Review of Judgments of Provost Courts.—It is observable that under the judicial scheme promulgated for the occupied area, there was no central authority provided, except that inherent in the Commander in Chief, and therefore in his personal representative at Treves, but which, as has been observed, was exercised only to a limited extent. No order was ever issued to remedy this situation beyond such as was contained in the order of March 5th, 1919, above referred to. In order that there might be harmony in procedure and an approach at uniformity in the imposition of sentences, it was necessary that some central authority be created. This was assumed by the Officer in Charge of Civil Affairs, Third Army, acting for and on behalf of the Commanding General. Technically, under the organization promulgated, the jurisdiction of this officer was confined to the Third Army area. In spite, however, of the specific provision that Division Commanders should be responsible for the administration of Civil Affairs in their respective districts, the Officer in Charge of Civil Affairs, Third Army, acting for the Commanding General under authority of the order of March 5th, 1919, assumed jurisdiction over the court procedure for the entire occupied area. All Provost Court reports made to the Officer in Charge of Civil Affairs at Treves were sent through the office of the Officer in Charge of Civil Affairs, Third Army, and were there inspected and reviewed, as far as this could be done upon the face of the record. Where technical errors appeared, the record was sent back for correction or explanation. When it appeared certain that a conviction was erroneous, such conviction was set aside by the Commanding General of the Army.

The procedure of the Provost Courts, was, of course, summary, and by the provisions of the order it was intended that it should be so. The limitation upon punishment was undoubtedly imposed in view of that fact. Because of this summary nature of the proceedings, some errors were bound to be made, some convictions were certain to be unjust. To meet this situation, a very liberal policy was devised for investigating all complaints in which it was claimed that there had been a failure of justice. Whenever such a complaint was filed, it received careful consideration. If it first reached the office of the Officer in Charge of Civil Affairs, Third Army, it was sent for investigation to the court where the case was tried, through the Officer in Charge of Civil Affairs in that jurisdiction, with instructions that the matter be thoroughly investigated and such action taken as the result of that investigation should seem to warrant, and that a full and complete report of the investigation be made to the Officer in Charge of Civil Affairs, Third Army. Ordinarily the action taken as a result of the new investigation was approved by the Officer in Charge of Civil Affairs, Third Army, and the action became final. The right was at all times reserved, however, to act contrary to the judgment of the investigating authority, and in many instances, sentences were set aside or modified, contrary to the recommendations of the local authorities.

Review of Judgments of Military Commissions.—Under the provisions of G. O. 48, W. D., 1918, the records of trial before Military Commissions were sent to the Acting Judge Advocate General for the American Expeditionary Forces and reviewed by him in the same manner as General Court-Martial cases. Regardless of this fact, however, the Commanding General of the Army claimed the right of control over the judgments of Military Commissions, and through his Officer in Charge of Civil Affairs investigated all charges of miscarriage of justice. In only two cases, however, did he issue orders in mitigation of the sentence imposed.

Procedure in Provost Courts and Military Commissions.—Procedure in Provost Courts, both Superior and Inferior, was required to be uniform. Written charges were filed. They were not required to be formal, but to state only the substance of the offense, the name of the alleged offender, and the place where and time when the offense was said to have occurred. It was required that every person tried by a Provost Court should be informed of the charges upon which he was to be tried, that he should be present in person at the trial, and that he should be confronted with the witnesses against him. It was further required that he should be permitted to be heard either in person or by counsel. Bail was denied the accused, but a speedy hearing was guaranteed him. All evidence was required to be under oath. Sentences were put into immediate execution without awaiting action of the reviewing authority.

No particular form of procedure was required by order for trial before Military Commissions. In practice, however, a definite and specific form of procedure was followed—namely, that of a General Court-Martial, in as far as the requirements of such procedure were applicable to trials by Military Commissions.

Proposed Draft of order Establishing Military Courts and defining their procedure.—As experience always carries suggestions for improvement, so has it been in the matter of court procedure. Specific discussion of this subject, however, does not seem to be called for. But it may perhaps be helpful to give a comprehensive draft of an order establishing military courts and defining their procedure, which would, possibly, remedy such defects and fill in such omissions as have been found in the order under which our courts have operated during the occupation. Such a draft will be found as Appendix No. 16, to this report. [Not printed. See FM 27-5, Military Government, paragraphs 20-32, on Military Tribunals.]

Method by which uniformity of court action was established.—It has been observed that the Commanding General of each area appointed a Superior Provost Court and thereby became the functionary to approve or disapprove its judgments, and that the Commanding Officer in each town or village occupied by troops likewise appointed an Inferior Provost Court and thereby became the proper officer for approval or disapproval of its judgments. During the full occupation of the territory, this system produced upwards of 200 Provost Courts. The tenure of detail of the Provost Court was uncertain. Sometimes an officer held such a detail for a week only, sometimes for several months. No means seemed to exist for bringing about uniformity of action by the various courts. The judgments of the courts showed clearly that there was a wide divergence of viewpoint. In one jurisdiction for a particular offense a 100-mark fine would be imposed, while in another jurisdiction for the same offense an imprisonment of three months would be imposed. The Inferior Provost Courts were entirely independent of the Superior Provost Courts. It was clear that what these courts needed was to come in personal contact and to discuss the problems arising before them, so that an approach to the same viewpoint could be attained and reasonable harmony of action would prevail. The constant changing of personnel of the courts, however, coupled with the fact that the majority of this personnel was performing other duties as well, rendered it well-nigh impossible to solve the difficulty.

After the troops had evacuated the Treves district the only military occupation of this area consisted of a company of Military Police at each Kreis center. Civil Affairs was then entirely divorced from the tactical organizations and a territorial organization was set up. An Officer in Charge of Civil Affairs, who was made directly responsible to the Officer in Charge of Civil Affairs, Third Army, was appointed for each Kreis. The Military Police reported to this officer. A Superior Provost Court was also appointed in the Kreis by each Officer in Charge of Civil Affairs. In this manner there were ultimately set up nine Superior Provost Courts whose tenure of office was reasonably permanent. The Superior Provost Courts in the remaining tactical divisions were also reasonably permanent, as they had been in the past. With this situation existing, orders were issued for weekly meetings of the Provost Courts of Regierungsbezirk Treves, in the city of Wittlich, and of the Provost Courts in the rest of the area, also weekly, in the city of Coblence. With these courts met also the legal officer of the Office of Civil Affairs of the Third Army, who thus formed a connecting link between the courts of the two areas. This officer sought to exercise no control over the courts. He made it entirely clear that the courts were in fact courts and that their decisions should in no way be interfered with, except as the reviewing authority should approve or disapprove their judgments. Full and free discussions were had of cases coming before the various courts and of conditions in the various communities which were likely to find their way into the courts. Differences of opinion were thrashed out. The total result was the bringing about of a great unanimity of opinion as to the best way of handling the various situations coming before the courts and a reasonably harmonious view with respect to the punishment which ought to be imposed. In this was proof that, with courts reasonably permanent, a very harmonious system of judical procedure could be arrived at, even while operating under orders stated in very general terms.

Vagrancy and Juvenile Courts.—As American troops in increasing numbers returned to America in the Spring of 1919, prostitutes, who had been plying their trade in France, began to enter the American zone in large numbers. The Office of Civil Affairs had recommended on several occasions that prostitution be declared a crime punishable by military tribunals. When the venereal rate rose to an enormous figure (See Table XI, Appendix No. 56) [Not printed.] as a result of the combined influence of the increased number of prostitutes and the lessened moral sense caused by replacing the men of the Third Army with volunteer recruits, the Army Commander decided to establish a Vagrancy Court. (The Order referred to is quoted in full in Chapter 7 on "Sanitation and Public Health"). The term "vagrancy" was broadly defined so as to enable us to try Bolshevist agitators and other similar persons found in the area without visible means of support, as well as prostitutes. The chart above referred to shows the effect of this court in removing risks of infection from diseased women, and shows conclusively that a military government, vested as it is with absolute power, can limit the dangers arising from prostitution if it is determined to do so.

The first object in trying these unfortunate women was to remove them as possible points of contact for venereal infection. But the broader grounds of humanity were also considered, and after conviction they were physically examined and, if found diseased, confined in a venereal hospital which the city of Coblence was required to establish for this purpose. German philanthropic women were encouraged to work among them and many were paroled in their custody. Those from outside the area were always deported, ordinarily after serving at least a portion of their sentences. Those residing within the area constituted much the larger class and were more difficult to deal with, as it would have been a great hardship on them to banish them from their homes. This type was usually paroled after a short period, in order to hold some sort of authority over them to insure that they would not have to be arrested again. From the establishment of these courts down to January 10th, 1920, 401 women and male vagrants had been convicted.

Juveniles in the early days of the occupation were a source of constant annoyance, as they committed many petty thefts of clothing or food, often, no doubt, at the instigation of their parents who were suffering from a lack of these necessities. With the establishment of a Juvenile Court, an effort was made to conform its practice to the best modern thought along this line, and where confinement was imposed, it was served in German reformatories.

Considered in the light of experience, the establishment of these courts marks a distinct advance in our system of administering justice with war tribunals. Where suitable officers are selected, they may be made of great value to an occupying army.

Statistics of trials.—The following table gives the total trials held in the American area and a classification of offenses committed from the beginning of the occupation to January 10, 1920, the date when the Rhineland Convention became effective:

Sale of prohibited alcoholic drinks	517
Unauthorized sale of light wine and beer	335
Sale or unlawful possession of U. S. property	1580
False statement to U. S. officers	68
Unauthorized assembly	314
Theft of U. S. property	781
Violation of circulation orders	2317
Failure to obey lawful orders of military authorities	556
Drunkenness and disorderliness	177
Unlawful possession of deadly weapons	302

Insulting language concerning U. S. army	143
Violation of sanitary regulations	918
Disrespect to American officers	81
Practicing prostitution while diseased	91
Discrimination in prices	16
Interference with U. S. troops (assaults of American soldiers)	357
Damage to or destruction of U. S. property	33
Vagrancy	401
Miscellaneous	712
Total	9699

Disposition of funds.—By a decision of the Assistant Comptroller of the Treasury in Paris, all funds collected by the American Forces in Germany had to be covered into the Treasury. Although reversing the precedents of a century and a half of Government experience, the decision was ordered by G. H. Q. to be carried out. It results that the United States has actually profited by the occupation, since the German government is required to pay for the cost of the maintenance of the armies of occupation by the terms of the armistice. Money received from fines and the sale of confiscated goods has been profit. The total amount thus received is as follows:

Fines	1, 847, 193. 37 Marks	
Forfeited bail	800. 00 Marks	
	1, 847, 993. 37 Marks	=$128, 274. 46
Sales of confiscated liquor	669, 782. 00 Marks	= $42, 147. 74

In addition to the foregoing, the Intendanture of the German VIIIth Army Corps was required to turn over 7,048,526.27 marks on account of the illegal sale (under presumed authority from Berlin) of German army property, in violation of the armistice. (See Appendix No. 22.) [Not printed.]

Abandoned enemy war material.—As we have never before in our history taken any profit from a military occupation, it is recommended that the total amount thus received from Germans be appropriated by Congress as a trust fund, and invested in U. S. bonds to be deposited in the Treasury, the interest on these bonds to be expended, under such regulations as the Secretary of War may establish, for the purpose of putting the best of American scientific and literary thought, as well as studies on political and sociological progress in America, within reach of German students. Suggestions from Germans have been invited as to the possible uses for such funds, and their views are set forth in Appendix No. 55. [Not printed.] From such investigation as has been made in the time available, it would appear that such funds could best be used to further the interests of the Amerika-Institut in Berlin and to establish and maintain an American library in Coblence. The former institution exchanges scientific literature with the Smithsonian Institution in Washington, and, if ample funds were available, the librarian of the American library in Coblence could be selected with a view to his occupying a chair in the University of Bonn. Precedents exist for such an investment of these funds, the American University in Pekin, China, being a case in point; while in all our dependencies, such funds were considered as held in trust for the native populations.

Superiority of territorial over tactical organization of courts.—A comparison of the operation of courts operating under tactical organization, and of those operating under territorial organization, shows conclusively the superiority of the latter system over the former. Probably it would, however, [not?] always be practicable to separate the Superior Provost Courts from tactical organizations and give them a territorial jurisdiction. It should be provided that the officer designated as Superior Provost Court should be attached to the Staff of the Commanding Officer whose jurisdiction included the particular territory involved, thus leaving to the Commanding Officer of that territory the power to approve or disapprove the judgments of these courts. With this situation, cooperation and unity of action by the Superior Provost Courts could be arrived at in the manner outlined above. The same result could be obtained from Inferior Provost Courts by giving to the Superior Provost Courts a certain supervisory power over the Inferior Provost Courts, such as would allow them to bring the Inferior Provost Courts into meetings similar to those suggested for the Superior. In this way, the entire court system, both Superior and Inferior, would be bound into a composite whole. To bring about this situation, the order defining the jurisdiction of Superior Provost Courts should, among other things, give to those courts the power to verify the records of Inferior Provost Courts and return them for the correction of any apparent irregularities; to advise the authority appointing them to set aside the judgments of Inferior Provost Courts and return the cases for new trial or for the dismissal of the accused, or to mitigate the sentence imposed by the Inferior Provost Court.

In what authority final control over judgments should be reposed.—The fact has heretofore been adverted to that the ultimate power in Civil Affairs in occupied territory was reposed in the personal representative of the Commander in Chief at Advance G. H. Q. at Treves. Experience developed the fact that this form of organization was not well suited to the existing situation, as far as military courts were concerned. By it final authority was placed too far away from the seat of activities at the Headquarters of the Third Army at Coblence, at the opposite end of the occupied area. There is where the business of the Army was transacted. There is where reports were received of the activities in all portions of the occupied zone. There is where the Commanding General and members of his staff received reports and discussed situations existing throughout the entire area. There is where policies outside of Civil Affairs were formulated and promulgated. And there is where final control over the judgments of military courts ought to have resided; for such control should be where the fullest knowledge is had of the affairs of the entire area. Whether, theoretically, it is good or bad organization that the Officer in Charge of Civil Affairs for the occupied area should function under the Commander in Chief of the forces, or as a member of the staff of the Commanding General of the Army, practice has demonstrated that in as far as his functions have to do with the military courts, these could not be properly performed at a distance far removed from the central activities of the Army of Occupation. So fully was this demonstrated that, as heretofore pointed out, the Officer in Charge of Civil Affairs in Occupied Territory issued, on March 5th, 1919, an order which gave to the Commanding General of the Army of Occupation the authority to revoke, mitigate or commute sentences adjudged by Provost Courts. This recognition of the practical requirements of the situation may well be accepted as a final determination of where the final power of review of the judgments of military courts should be, namely, in the Commanding General of the occupying forces.

Acts which constitute punishable offenses should be clearly specified.—The proclamation to the inhabitants of Germany issued by General Pershing on December 9th, 1918, attempted within a short space to regulate the entire relations between the inhabitants and the army. This was largely added to from time to time by the issuance of additional orders at Advance G. H. Q., at Headquarters of the Third Army, and by Commanding Generals of each of the Divisions and Corps, within the limitations pointed out above. These additions may largely be regarded as having been made necessary by special situations which could hardly have been foreseen, and most of which would probably not arise in the military occupation of other or different areas. It is probably true that by process of construction, the original proclamation could be made to reach most offenses against which the army needs to protect itself. Could the courts be made permanent, and could it be depended upon that men of keen sense and sound judgment would always be detailed as Provost Courts, it would be entirely safe, from the standpoint of the army, to rely upon definitions of offenses couched in general lan-

521572—43——8

guage rather than in a long series of specifications. The fact remains, however, that in practice, courts cannot always be composed of men of the characteristics mentioned; furthermore, as it is desirable that offenses should not be committed, wisdom, on the one hand, and fairness to the civil population on the other, require that acts that are to constitute punishable offenses should be specified so clearly that the civilian inhabitants would certainly know what conduct is expected from them. Special Rule (g) of the proclamation of the 9th of December provides that whosoever "commits any act whatsoever injurious to the American army or the obstruction of the military government, will be punished as a military court may direct." To the civilian population, this language meant little in particular. To the various courts it has meant everything in general. Under it, men have been convicted for filing a claim against the army based on the destruction of property, because they asserted that the property destroyed was worth 300 marks, whereas the court found it to be worth only 200 marks. Under it, civilians have been convicted for disrespect towards American officers, non-commissioned officers and privates. In some instances, a conviction was undoubtedly warranted, but in other instances, there is no question that this broad language was so interpreted as to result in injustice. Examples of convictions under this section for offenses which were petty (if offenses at all), could be given in considerable number. But probably enough has been said clearly to demonstrate that it is unsafe to put an order couched in such general terms as the one in question into the hands of a great number of Provost Courts, whose members are unlearned in the law, and whose experience in such matters is most limited. Ultimately, justice can be and in this zone has been done through the power of the Commanding General of the Army to set aside judgments erroneously made; but in the meantime an injustice has been done, and the admission of that injustice which comes with the reversal of judgments does not fail to have the effect of lessening respect for our courts upon the part of the civilian population.

In this connection, the view is often expressed and seriously argued, that we are dealing with a defeated enemy and that we need not overtrouble ourselves as to the treatment accorded. A mature reflection, however, must convince even the most radical that the question involved is really not what is due the inhabitants of the defeated country, but what is owed to the victorious country by the army which represents it. An occupying army in a defeated country is making history which is bound to be written. As that army conducts itself, so is the world largely to regard the country which it represents. If its army is dishonorable in its relations with a fallen foe and treats the population with injustice and subjects the people to a rule more harsh than is necessary for the preservation of order and the establishment of proper decorum and respect, that army and the country it represents are bound to stand in disrepute before the civilized world. The proper viewpoint of the soldier within an occupied area is beautifully and forcibly expressed in General Orders No. 100, Adjutant General's Office, 1863. Every officer and soldier entering an enemy's country as part of an occupying force should have its words indelibly impressed upon his memory:

"Military government simply is military authority exercised in accordance with the laws and usages of war. Military oppression is not military government; it is an abuse of the power which the law of war confers. As military government is carried on by military force, it is incumbent upon those who administer it to be strictly guided by the principles of justice and humanity—virtues adorning a soldier even more than other men, for the very reason that he possesses the power of his arms against the unarmed."

Proposed draft of orders to the inhabitants of an occupied territory.—As far as has been possible to ascertain, the army of the United States has never promulgated a code of laws for an occupied territory which is in any respect full or complete. Other countries whose armies have more frequently been upon foreign soil have promulgated orders more or less complete in detail, but it appears that not one of them has covered the ground so fully as to eliminate the objections raised above. With this fact in view, it seems desirable to put into one document, and leave for the benefit of armies that shall come after, a draft of orders, based upon our experience in occupied Germany, which seems to fairly well cover all general situations that have arisen, and that would seem likely to arise elsewhere. Particular situations are sure to be found everywhere and these, of course, will have to be met by particular treatment. These proposed orders appear in Appendix No. 16, to this report. [Not printed. See FM 27-5, Military Government, Appendix V, Ordinances.]

Orders defining offenses should emanate from one authority only.—Closely allied to the question just discussed is that of determining the source from which orders for the control of the civilian population should emanate. It has heretofore been pointed out that the civil government of this area was, by G. O. 225, very largely decentralized, and that orders for the government of the civil population of American occupied territory emanated from ten different sources. Experience has shown that such a system constitutes a large impediment to the following-out of a general policy by central authority. Probably without legal right, yet with recognized authority, the Commanding General of the Army of Occupation has in fact scrutinized orders issued by the Commanding Officers of the various Divisions and Corps and has not hesitated to order their revocation when he has considered them to be in derogation of orders issued by the central authorities, or in contravention of his policies with respect to the government of the area.

The government of the people of an occupied territory must be a strong government. The respect of the people for the authority of the army must be acquired. If that authority wavers, such respect will not be acquired. The issuing of an order today and its revocation tomorrow can produce but one result, namely, a lack of respect for the issuing authority. That the government of the civil population may be strong, that it may be unwavering, that it may produce respect for its own authority, it is absolutely essential that it shall emanate from one source, and from one source only. Whether that source shall be the Commanding General of the Army of Occupation, or other higher authority not otherwise participating in the occupation, is a subject about which there may well be difference of opinion. But experience has shown that there is no basis for differences of opinion as to whether that authority should be centralized or decentralized. The greatest efficiency will be acquired through centralization of policy and decentralization of execution.

CHAPTER No. 6

THE "ANORDNUNGEN" (ORDINANCES) OF DECEMBER 9TH, 1918, AND THEIR INTERPRETATIONS AND LATER MODIFICATIONS

Anordnungen; Identification and Circulation; Arms and Ammunition; Alcoholic Drinks; Assemblies and Meetings; Publications; Post, Telegraph and Telephone; Pigeons; Photographs; Special Rules.

ANORDNUNGEN

> The following rules and regulations are published for the guidance of the inhabitants of the district of Germany occupied by the American forces and will be strictly observed:
>
> *　　*　　*　　*　　*　　*　　*
>
> Every person must familiarize himself with these regulations and all others hereafter made by the American military authorities. Ignorance of these regulations will not be accepted as an excuse for their violation. In case of doubt as to the requirements of the regulations, inquire at the Rathaus.
>
> BY COMMAND OF GENERAL PERSHING:
>
> JAMES W. MCANDREW,
> *Chief of Staff.*
>
> OFFICIAL:
> ROBERT C. DAVIS,
> *Adjutant General.*
>
> 9 December, 1918.

In past occupations, the orders of the invading forces have usually been published to the civil population as proclamations, posted conspicuously in all the towns and villages of the territory falling under the temporary dominion of the conqueror. This was the case in Mexico in 1848, in the South during and just after the Civil War, and in later American occupations of Cuba and the Philippine Islands.

Marshal Foch in the late autumn of 1918, when the invasion of enemy territory appeared a distinct possibility, began the preparation of a proclamation which should serve as a model to the commanders of the Allied armies in case occupation of Western Germany actually should occur. This proclamation was distributed among the armies shortly after the armistice, when it became known that the entire German civil administration would be placed at the disposal of the Allied armies.

The "model proclamation," while outlining a general conception of the occupation as desired by the Allied General Staff, left ample opportunity for each nation to carry out details in accordance with its national characteristics and feelings. Such a policy proved wise in the long run, for in many matters, such as the payment of claims and the recognition of the Rhenish Republic, the attitude of England and America on the one hand, and that France on the other, were so contrasting that any single authority would have met with insuperable difficulties in enforcing a uniform compliance throughout. The "model proclamation," however, was never meant to be more than a guide, and in such a light was it considered by American G. H. Q.

As has been mentioned in a preceding chapter, while the proclamation was intended to be issued simultaneously with the entrance of our troops into Germany, its preparation was so delayed that it was not until Army Headquarters were installed at Coblence, that it was made known to the civil population through the press or by posters. In the meantime, the army had governed the population without publication of any definite code of law, relying solely on the authority granted by the laws of war.

However much the proclamation of December 9th, 1918, may be criticised on the score of impracticability, it is scarcely feasible, in the seclusion of a headquarters on friendly soil, to compose a code of laws for the government of a hostile people. Knowledge of the customs of the country is vital to such a task, even acknowledging the right of the conqueror to make what ordinances he will. Two-thirds of the trouble which the army experienced and half of the prosecutions before its provost courts were in regard to offenses which constituted a departure, for the Germans, from the customs of centuries. Little or no difficulty was encountered with the civil population in enforcing purely military regulations. The reason for this is to be found in the readiness of the average German to understand the necessity of war measures, and his innate discipline in accepting them. The American regulations, under their German title, the "Anordnungen," which will be discussed in detail in the course of this chapter, cover a wide variety of subjects and circumscribed the daily life of the civilian with so many restrictions and inconveniences that it is not surprising that violations continually occurred. Considering that the German peasant is hardly as intelligent as the average American, it is also not to be wondered at that he could not grasp the continued state of war, which obtained despite his homecoming. It seemed impossible for him to comprehend why, when to all appearances peace had come, he was not permitted to travel to a neighboring town without an array of passports, or why the American military authorities were interested in preventing his enjoyment of a glass of wine with his supper at eight o'clock in the evening. However necessary these restrictive regulations may have seemed to the security and welfare of the American army, they struck deeply at the customs of the people and stored up trouble immeasurable for the troop commanders, provost marshals and army courts. Desirable as law may have been from an idealistic viewpoint, it was bound to be modified, if prepared without a thorough knowledge of the psychology of the population which the army was expected to govern. We have only to point to the failure of the German laws in Belgium to attain even a measure of success as an example of the truth of this statement. Their very harshness, their failure to appreciate the patriotism and the habits of life of the Belgians, reacted in the end against them and brought together a far from homogeneous race into a united and defiant nation.

Were we to consider the Anordnungen as a hard and fast finality, the results of more than a year of occupation would have to proclaim them a failure. Their provisional character was, however, always realized. The Anordnungen, the basis of all orders of the American military government, soon ceased to be recognizable, so many were the modifications which proved to be necessary, but, in a large sense, they still obtain as the law of the American portions of occupied Germany.

The various regulations of this proclamation will be taken up one by one in the following pages, and their modifications, as made necessary in the light of practical experience, pointed out separately. It is hoped in this way to emphasize the importance in the future of having a thorough knowledge beforehand of the customs of countries to be occupied.

IDENTIFICATION AND CIRCULATION.

> *Identification.*—Every person above the age of twelve must carry at all times an identity card bearing his signature and address. Such card will be issued and stamped by the appropriate civil authority.
>
> A change of address must be immediately notified to the appropriate civil official and indorsed by him on the identity card.
>
> The head of each household must keep posted on the inside of the outer door of the building a list showing the name, nationality, sex, age and occupation of every person of his household residing in the building.
>
> *Circulation.*—Circulation will be controlled by the American authorities. Burgomasters, under the direction of the American authorities, will regulate travel within the American zone, and will be held responsible for a strict compliance with all regulations. Authority to leave the zone will be granted only by a division or higher commander.

These regulations were promulgated to assure the safety of the army. In the months following the armistice, one could never feel quite certain that hostilities might not recommence. Spies were therefore to be reckoned with, and, incidentally, the many radical agitators who travelled from one city of Germany to another. A strict control of circulation was therefore believed to be necessary.

Contrary to expectations, however, disorders did not arise, and the people of occupied territories eagerly turned from four years of war to the all-absorbing task of revivifying their decayed industrial life. The American army was therefore confronted with a situation to which the Anordnungen were never meant to apply. However necessary restrictions of circulation may be in war, they hardly contribute to restore communities to their ordinary existence. Each week of the occupation, requests of business men to travel from zone to zone and from occupied to unoccupied territories became more and more numerous. They finally became so numerous that the American circulation personnel was not nearly sufficient to investigate the thousands of individual requests for passes. The stamping of passes became a perfunctory performance, irritating to the Germans, who were compelled to wait in line for hours in all sorts of weather, and without corresponding benefit to the United States. While fully alive to such conditions, the American army felt bound to proceed in this matter in accordance with the desires of its Allies. A different set of rules for circulation in each of the Allied zones would have been even more disastrous to the economic unification of occupied territory than were the severe restrictions of the Anordnungen. Alteration of the circulation regulations was therefore primarily a subject for conference and mutual agreement among the Allies. Such conferences were held monthly at General Payot's headquarters at Wiesbaden, and formed a part of the regular proceedings of the *Direction Générale de Chemins de Fer du Ravitaillement en Allemagne* (Direction of Railways for Food Supply in Germany). The severer restrictions were gradually removed, first by modifying the requirements for travel by a civilian between zones, and later to a very considerable extent between occupied and unoccupied Germany. The adoption of a tri-lingual circulation pass, valid in all the occupied zones, greatly simplified matters.

The demands on the American circulation personnel, however, were diminished only slightly by these modifications. By August, 1919, and with the return of the majority of the army to the United States, it became necessary to employ Germans under American supervision. It is to be clearly emphasized, however, that the employment of German civilians in the American pass office did not place the control of civil circulation in German hands. The army circulation rules remained in effect as long as American military government continued in undisputed control.

With the publication of the ordinances of the High Commission as military orders of the Commanding General, American Forces in Germany, on January 10th, 1920, the provisions of the Commission in regard to circulation naturally supplanted previous military regulations. Germans were thereafter permitted to circulate freely in occupied territory, subject only to their possession of an identity card and compliance with the rules of their own government.

Up to January 10th, 1920, 2317 German civilians were convicted of violations of the American circulation regulations.

> ARMS AND AMMUNITION.
>
> The carrying of arms or deadly weapons is forbidden except by the local police.
>
> Every person in possession of arms and ammunition of any kind must deliver them to the American authorities at such time or place as may be appointed. A receipt for each weapon will be given at the time of delivery and the weapon tagged with the name of the owner.

Precedents for forbidding the civil population to carry arms and ammunition during a military occupation are so numerous that the publication of such regulations by the American army in this instance cannot be considered a severe restriction. During the march to the Rhine, the Commanding General of the Third Army had recognized the necessity of a regulation of this nature, and in Memorandum No. 4 had instructed unit commanders to forbid civilians to carry deadly weapons. This regulation was acknowledged to be a temporary expedient, but its immediate application was considered none the less desirable. General Pershing's proclamation, when received, definitely excluded the civil police from the necessity of turning in their arms, which was a measure that practical experience during the advance had already shown to be advisable. The army, in fact, had gone even further and had temporarily permitted duly appointed government foresters to carry light hunting weapons. This permission was revoked by General Pershing in the Anordnungen, the police alone being authorized to carry arms.

While the proclamation was definite enough in designating the class of people excepted from its operation, the manner of depositing arms was not sufficiently clear and further explanatory instructions seemed advisable to the army. The primary purpose of the order was to make sure that arms and ammunition were under proper guard and not at the disposal of a revolutionary or enemy group. At the same time, although it was realized that the owners of arms should receive proper receipts for their property, the army did not wish to accept responsibility for their return. Memorandum No. 5, Third Army, January 5th, 1919, therefore ordered that arms and ammunition should be turned in to the German Burgomasters of towns specifically designated in every Kreis. These officials would be required to issue receipts to the owners and would thereafter be held responsible for their safe keeping. Guards over the arms, however, were to be furnished by the American army. This procedure relieved local officers of a large amount of "paper-work", and incidentally placed the responsibility upon the Germans for the numerous losses which took place. When in June, 1919, claims against the United States for losses of firearms began to reach Headquarters Third Army, in large numbers, the Office of Civil Affairs specifically informed the Oberpräsident of the Rhine Province that we had never accepted responsibility for the safekeeping of arms, and that the owners must seek reimbursement from their own government.

Modifications of this paragraph of the Anordnungen during 1919 were fairly numerous, due to the army policy of placing more and more responsibility on local officials. Another reason for the many modifications was our recognition of the inadvisability of restricting hunting indefinitely, which sport was, in fact, rendered impossible by the Anordnungen. The army, as has been pointed out, had in December temporarily authorized foresters to carry hunting arms. Foresters are in fact "rural police", whose duties require them to protect the forests and fields and particularly to prevent poaching. This privilege was restored to them in May, under the provisions of General Orders No. 42, Third Army, which specified four classes of officials who were to be entitled to carry arms:

(a) National gendarmes;
(b) Municipal police;
(c) National forestry officials;
(d) Municipal forestry officials.

In July, 1919, following the signing of peace at Versailles, permission was granted to other groups of government officials to carry arms while on duty. These were the personnel of the railway and customs services. The reduction of the American Forces from seven to two divisions had rendered it impossible for the Commander in Chief to spare the former number of men for guard purposes. The German government showed no evidence at this time of renewing hostilities, and danger was to be feared only from Sparticists or fanatics. Against men of these classes the German railway guards appeared to offer sufficient protection. The American army had also by this time come to the conclusion that it would be desirable to permit Germany to reestablish customs posts on her western frontier. Without a customs service, fully operating, there appeared to be small chance to restore the economic life of Germany, which was absolutely necessary if she was to pay even a part of her indemnities to France and Belgium.

Customs guards in American occupied territory were therefore on July 28th, 1919, accorded the right to carry arms while on duty. It is interesting to note that when the western Kreise bordering Luxemburg and Lorraine were ceded to the French in August, 1919, this privilege was revoked, and the German customs-frontier thereafter remained inoperative until re-established by order of the Interallied Rhineland Commission on November 15th, 1919.

In regard to hunting, the Office of Civil Affairs fully appreciated the burden of the restrictions placed on that sport. Hunting privileges are in Germany a source of income to the individual or community owning the land in question. Hunting rights are auctioned by the village or by the property owner once a year to the highest bidder. Our regulations on arms and ammunition therefore deprived many towns of a considerable proportion of their annual income. As early as March, 1919, letters reached the Office of Civil Affairs from many different sources, begging that hunting weapons be returned and that orders forbidding the sport be revoked. In not a few localities, these orders seem to have resulted in real hardship to the farmers, as wild boars had multiplied and were ruining fields and crops. The Commanding General had for a short time in January permitted civilians to draw their arms from repositories on receipt, but the results had been unsatisfactory and had led to general abuse of the privilege.

In August, however, the Officer in Charge of Civil Affairs instituted a thorough investigation of German hunting laws and customs, with a view to restoring hunting privileges to persons duly authorized under their own laws. This investigation disclosed the suspected fact that unnecessary financial hardships had been placed on many communities by the American regulations. Paragraph 14 of Civil Affairs Bulletin No. 14, August 16th, 1919, accordingly authorized the return of sporting arms and ammunition to their owners and authorized hunting in accordance with German laws.

Violations of our regulations pertaining to arms and ammunition were fairly frequent during the whole occupation, and were severely punished when brought to the attention of provost courts. Up to January 10th, 1920, there had been 302 convictions for such violations, in a large number of which cases imprisonment was awarded.

Our policies, in regard to both arms and ammunition and hunting, were very similar to those pursued by all Allied armies. Both England and France authorized hunting at the same time and in the same manner as did the American army. They, however, at all times permitted officers and soldiers to hunt, which privilege for a long time was denied to Americans. When it was finally authorized, in the Autumn of 1919, a provision of the order required officers and men to comply with the German civil regulations.

During the latter part of 1919, the only material change in our policy as to arms and ammunition was to permit members of the civilian police reserves, which had been organized under American supervision, to carry pistols.

The Ordinances of the High Commission, issued on January 10th, 1920, as orders of the Commanding General, A. F. G., made very few changes necessary in established principles. Though a few more restrictions were placed on hunting, they were unimportant.

ALCOHOLIC DRINKS.

> The sale or gift of all alcoholic drink, except light wines and beer, is forbidden. Alcohol for medicinal or industrial purposes is not within this prohibition. The sale or gift of light wines or beer is prohibited except from 11:00 A. M. to 2:00 P. M., and 5:00 P. M. to 7:00 P. M. A violation of these orders respecting drink will render the offender liable, in addition to other punishment, to confiscation of his stock and the closing of his business.

The Moselle Valley, long celebrated as the center of Germany's wine industry, lay in the heart of American occupied territory. The civilian population had been accustomed to regard wine much as Americans regard water. They were of course entirely ignorant of the strong feeling against alcohol and wines which had come into prominence in the United States since the commencement of the war. They would never have understood any order prohibiting the sale of wine, and it is inconceivable that any military force could have stamped out the traffic in either wine or beer. On the other hand it was realized that a double set of regulations for soldiers and civilians would have been equally impracticable for the time being. Illicit sales to soldiers could hardly have been prevented under such conditions, and it is quite possible that much ill-feeling would have been stirred up among the soldiers at the sight of the conquered enemy enjoying privileges that were denied the conqueror. The solution of the problem, reached by General Headquarters and published in the Anordnungen, attempted to find a medium between the desire of the army and the necessity of recognizing the use of wine and beer by the civil population. Two general principles were laid down and thereafter served as a guide to future commanding officers in promulgating further regulations on the same subject. These were:

1. An absolute prohibition of the sale or gift of strong alcoholic drinks, sale of beer and light wines alone being permitted.
2. Police hours for public hotels and cafes, in order that the consumption of wines and beers might always be supervised by the American military authorities.

The first principle required further interpretations as to the phrase "light wines."

Letter of Instructions No. 9, Third Army, January 30th, 1919, directed that sherries, ports or maderias were not to be considered as "light wines" within the meaning of this order, and on February 24th, the same distinction was extended to include German and French champagnes.

It is needless to remark that these regulations, lenient though they were, were not looked upon kindly by the civil population. The German pointed to the fact that his armies in Belgium had never issued any such regulations, and realization of the fact that Americans held different views in regard to alcohol than his own only gradually made itself felt. Experience has shown that military regulations in an occupied territory are extremely difficult to enforce if they run counter to long-established customs. Our experience in the Rhineland in enforcing the regulations, both as to circulation and alcohol, amply confirm this statement. Up to January 17th, 1920, there had been 863 convictions of German civilians for violations of regulations on alcohol—524 for selling cognac, champagne, etc., and 339 for sales out of legal hours. The number of civilians convicted was exceeded only by the number for two other offenses, viz: violation of circulation orders and sale of United States property. A large majority of the trials for violation of this regulation, however, was for sales to members of the military forces. The civilian was never wholly to blame for sales of this character, as the soldier in his eagerness for cognac offered such high prices that the profit to liquor dealers was large enough to risk imprisonment.

However as the occupation wore on it became more and more clear that the regulation should conform more nearly to local customs. In February, 1919, army orders directed that the evening closing hour of restaurants and cafes be extended to 10:00 P. M. The change was largely due to the fact that the German evening meal was usually not before 8:00 P. M. The former regulation had therefore made it impossible to serve wine with meals in hotels, clubs, private houses, etc. The new regulation, while pleasing the civil population, was not agreeable to some organization commanders. Drunkenness among the soldiers promptly increased, and commanders on their own initiative, reestablished the former police hours, or, in some cases, set new hours, in order to cope with their particular situations. However desirable such a course may have been from the organization commanders' viewpoint, the result was disastrous to uniformity as to military government. In one town, for instance, cafes were not allowed to sell even wine or beer; in the neighboring village, the local commander permitted them to remain open until 7:00 P. M., while in a third village the army closing hour of 10:00 P. M. remained in effect. The Army Commander, appreciating the impossibility of permitting each commander to make his own regulations, directed commanding officers in Letter of Instructions No. 15, March 13th, 1919, not to extend nor restrict the police hour as established by the Army.

When peace had been signed at Versailles, it became desirable to take an even further step, and, by removing unnecessary restrictions, restore the normal life of occupied territory to the largest extent compatible with the security of the army. The Regierungspräsident of Coblence was requested to submit such recommendations as he desired, with a view to modifying the standing army regulations on alcohol. Much to the surprise of the military authorities, he informed them that he did not favor removal of the restrictions on cognac and strong liquor, because he feared friction between soldiers and civilians as a consequence. His letter merely asked that cafes be permitted to remain open an hour longer on week-days and until midnight on Sundays and holidays. The Commanding General decided that this request was both moderate and wise and accordingly authorized the hours requested in a letter to the Oberpräsident of the Rhine Province, dated August 22nd. The police hours thus established remained in effect as far as civilians were concerned, until the close of the military government.

The return of the regular divisions to the States in the Summer of 1919 and the arrival of untrained replacements from America, rendered the success of the new closing-hours, from the military point of view, very questionable. It had been a cardinal American principle from the first days of the occupation, that soldiers and civilians, as far as service in public-houses was concerned, should be treated alike. As the court-martial statistics rose constantly, without corresponding increase in offenses by civilians, it was realized that a departure from former principles must be made and one set of hours fixed for soldiers and another for civilians. To have punished civilians for the misdeeds of American soldiers, would have been an act unworthy of the ideals with which the army entered the war. Sales of wine and beer by cafes, restaurants, etc., to soldiers were therefore restricted to the hours from 5:00 P. M. to 9:30 P. M. The civil population was however warned, that as the sale of forbidden cognac and other strong liquor to soldiers showed no signs of diminishing, the punishment in every case would in future include confiscation of the owner's stock of liquor of the brand illegally sold. (Civil Affairs Bulletin No. 38, November 4th, 1919.)

On December 23rd, 1919, a new regulation of the Army Commander (Civil Affairs Bulletin No. 47), interpreted the term "light wines" to include sparkling wines, such as champagne, sparkling Burgundy and "Schaumwein". Sherry, port and maderia were still left under the ban. This was the last order issued on the subject prior to the publication of the Ordinances of the Inter-Allied Rhineland High Commission on January 10th, 1920. In the meantime, however, a letter had been sent the Commanding General by the Deputy French Commissioner of the Inter-Allied Rhineland Commission, requesting that the American authorities permit the sale of cognac, etc., to civilians. The Commissioner stated that his government viewed our regulations as a serious detriment to French foreign trade, and requested that fines collected by our provost courts from civilians for offenses involving purchases of French cognac, be revoked. The Commanding General felt that the sale of cognac was a subject in which the German officials were primarily interested and that it was entirely proper in this case to consult them. Both the Oberpräsident and the Regierungspräsident maintained their previous position, that the unrestricted sale of alcohol for the time being was clearly inadvisable. The request of the French Deputy Commissioner was therefore refused.

The Ordinances of the Inter-Allied Rhineland High Commission went into effect on January 10th. By the provision of Article 28 of the Ordinance regarding criminal and civil law, traffic in cognac, etc., among civilians was no longer forbidden, although sales to soldiers were made punishable offenses.

ASSEMBLIES AND MEETINGS.

> All gathering of crowds is forbidden.
>
> No meeting or assembly of persons shall take place without authority of the local military commander. Sessions of courts and councils, schools and religious services may be held as usual.

The restriction and supervision of the right of public assembly in the Rhineland was a military measure which had many precedents in previous occupations. In the Autumn of 1918, Germany's expressed desires for peace were viewed with considerable suspicion, and rigid regulations were believed necessary to assure the safety of the invading army. The advisability of promulgating some such regulation had been foreseen as early as December 1st, and Memorandum No. 4, Third Army, had directed unit commanders to forbid crowds and meetings in towns along the projected line of march where troops might billet over night.

A resumé of the later modifications of this provision of the Anordnungen may be said to be similar to the trend of modifications of many other ordinances. The continuation of a rigid military government, with hard and fast restrictions, becomes in the end as burdensome to the army as it is oppressive to the civil population. Severest measures were proper while hostilities were in progress, but there existed no justification for retaining them when once the occupied territory had become peaceful and orderly.

The right of assembly was therefore gradually restored to the civil population of the American area. When restriction was removed, however, it was so safeguarded that no increased danger was feared in consequence for the troops from riots or hostile demonstrations.

Every facility was given to the people, even in January, 1919, to discuss the platforms of the political parties which were engaged at that time in the campaign for supremacy in the national elections. As a matter of fact, requests for permission to hold meetings were granted at once under ordinary circumstances. If the meeting appeared suspicious, an American representative was directed to be present and report or check, as the case warranted, inflamatory speeches or disorder. The moderation with which regulations were enforced, may be judged from the fact that street parades and celebrations were permitted on May 1st in honor of the German revolution. The many religious processions on church holidays, so common in the Rhineland, were always permitted, although display of the national colors was stringently forbidden.

The return of the German prisoners of war from English and American captivity towards the close of summer was a factor which gave rise to apprehensions by the military authorities of possible public disorder. It was practically certain that the families of the prisoners would congregate at railway stations to greet them, especially as many prisoners had been absent from their homes for more than four years. The American authorities did not intend to forbid legitimate welcome of relatives, but they feared that the legitimate welcome of a family might easily be turned by agitators into a hostile demonstration against the occupying forces. Orders were therefore drafted by the Office of Civil Affairs rigorously forbidding processions or celebrations in honor of returning prisoners of war and incidentally prohibiting the display of the German national colors. However, on the special plea of the Oberpräsident of the Rhine Province, this order was never published in the local press. His request was based on the fact that the publication of such an order might be interpreted as forbidding the legitimate welcome of the prisoners' families, and, if such were the case, would be certain to give rise to ill-feeling. He begged that he himself be permitted to transmit our instructions to subordinate officials, and said that if this were allowed, the civil officials would accept responsibility in case of disorder. Although the danger of leaving matters of such importance to the discretion of the civil authorities was realized, the request was granted, largely in view of the Oberpräsident's correct attitude in carrying out all previous demands of the army. However, although the same cooperation from the Oberpräsident was received in this matter as in others, an official in the Regierung of Coblence failed to transmit the instructions of the Oberpräsident to the Landrat of Mayen, and there was an unnecessary delay in informing the Landräte of the other Kreise. It was therefore not unnatural that, during the last week in September, our orders should have been violated in the city of Mayen (Kreis Mayen), at that time without an American garrison. The national German colors were flown from a number of homes in the city, and a celebration seems to have been held by the populace generally in honor of the returning prisoners of war. No effort was made by the civil authorities to check the celebration. While, in this instance, the Regierung of Coblence seems mainly to have been at fault, the American policy in regard to the display of the German flag had been so repeatedly announced at the time that permits for religious processions were granted, that the local Mayen officials could not escape a share of the guilt. Both the Burgomaster of Mayen and his chief of police were accordingly suspended from office for thirty days. The responsible official in the Regierung was also suspended from office for a similar period.

Reference has been made in these pages to the fact that supervision of public meetings placed a heavy burden on the shoulders of the American personnel detailed to that work. In Coblence particularly, there were hundreds of meetings weekly, and the Second Section of the General Staff, whose duties included supervision of such meetings, could cover only those of a political and industrial character. In the course of the year 1919, it became evident that no attempt was being made by the German Government to stimulate disorder in our area. There was nevertheless a certain danger from radical and revolutionary agitators, which, however, the Oberpräsidium and the Regierung had an even more lively interest than ourselves in checking. Accordingly, on August 26th, control of meetings was turned over to the civil authorities with the following reservations: That notice of meetings of an industrial or political nature should be given by the civil authorities to the Second Section of the General Staff forty-eight hours in advance. This enabled intelligence agents of the army to cover any meeting likely to arouse suspicion.

It must be recorded that the Oberpräsidium, the Regierung and the Landräte entered into the new arrangements in an entirely proper spirit, and referred all doubtful cases to the American authorities.

> PUBLICATIONS.
>
> A copy of each newspaper or other periodical publication will be delivered to the local military commander immediately on issue. The appearance of any matter reflecting on or injurious to the American military government will render the publication liable to suspension or suppression. Excepting the periodical press, no printed matter will be published without permission from the local American military authority.

The moderation of the restrictions imposed by the American military government on the press proved in the end to be a wise course. Not only did the United States desire a free discussion of the political issues which, in December, 1918, were agitating Germany, but it desired in particular that the Rhenish people should have a share in the foundation of the new national government. Criticism in the German press of the Allied military governments was of course objectionable, in view of the danger that such criticism might incite disorder. This fundamental principle governed all actions in regard to censorship, and though criticism of the military government was later so defined as to include criticism of the policies of the Allied peace delegates at Paris, it was aimed solely at constituting a safeguard for public order. Newspapers were at all time permitted to discuss the domestic issues before Germany, without interference or pressure on our part. Not once did the American authorities instigate the publication in a German paper of an article of a propagandist nature. Within the limits set by the Anordnungen, the right of free speech was safeguarded to the population of occupied territory.

Coblence was not a center of the printing trades, and the authorities were never forced to censor books or periodicals other than the local press. The importation of magazines, etc., from unoccupied Germany was an entirely different matter, and one which called for military action. Censorship of periodicals originating elsewhere was, for a while, very irregular. Several magazines were forbidden sale in the American zone,—*Kladderadatsch*, for instance, which printed matter derogatory to the United States. It was felt, however, that entirely to exclude German periodicals published on the other side of the Rhine was to impose unnecessary hardships on the civil population, the weekly and monthly press on the left bank being unimportant in quantity and quality.

Finally on September 27th, 1919, booksellers were made responsible that no publications on sale in their shops contained matter offensive to the military government. This proved a satisfactory solution of the problem, and thereafter but very few complaints on violation of orders reached the Office of Civil Affairs. A certain number of suspensions of the local press for short periods are of record, but in no greater number than one would except, in view of the prevailing disturbed state of public feeling. No periodical was ever permanently suspended.

Publication of all rules in regard to censorship was decided to be, as early as January, 1919, a function of the Office of Civil Affairs at Advance G. H. Q. Enforcement of the rules so laid down was the duty of the Second Section of the General Staff, and at all times exercised by it.

POST, TELEGRAPH AND TELEPHONE.

The mail is subject to the censorship of the American military authorities.

The use of the telegraph or long distance telephone is forbidden except by permission of the local military commander.

The use of aerial wireless apparatus is forbidden. All private telephones or telegraph apparatuses, ground or aerial

> wireless apparatus, must be reported to the military commander at once.
>
> No person may, without authority of the local military commander, transmit any message or communication to any person outside the territory occupied by American troops except through the post office.

The above regulations certainly placed severe restrictions on the civil population, but as they were prepared to govern an enemy country in a state of war, the necessity for a tight rein on all channels of communication must be recognized.

On January 10th, 1919, the probabilities of further war having been dissipated by the rapid demobilization of the enemy's army, Advance G. H. Q. authorized resumption of telegraphic and telephonic communication, both within occupied territories and with unoccupied Germany. Provision, however, was still made for a rigid censorship. Messages in regard to private affairs were permitted only in urgent cases. Telephonic or telegraphic messages of a trivial or unimportant nature were still forbidden.

On April 5th, Orders No. 24, Advance G. H. Q., still further modified our instructions by permitting nearly all kinds of commercial messages to be exchanged with unoccupied Germany. It also announced the resumption of mail service between occupied territories and Allied countries. On May 9th, the German parcel post system was even authorized to resume operation between the Rhineland and the rest of Germany. All packages were, however, rendered liable to censorship.

On August 21st, two months after the signing of peace at Versailles, and following an agreement among the Allied armies, the Oberpostdirektor of Coblence was informed that mail, telegrams and telephone messages might follow their normal routes without restriction. All messages were still subject to censorship, but such censorship was to be conducted at irregular intervals, in accordance with the "Sondage" system. Groups of American censorship personnel were to go from one town to another of our area and censor telegrams, telephone and post during their stay.

The Ordinances of the Inter-Allied Rhineland High Commission (G. O. 2, A. F. G., January 10th, 1920) practically restored the mails, telegraphs and telephones to their peace basis. The High Commission, however, reserved the right to censor or seize any package or message whose appearance or nature aroused its suspicions.

> PIGEONS.
>
> The use of carrier pigeons is forbidden. Owners of carrier pigeons will make an immediate report of their pigeon cotes to the military commander, with a list of the marks of their pigeons. Pigeon houses must be kept open day and night.

The above regulation was applicable primarily and solely to a state of war. As soon as civilians were permitted to circulate between occupied and unoccupied Germany, any use of pigeons to transmit information concerning the Allied armies was, in comparison, cumbersome and superfluous. Such circulation was never prohibited, and in consequence, almost as soon as the proclamation was posted throughout the American area, the prohibition became a matter of no importance. Attempts by the Provost Marshal and the military police to enforce the order strictly would have been merely so much waste effort, and were consequently abstained from. Throughout the whole period of military occupation, not a single prosecution was conducted for violation of this ordinance. It may be stated, however, that the pigeon cotes in the American area were almost empty, the vast majority of carrier pigeons throughout Germany having been requisitioned during the war for the use of the armies in the field. So slight was the interest of the civil population in this restrictive regulation that no appeal ever reached the Office of Civil Affairs from German sources for its modification or revocation, even when the military authorities were perfectly willing to have taken such a step.

PHOTOGRAPHS.

> The taking of photographs out of doors, except by permission of the local military authority, is forbidden.

This was another instance in which the proclamation took steps to guarantee the safety of the army, foreseeing the possibility of resumption of the war. The civil population of course obeyed this order, but as the country returned gradually to an appearance, at least, of peace, they deemed it an unnecessary hardship. No particular efforts were ever made by the military police to enforce this paragraph, and it gradually lapsed. There is not even a single record of trial of a civilian for its violation. The use of cameras finally became so common that a formal revocation of the order would have only called attention to the previous unchecked violations of it and it was believed best to let it remain a dead letter.

SPECIAL RULES.

> Whosoever:
>
> (a) Attacks, disturbs or impedes any American troops or officer thereof * * * will be punished as a military court may direct.

Such a regulation was of course absolutely necessary to the occupying forces. Brawls, fights and assaults constantly occurred between soldiers and civilians as a result of over-drinking, and in at least half of these, civilians were the aggressors. Whoever the offender in the matter, however, physical violence by a civilian towards a soldier, would have gone a long way towards weakening the army if permitted to go unpunished. This regulation did not of course make it an offense for a civilian to defend himself from unprovoked attack by a soldier, but it did protect the soldier from assaults by civilians, and our military policemen from any physical attempt by the wrongdoer to escape the consequences of his offense. As will be noted in a succeeding chapter, dealing with relations of the army with the people, physical encounters in which soldiers and civilians took part were by no means uncommon occurrences during the occupation. Soldiers were constantly being brought to trial by court-martial, and civilians by the provost courts. Most of the cases, however, were mere brawls, whose origin could be traced to excessive consumption of alcoholic beverages, rather than to defiance of the authority of the military government. Up to November 11th, 1919, 315 civilians had been convicted for violations of this regulation. The subject of the punishments accorded soldiers so involved is treated in chapter 11 of this report, entitled, "Relations Between Army and Civil Population in Occupied Territory."

> Whosoever: * * *
>
> (b) Destroys, damages, or disturbs any railway, telegraph or telephone installation, lighting, water or power system, or any part thereof, * * * will be punished as a military court may direct.

This was still another instance in which the proclamation provided suitable ordinances to protect the security of the army. As Germany every day showed a more and more evident intention of abiding by the terms of the armistice, self-destruction of public works became more harmful to her own interests than to those of the armies of occupation. This paragraph therefore became superfluous and no violation of the regulations was ever recorded.

> Whosoever: * * *
>
> (c) Destroys, damages, steals or secretes any property of or in possession of, the American army, * * * will be punished as a military court may direct.

Up to November 11th, 1919, twenty-six civilians were convicted by provost courts for destroying government property, and 724 for stealing the same. The great number of Germans who appear to have stooped to theft during the occupation would at first sight give cause for reflection, and possibly furnish grounds for the statement that the Rhinelander is not honest. There can be no question, however, that, in view of the dearth of many articles of food and clothing, a certain amount of theft from government stores had to be accepted as normal under prevailing conditions. The *morale* of the nation had been shaken by four years of suffering and want, and the vast stores of army provisions, blankets, etc., offered a tempting prize to people who lacked these very articles. The protection of these supplies was usually insufficient, and arrests were ordinarily made only when the stolen goods were found in possession of the thief. Thefts appeared therefore a simple matter to the lower class German, and as one incurring slight risk of detection. It should be noted, in connection herewith, that many civilians were employed in army depots and repair establishments, which rendered them peculiarly liable to temptation. 781 convictions of German civilians are of record for these offenses up to January 10th, 1920.

> Whosoever: * * *
>
> (d) Purchases, receives in pawn or has in his possession, articles of clothing, equipment or rations furnished to American soldiers or belonging to the American armies, * * * will be punished as a military court may direct.

This regulation was at all times difficult to enforce. More than 1,500 convictions of German civilians for violations thereof are recorded. Despite these convictions, army property still continued to find its way into unauthorized hands, affording ample evidence that our preventative measures were not effective. Lack of food and clothing, which the army possessed in abundance, is, of course, sufficient explanation of the illegal traffic. Clothes, soap, cigarettes, and many articles of foodstuffs could be procured by the Germans from their own sources only in limited quantities. Added to this fact, it must be borne in mind that the prices which had to be paid by a civilian for shoes, clothing, meat, etc., were enormous, particularly in the winter of 1918. Even the most necessary articles became prohibitive luxuries to the laboring man. It is therefore not surprising that soldiers were often approached by business men of the "Schieber" variety and offered tempting bribes to transfer these luxuries *sub rosa* to civilian ownership. While exhorbitant bribes were no doubt in many cases to blame, the soldier cannot but be accorded a share in the guilt. As long as government property was not authorized to be disposed of to the civil population, offenses could be readily detected and the civilian concerned punished; but in the spring and summer of 1919, the situation became more complicated, and thefts more difficult to detect. The Liquidation Commission proceeded about this time to dispose of much of the salvaged clothing and surplus horses and motor transport belonging to the army. The army itself and, to some extent, the Hoover Commission also, undertook at the same time to supply large quantities of food and tobacco to the Rhine Province. Such sales increased the chances for the average German to escape detection when he bought sales-commissary articles from soldiers. Purchases of cigarettes by soldiers at the sales-commissary indeed increased so phenominally that the army finally found it necessary to ration such purchases. This measure to some extent checked the transfer of foodstuffs and cigarettes to German hands.

The problems of preventing the illegal disposal of United States property to enemy civilians proved so difficult during the occupation of Germany, that it warrants close attention, in view of the possibility of a future occupation of the territory of an enemy.

> Whosoever: * * *
>
> (e) Destroys, damages, pollutes or secretes any fodder, water or other thing useful to the American army, or
>
> (f) Acts as a spy or commits war treason against the American army, * * * will be punished as a military court may direct.

No necessity ever arose to prosecute civilians under either of these statutes. The comparative peace which the Rhineland enjoyed during the occupation, in contrast to the anarchy in unoccupied Germany, accounts in no small measure for the absence of a feeling of hostility by the civil population towards the American forces. The Rhinelander was heartly sick of the war. He had not only done his share of the fighting but had been called upon in addition to see the armies of Germany and the Allies march across his territory. He was in no mood to bring further burdens upon himself by useless friction with the occupying troops.

> Whosoever: * * *
>
> (g) Commits any act whatever injurious to the American army or in obstruction to the military government, or in violation of these regulations or any regulations or orders hereafter made by the American military authorities, or refuses to comply with a requisition; or
>
> (h) Attempts to do any of the above things, or advises or assists anyone else to do or attempt to do any of them;
>
> will be punished as a military court may direct.

This regulation naturally demanded explanation from time to time, as offenses against the military government took place. The Army Command and the Office of Civil Affairs did not believe it wise to create punishable offenses under the terms of this article without informing the population of their action. Additional orders were therefore issued from time to time on a wide variety of subjects. The list of acts thus made punishable offenses was very numerous, and there were consequently a great number of persons tried by provost courts.

Some of the more important of these offenses up to January 10th, 1920, are noted in the following table, with the number of convictions for each offense:

False statements to personnel of United States army	68
Failure to obey lawful orders	556
Drunkenness and disorderliness	177
Violation of American sanitary regulations	918

Disrespect to officers	81
Prostitution while diseased	91
Discrimination in prices against Americans	16

Miscellaneous.

The term civil authority, as used in these regulations, means the Burgomaster or Gemeindevorsteher or the head of the local community.

Every person must familarize himself with these regulations and all others hereafter made by the American military authorities. Ignorance of the regulations will not be accepted as an excuse for their violation. In case of doubt as to the requirements of the regulations, inquiry should be made at the Rathaus.

No comment is necessary on the concluding paragraphs of this proclamation. The Anordnungen must be regarded as a remarkable document, particularly for the comparatively moderate restrictions therein imposed on the civil population. Coming as the occupation of the Rhineland did on the heels of a struggle distinguished for bitterness and hatred, one must recognize the judicial quality of the proclamation, and the complete absence of harshness and pettiness. Any other policy would have been unworthy of the ideals for which America had struggled. A careful reading of the proclamation must lead the reader to a clearer appreciation of the spirit with which the army of occupation carried out the task imposed on it.

Chapter No. 7

SANITATION AND PUBLIC HEALTH

Administration of Sanitary Laws; Duties, etc., of Kreisarzt; Supervision of Medical Practice; Health Laws; Statistics of Infectious Diseases; Street Railways; Sewage Disposal; Prevention of Nuisances; Street Cleaning; Removal of Waste; Milk Supply; Dwellings; Water Supply; Prostitution; Recapitulation; Population.

Administration of sanitary laws.—Administration of sanitary laws and of matters concerning public health of the Kreis, are in Prussia largely exercised by, and dependent upon, the Kreisarzt, who is the government official of sanitation and public health for the Kreis. In this capacity the Kreisarzt acts as advisory official to the Landrat in the Landkreis and to the police in the Stadtkreis, and is also the medical legal official of the district in which he serves.

Appointment of the Kreisarzt is made by the Minister of Public Health. He is sworn in by the Regierungspräsident and assigned to duty by the Regierungs und Geheimer Medizinalrat. After ten years of service, he may be promoted to the fourth class of officials, thus becoming a Medizinalrat; after twenty-two years of service he may become a Geheimer Medizinalrat. He may resign his post at any time subsequent to appointment and assignment to duty, provided there be no charges against him, and he may be pensioned if disabled and unable to perform his duties. He may also be removed from his post for the good of the service.

The immediate supervisor of the Kreisarzt is the Regierungspräsident and, above the latter, the Minister of Public Health. All communication between the Kreisarzt and other officials, except in case of emergency, are required to pass through the office of the Regierungspräsident, and the Kreisarzt is required by law to make no disclosures to a third person,—a law which, however, is more honored in the breach than in the observance. Three times each year the conduct of the Kreisarzt in his official duties is investigated by the Regierungs und Geheimer Medizinalrat.

As an advisory official, the Kreisarzt advises the Landrat in the issuing of orders to the police for the control of public health; keeps the Landrat informed of the sanitation and public health of the Kreis, and transmits his reports and communications through the Landrat to the Regierungspräsident. On the other hand, any reports or comments concerning public health matters originating with the Landrat, and addressed to the Regierungspräsident, must first be submitted to the Kreisarzt, after being forwarded to the Regierungspräsident by the Landrat.

In each Bezirk is a supervisory official, known as the Regierungs-und Geheimer Medizinalrat, who is responsible to the Regierungspräsident and who supervises the activities of all Kreisärzte. All his knowledge concerning their actions and concerning sanitary conditions in the various Kreise, come through the Regierungspräsident. All these officials are appointed by the central government in Berlin, and apparently hold their offices indefinitely, except in case of very gross inefficiency or of neglect of duty.

In the whole system of the prevention of infectious disease and supervision of public health matters, two extreme conditions might occur. Under supervision of an efficient Geheimrat, the action and methods of the Kreisärzte would probably be similarly efficient. On the other hand, under an inefficient, inactive Geheimrat, the result would be that each Kreisarzt would perform his duty, if he performed it at all, without any real effectiveness. It is possible to imagine all degrees of efficiency between these two extremes. In Regierungsbezirk Treves, in American occupied territory, there was a fairly effective management of public health matters, but in Regierungsbezirk Coblence there was an ineffective management, with different methods prevailing in every Kreis. In this latter Bezirk, it was sufficient for the Geheimrat to know that infectious diseases had been reported to the Kreisarzt. Whether the Kreisarzt took any action, or made any investigations, was a matter of apparently no concern to the Geheimrat. This condition led to his early dismissal from the public health service and to the appointment of another Geheimrat in his place. The Kreisarzt of Ahrweiler was of the same inefficient and inactive type. As long as physicians made some kind of report to the police authorities, and as long as these reports were transmitted in turn to him, he was perfectly satisfied. Investigations were practically never made to determine the correctness of diagnosis, or whether all cases of infectious diseases were reported. His records of the cases reported to him were always easily accessible and well kept, but the effectiveness of his public health administration was practically *nil.* The only area in the American occupied territory which has caused any great concern to the occupying forces with regard to prevalence of typhoid fever therein, was this particular Kreis. In the town of Altenahr, more than 10% of the population fell ill with typhoid fever within a period of two months, due entirely to the inactivity and lack of vigilance on the part of the Kreisarzt. He was therefore removed from his post and a very able and efficient official put in his place.

Duties, etc., of Kreisarzt.—In Stadtkreise, the Kreisarzt has the same position with regard to the police of the town as has the Landrat in Landkreise. He is vested with police authority and with the power to issue orders to the police in sanitary matters. In fact the Kreisarzt is the medical export whose opinion is required at any time on any matters pertaining to the public health. He is also required to give expert medical testimony in court concerning Kreis matters. He is required to maintain relations with the private physicians of the Kreis and to make an investigation in case of infectious disease, without first inquiring whether or not the case in question already has an attending physician, and if it has, to inform the physician, so that he may be present at any investigation made.

In the matter of salaries, Kreisärzte fall into two classes, one class being employed on full time, and not permitted to assume any private practice whatever, the second class being employed on part time and permitted to have a certain amount of private practice, provided it does not interfere with their official duties. Both classes are paid for the necessary expense of carrying on their duties and are provided with an official residence. During the first ten years of service the Kreisarzt may be retired on a pension, if disabled by reason of the occurrence of disease or of injury acquired in service, if this is not due to a fault of his own. After ten years he may be retired on pension by reason of any total disability, no matter how acquired, and after reaching the age of sixty-five he may be retired on pension, whether disabled or not. The Kreisarzt is required to possess a knowledge of the climate of the Kreis, the habits of its people, the state of the drinking water, and the sanitary conditions in houses and factories. He is also required to make any simple physical, chemical, or microscopical tests incidental to his work; but any more important investigations of a chemical, microscopical or bacteriological nature may be referred to an official laboratory. These latter include such examinations as may be necessary of faeces, urine, blood and water.

Supervision of medical practice.—The Kreisarzt is required to keep a register of all medical or quasi-medical persons within the Kreis, such as physicians, chemists, dentists and midwives. He, and other physicians as well, are required to take cognizance of unqualified practitioners of all kinds, and to adopt such measures as may be necessary under the law, to prevent the practice of quack doctors.

Apothecaries.—He is further required to exercise supervision over apothecaries, in relation to the condition of their shops and drugs, and to make an inspection of each shop and the business carried on therein at least once each year. Besides supervising trade in drugs in apothecary shops, he must supervise sale of drugs by persons other than apothecaries and in places other than their shops.

Midwives.—The Kreisarzt is further required to supervise the activities of all midwives in the Kreis. This includes an examination into their fitness, personal inspection of their recordbooks, and an investigation of their work in their own localities at least once every two years. Heavy penalties are provided for infractions of the laws regarding the practice of midwifery. These laws are enforced by the Kreisarzt. On the other hand, a curious and benevolent provision stipulates that the Kreisarzt shall protect midwives from accusations of an unjust nature, or from accidents of any kind which are not due to their own neglect or fault. Before being examined in regard to their professional fitness, applicants for license to practice midwifery must present a certificate from the police regarding their personal qualifications. Except under exceptional circumstances, applicants must be between twenty and thirty years of age, and must have had a common-school education. At least once every two years the Kreisarzt must reexamine all midwifes. In the event of this subsequent examination proving unsatisfactory, midwives may be reexamined after six months. Four weeks notice of examination must be given the applicant. The Kreisarzt is also required to see that a sufficient number of qualified midwives are provided in the Kreis. Persons practicing without a license are subject to prosecution and punishment.

Disinfectors.—These officials are appointed by the Kreisarzt after an examination in bacteriology, means of prevention of infectious disease, methods of disinfection and knowledge of disinfecting apparatuses. They are reexamined every three years, and in case they fail to pass, reports are made to the Regierungspräsident and also to the people on whose recommendation they were appointed. A report is also made every year to the Regierungspräsident concerning the efficiency of these officials. In the law regulating disinfection, it is not clear as to who bears the responsibility of checking the work of disinfectors, and in the actual practice of disinfecting there seems to be no direct knowledge or supervision thereof.

When the coroner for any reason is not available, the Kreisarzt may undertake his duties.

All towns within the Kreis must be inspected at least once every five years by the Kreisarzt, according to a plan laid out by the Landrat. Reports on results of these inspections are made to the Regierungspräsident and include both statements of the conditions as found and of the measures taken for the protection of public health.

The Kreisarzt is required to maintain cleanly conditions both within and without all dwellings, and to see that all dwellings are provided with sufficient light and air. He is required to report in the absence of proper light and ventilation to the building officials. He is likewise required to prevent the occurrence of any condition which might promote the occurrence or spread of infectious diseases. In spite of this provision of the law, several Kreisärzte have stated that they are often not permitted by the occupants to make an inspection, or to enter a building unless some disease necessitating investigation has occurred therein. Approval of the Kreisarzt is required for all building regulations and ordinances in so far as they concern light and ventilation, water supply, drainage, removal of refuse, or sanitary matters in general.

Water.—The Kreisarzt is required to maintain a supervision of the water supply of the Kreis and to make examinations and investigations as often as may be necessary to ensure its purity. The frequency of examination is determined very largely by the condition found on the previous examination, or by the occurrence of drought or flood. In general, the water supply is required to be examined within a period of from one to two years. The Kreisarzt must, according to law, be guided more by an investigation of the sources of the water supply than by the results of chemical or bacteriological investigation. This latter provision is a very sensible one, but it is also a fact that the provision regarding reexamination every one to two years has been understood by many Kreisärzte to indicate that they need make no investigations of water supply that has already been examined within a period of two years, except in the case of an epidemic of typhoid fever. This, however, is manifestly not the intention of the regulation.

In the case of installation of a new general water supply, the plans must be submitted to, and approved by, the Kreisarzt before the project may be carried out. This official is also required to maintain a vigilant watch to prevent the existence of nuisances at the sources of water supply.

The Kreisarzt must also see that sewage and all refuse is properly disposed of, whether by drains or by carting it away. In case he needs expert advice, he must apply to the Experimental and Investigating Institute for Water Supply and Waste Water (Versuchsund Prüfungsanstalt für Wasserversorgung und Abwasserbeseitigung) in Berlin.

Food.—The Kreisarzt is required to supervise the sale of food, not only with regard to its physical condition, but also its food value and price. He is required especially to supervise the milk production, because, as the law states, milk forms a food for infants and children. The fact that it is a culture medium for many pathological bacteria and is frequently infected by handling, appears to have been considerably overlooked. This, however, is explained largely by the reason that the German people almost always boil milk before using.

With regard to meat, the Kreisarzt must exercise vigilance to prevent the conveyance of parasitic disease, and is also required to supervise cattle slaughtering. Meat control, however, is in general under direct care of the Tierarzt, and the work of the Kreisarzt is mostly of a supervisory nature and from the standpoint of protection of public health.

Manufacture of mineral waters is under the supervision of the Kreisarzt, who must maintain in a clean condition all places where mineral water is traded in. He must also verify the label and see that it is not false nor misleading in appearance.

Alcohol.—The supervision of the sanitary law regarding supervision of sale of alcohol is so peculiar that the text is quoted in part:

> "The Kreisarzt must observe any tendencies towards excessive use of alcohol in his Kreis, and institute and support measures to check the abuse of alcoholic beverages. He must decide whether local restrictions in granting of licenses for retailing beer, wine, or spirits, police orders concerning opening and closing hours of public houses, and establishment of consultation rooms and medical stations for victims of alcohol, appear advisable. The Kreisarzt must supervise retailing of non-alcoholic drinks in public houses, and is expected to lend his support to societies whose object is the prevention of excessive use of alcohol. Further, the Kreisarzt will seek to enlighten the people as to the dangers from both a sanitary and a social point of view, of excessive use of alcohol.
>
> "Suitable measures will be taken to this end, such as the holding of public meetings, distribution of pamphlets and discussion with school teachers at official conferences on educational work among their pupils."

The sanitary law requires the Kreisarzt to investigate all outbreaks of infectious diseases when reported to him by the police or even if not so reported. Not only is he required to make these investigations in established cases of diseases, but also when they are only suspected. He must make every effort to determine the nature of the disease, including, if necessary, bacteriological investigation or autopsy. The Kreisarzt communicates the result of his investigations to the police and to the Health Commission.

In cases of infectious diseases among animals, the Kreisarzt makes an examination in cooperation with a veterinary surgeon.

Three classes of persons are subject to examinations:

1. Those suffering from a known infectious disease;
2. Those suffering from a malady whose nature has not been determined;
3. Those suspected of being carriers or of having been exposed to infections.

The Regierungspräsident may order all persons coming from seats of epidemics of cholera, plague, typhus, typhoid, yellow fever, trachoma, or smallpox, to report to the police on arrival in a certain district. The following are subject to observation:

1. Those suffering or suspected to be suffering from trachoma, glanders, relapsing fever and typhoid;
2. Those suffering or suspected to be suffering from, or infected with, cholera, typhoid, yellow fever, plague and smallpox;

3. Prostitutes suffering from syphilis, gonorrhoea or chancroid;
4. Persons suffering or suspected to be suffering from rabies, or who have been, or are suspected to have been bitten, by a rabid animal.

In the case of epidemics of certain diseases, the Sanitary Police may keep places suspected of being centers of infection, under close observation, may close such factories, schools, etc., as may be necessary, and may forbid all assemblages of people.

In any place in which cases of cholera, typhus, yellow fever, plague, smallpox, dysentery or typhoid have occurred or are feared, as well as in surrounding districts, the use of wells, lakes, brooks, water pipes, water supplies, public baths, swimming pools, wash places and lavatories, may be forbidden or restricted.

Infected places may be disinfected and infected materials destroyed as the case requires. The Kreisarzt must see that places of isolation, medical care, medicines and suitable means of disinfection are provided for victims of infectious diseases, even to the extent of establishing a public steam disinfecting plant in large communities. He must either personally vaccinate the people, or appoint a physician for that purpose, and supervise the sale and manufacture of vaccine. In case of prejudice against vaccination, he is expected to inaugurate and carry out an additional campaign. He is required to keep a complete record of persons vaccinated and to forward a yearly consolidated report to the Regierungspräsident.

With reference to industrial hygiene, the Kreisarzt is required to maintain a supervision over dangerous and offensive trades, for the purpose of protecting the workers as far as possible from conditions which endanger life or are detrimental to health, and also to protect those living in the neighborhood against detrimental results of the carrying on of such trades.

The Kreisarzt is responsible for supervision of school hygiene and for proper construction of school buildings, and is required to approve their plans, to propose constructions, to provide for proper ventilation, lighting, seating, water supply, school baths, etc. His supervision of school hygiene is carried on in cooperation with the school directors and school physicians. He may close any school for the protection of public health, under certain specified conditions which are: When one or more cases of any kind of infectious disease, ranging from mumps and whooping cough to smallpox, occur in a school; in case of an outbreak of an infectious disease in the village or town where the school is situated, or in the neighborhood. Before a school which has been closed is re-opened, the Kreisarzt must assure himself that all precautions, including disinfection, have been taken to prevent the spread of disease.

The Kreisarzt is required to enlist the interest and support of school teachers in order to improve school hygiene and the physical condition of school children, to urge the necessity for outdoor classes and exercises, and to improve the condition of physically undeveloped children.

Baby Farming.—The Kreisarzt is in charge of all places where babies under six years of age are boarded and cared for. It does not appear that the law requires a specific permit to board and care for infants, but it is evident that the Kreisarzt has the power to refuse permission to carry on this business. Not only is he required to supervise the conditions of places where children are boarded, but he must supervise the conditions of the children themselves, and, in the event of a death, to investigate its cause, in order to determine the responsibility, if any.

A further duty of the Kreisarzt is, to take all means possible to reduce infant mortality and to disseminate any valuable information by means of pamphlets, calling attention particularly to the value of feeding of infants. He must provide rest-rooms for mothers in factories and other working-places, and see that their hours of labor are so arranged as to maintain them in a healthy condition.

The Kreisarzt is still further required to supervise the operation of benevolent institutions, to see that medical attention is given to the needy, and employment offered them after illness. He makes a yearly report of conditions in institutions, including the sufficiency and efficiency of the medical staff and nurses, and, if examination has revealed unsatisfactory conditions, to have them corrected. Asylums for epileptics and insane are also under his supervision, and plans for new hospitals and extensions of old buildings must be submitted to him for approval. He investigates plans of proposed private hospitals and maintains a supervision of them after they have been opened.

The Kreisarzt must further keep himself informed through various people having knowledge of existing conditions, of the existence of crippled children, in order that they may receive proper medical attention, and, if cures are impossible, that they may be trained for certain trades.

Administration of Kreis laws for the care of imbeciles, epileptics and idiots is carried out by him, and he must render medical assistance in case of need.

He is required to urge the establishment of public baths and bathing places and the construction of shower baths in schools, and is responsible for the sanitary condition of these places, and for provision at swimming pools in case of accidents. So-called healing springs, watering places and spas are under his care and supervision. He is required to investigate all places of this kind at least once a year and to make a report to the Regierungspräsident, stating the sanitary conditions, the number of patients and the improvements made or contemplated.

The general care of dead bodies and investigation of cemeteries is still another duty of the Kreisarzt. Certain confidential work is supposed to be carried out by him, upon order of the Medizinalgeheimrat or the Regierungspräsident, but he is not required to perform this duty except upon such orders. This confidential work is as follows:

1. Investigation of the sanitary condition of jails, prisons, correctional institutions of all kinds, and the care of prisoners during transportation;
2. Examination of, and report on, condition of health of public officials and teachers in public schools, and the examination of candidates for public school teachers;
3. Examination of such soldiers and non-commissioned officers of the 1870–1871 war, or of a previous war, as are completely incapacitated for work, in order that they may be permitted to share in the benefits of special funds;
4. Examination of recruits and determination of their fitness for military service;
5. Examination of applicants for benefits under health insurance act;
6. Examination of applicants for benefits under Workers Mutual Aid Societies;
7. Giving of confidential advice to the water supply authorities.

Instructions are given in utmost detail as to the forms of reports, including headings, and a special prohibition is made against the use of foreign terms.

The yearly report is made to the Regierungspräsident. It must contain a statement of the activities of the Kreisarzt and the conditions of public health in his Kreis.

His official correspondence is forwarded free of charge, if officially sealed.

Health laws.—The following diseases must be reported to the police authorities of the locality where the patient lives or has died: leprosy; Asiatic cholera; typhus; yellow fever; Oriental plague; small pox; diphtheria or croup; epidemic cerebros spinal meningitis; puerperal fever; granuloses; trachoma; recurrent fever; dysentery; scarlet fever; typhoid fever; anthrax; glanders; hydrophobia (rabies)—all cases of injuries by mad dogs, or by dogs suspected to have been mad; meat, fish or sausage poisoning; trichinosis; each death from tuberculosis of the lungs or larynx.

Further, the name, age, place of residence of patient and name of disease must be reported to the police authorities within twenty-four hours. It is apparently possible for the patient to change his residence, provided the police authorities of both the former and the new place are notified of the change. This is obviously a bad principle.

521572—43——10

A curious provision of the public health law is to the effect that if a patient suffering from an infectious disease is already under treatment at the time of the report, the official surgeon or Kreisarzt is not permitted to see him, if the attending physician declares that a visit might endanger the patient's life. The attending physician must be given an opportunity to make this declaration if he so wishes. In cases or suspected cases of puerperal fever, the consent of the head of the family must also be obtained. As a practical result of this provision, the Kreisarzt seldom attempts to visit a case of infectious disease reported.

In cases of suspected typhoid or glanders, an autopsy may be ordered by the police on recommendation of the Kreisarzt, in order to obtain a correct diagnosis.

In cases of trachoma, diphtheria and scarlet fever, only the first case in a family need be reported. All preventative measures against spread of infectious diseases and all measures of isolation are under control of the police authorities, who are, in general, permitted by the Kreisarzt to perform this duty with only slight directions and supervision. Details of particular interest in regard to special diseases are:

Diphtheria (*Croup*).—Removal to a proper hospital for treatment cannot be ordered if either the official or attending physician believes that sufficient isolation may be obtained in the dwelling of the patient. Diagnosis of the disease is as a rule not confirmed by bacteriological examination, nor is such bacteriological examination, for the purpose of definitely determining the duration of the infective period, usually made. No systematic effort is usually made by the public health authorities to place the antitoxic treatment of diphtheria within the reach of all cases of diphtheria, and, as a matter of fact, the majority of cases of diphtheria do not receive antitoxin.

Disinfection is under direction of the police, and is seldom directly supervised by the Kreisarzt. In general, the report of the police that disinfection has taken place is all that is required.

Puerperal Fever.—Physicians or other professional medical practicianers are required in each case of puerperal fever in which they are consulted, to give immediate notice of the disease to the midwife or child-bed nurse who is in attendance on a patient. No midwife nor child-bed nurse in attendance on a patient having puerperal fever, is permitted to attend any other person until a week after the end of such attendance, and then only after a thorough cleaning and disinfection of the body, clothing and instruments in accordance with instructions of the Kreisarzt. However, they may be allowed to take up their business again within the period of a week if the Kreisarzt so permits.

Typhoid Fever.—Cases must be isolated, and houses in which the disease exists placarded. The Kreisarzt is required to make an inspection and to submit a detailed report regarding the possible source of infection and the means taken to prevent the spread of the disease. The report of Bezirk Coblence is required to be made out on a special form. At the beginning of the American occupation, while these reports were regularly made out, they seldom gave any valuable information. It was not until the Officer in Charge of Civil Affairs required that all such reports should be submitted through his office that they improved.

Careful examination of public health conditions in American occupied territory led to the conclusion that all statistics of typhoid fever were misleading, because:

1. Many cases of so-called influenza are typical cases of typhoid fever;
2. Many cases of typhoid fever are not recorded as such, for the reason that laboratory examination of faeces does not immediately record the presence of typhoid bacilli, or the first Widal test may be negative;
3. Many cases of mild typhoid fever escape notice.

This is particularly true of mild or irregular cases among children. Two serious epidemics of typhoid fever occurred in the Ahr Valley during the period of American occupation, both due to disregard of ordinary preventive measures. In both instances, the epidemics were terminated only by the action of the American authorities in administering anti-typhoid vaccine to the whole civil population.

The Brück Epidemic.—Beginning in December, 1918, and running through January and February, 1919, a large number of cases of typhoid fever occurred in the town of Brück in the northern part of Kreis Adenau, on the Ahr river. Prior to the epidemic, the population of the town, numbering about 450, had obtained its water supply from three public sources and three private wells. One of the public sources was a pump over a well in the main street, not more than three meters distant from the privy-vault at the railway station. The close proximity of the privy-vault apparently caused no concern to any public health official, the assumption being, that because the privy-vault was constructed of cement, no pollution of neighboring ground could result therefrom. It was not until some returning soldiers had actually deposited typhoid bacilli in the privy-vault, the well-water had become infected with these organisms and the population of the town seriously infected with typhoid fever, that an examination of the water and the privy-vault was made. It was then discovered that the privy-vault was cracked and leaking and the water in the well grossly polluted. Even then, the only precaution taken was to issue a proclamation to the inhabitants of the town, warning them against the use of this water for domestic purposes. A few weeks later, on demand of the military authorities, a chain and padlock were placed around the pump and pump-handle, and when this measure proved ineffective the pump was dismantled entirely in January, so that no further use could be made of the water. About this time, another well in the public street, about 200 meters from the one just described, which had not been in use for a long time, was opened to the public. The water of this well was not examined chemically or bacteriologically before being placed in use, and it was not until demand was made by the military authorities that such examination was made. The water was then found to be grossly polluted. The third supply was a very small one, brought by pipe from a neighboring hill, and was found to be unobjectionable. This supply, however, was so small as to be utterly insufficient for the needs of the population. The pump over the well of the second public water supply was dismantled by order of the military authorities in spite of strenuous protests of the German civil authorities, both of the town and of the Kreis. They asserted that if this water supply were put out of use, the populace would be compelled to use the Ahr river water, which was even worse. As the Kreis and local officials seemed helpless in the matter, it remained for the military authorities to develop a water supply found in the neighboring hills and to bring it into the town by pipes. During all this time, however, the town had become so seriously infected with typhoid fever, that by February, 1919, more than 10% of the total population had had, or were having, typhoid fever. At this time it was proposed to the Geheimrat of Bezirk Coblence, that the whole civil population be immunized against typhoid by injection of typhoid vaccine. The Geheimrat stated that this had been tried in several places in Germany but without success, inasmuch as the population would never submit to the second and third administration of vaccine, after having had one. It was explained to him that the vaccine in use in the U. S. Army required only one injection; but this called forth further objections, because of the large number of organisms to be injected at one time by this method, which, he feared, would produce serious results. Despairing of any cooperation on the part of the civil authorities of either the Bezirk, Kreis or town, the O. C. C. A., Third Army, issued an order to the Regierungspräsident to assemble the whole populace on a certain day in March for the purpose of being vaccinated. The vaccinations were performed by the Kreisarzt, under supervision of a military officer. Within three weeks thereafter, the epidemic came to an absolute end. During the whole time of the epidemic, approximately 200 soldiers were quartered in this small town. These soldiers had of course been immunized against typhoid. Their drinking water was brought in carts from a neighboring and safe supply, and carefully chlorinated before being used. There were no cases among the soldiers.

The Altenahr Epidemic.—A second epidemic occurred in the town of Altenahr in the Ahr Valley, in the southern part of Kreis Ahrweiler. Altenahr contains a population of approximately 700, more than 10% of whom had acquired typhoid fever before the epidemic was brought to an end. The existence of typhoid in this town first came to the notice of the military authorities when four soldiers of the occupying troops were taken ill at one time. An investigation made by the Sanitary Officer in the Office of Civil Affairs, in conjunction with the representative from the Chief Surgeon's office, at once revealed the fact that there were at least twenty cases of unrecognized typhoid fever in the town. All these cases had been diagnosed by the local physician as influenza. The Kreisarzt of Kreis Ahrweiler was utterly ignorant of the public health conditions in Altenahr, explaining that he had received no reports to indicate the necessity for investigation on his part. It further developed at this time that it was not the custom of this Kreisarzt to make investigations even when reports were received. It may be interesting to record the questions asked of the Kreisarzt, and his answers with reference to his activities in public health work:

Q. Do you know anything concerning the water supply of Altenahr?

A. No; I have not caused it to be investigated.

Q. Do you know whether this water supply has ever been investigated and analyzed?

A. No; but it must have been, as the law requires such investigation at least every two years.

Q. Have any reports concerning this water supply been brought to your notice during the past few years?

A. No; but they must have been made, as the law requires it.

Q. Have you ever made a personal investigation of the sanitary conditions of Altenahr?

A. No; there have been no reports requiring such investigation.

Q. As a matter of fact, when reports of typhoid fever are received by you, do you make a personal investigation to determine possible sources of infection?

A. No; not since the war, as I have no transportation.

This Kreisarzt, by order of the military authorities, was immediately removed and replaced by an efficient and able executive. The amazing fact in regard to the matter is, that in September, 1919, with a full knowledge of the inefficiency shown by this Kreisarzt, and his lack of initiative and ability in public health matters, the acting Regierungspräsident requested permission to reinstate him, stating that this request had been made from Berlin and was submitted through him. At this time, reliable information was received that the request had come from Berlin as stated, but that it was made in response to a previous request sent to Berlin by the acting Regierungspräsident. The request was denied, and other permanent Kreisarzt was subsequently appointed by Berlin. An investigation made by the military authorities in the town showed that the main water supply used by the troops was obtained from a well in the cellar of a hotel. The top of this well was at least one metre below the level of the bottom of a large cesspool in the court-yard of the hotel, and not more than four metres distant from the well. The water was drawn from the well through a pipe, which led to the pump immediately above the cesspool, so that the most casual observation of the water supply could not fail to reveal the presence of the cesspool. The latter received the entire drainage from the hotel, which sometimes housed fifty persons. An examination of this water showed it to be grossly polluted and unfit for use, as was likewise the case with another well. The pumps over both these wells were immediately dismantled. Another water supply was in existence, but had been used very little. This was a general supply obtained from a well near the Ahr river, pumped to a reservoir above the village and distributed through pipes. The reason for failure to use this supply, which was unobjectionable, was the lack of gasoline for driving the pump in the pumping station. The necessary gasoline was supplied by the military authorities and the water supply brought into use. By that time, however, the disease was spreading by actual contact. As in the case of Brück, the entire civil population of Altenahr was immunized with typhoid vaccine and the epidemic thus brought to an end, no further cases developing after a period of three weeks among those who had been vaccinated. It is a noteworthy fact that chlorinization of the water supply used by the American forces had not been carried out properly, and that the soldiers had been permitted to use the well water from the hotel. An officer's mess was conducted in the hotel at the time of the occurrence of the epidemic, and the water used on the table of this mess was from the well of the hotel. One of the mess attendants was found to be a typhoid carrier, and may possibly have been the cause of the infection of the well water in the hotel.

Rabies.—The sanitary law requires a report on all cases of rabies, all cases of suspected rabies, and all persons bitten by mad dogs, or dogs suspected to be mad. The curious custom, however, prevails in this district, of immediately killing a dog that has bitten any person or persons, and sending the head or brain to Berlin for pathological examination. In the opinion of all persons having had experience with rabies, the first action to be taken in the case of any one who has been bitten by a dog, is to confine the dog, keep him alive and under observation, thus to determine whether or not he is rabid. The sending of the dog's head and brain to laboratory for examination as a routine procedure, leads to great delay. Frequently, on account of the length of time required for the head to get to the laboratory, decomposition has taken place to such an extent as to preclude the possibility of any diagnosis whatever. In many cases the diagnosis is received entirely too late to be of any value in connection with the beginning of Pasteur treatment. In American occupied territory several weeks almost elapsed between the occurrence of the dog-bite and the receipt of the diagnosis from the laboratory. Attention may be drawn to the following points:

1. While the muzzling and leashing of dogs is provided for by a German ordinance, and a penalty imposed on dog-owners for failure to comply with the ordinance, no provision is made for the collection and destruction of stray and ownerless dogs. This collection is however of utmost importance during a prevalence of rabies.

2. The assumption that the period of incubation of rabies may be as long as three months.

During July, August and September, 1919, a few cases of rabies were reported in American occupied territory, and strenuous efforts were made by the German civil authorities at this time to have an Army order issued, requiring the same muzzling and leashing for dogs belonging to Allied officers and men, as was required for dogs owned by Germans. The necessity for such action not being apparent to the medical officers of the occupying forces, the order was not issued.

Statistics of infectious diseases.—Statistics have been collected for three periods in each disease of importance: The pre-war period of five years, 1909, 1910, 1911, 1912, 1913; the war period, 1914, 1915, 1916, 1917, 1918; the period of occupation from January 1st to June 30th, 1919.

Diphtheria.—A study of these statistics with regard to the incidence of diphtheria, appears to indicate that in Regierungsbezirk Treves there was a distinct increase in the war period over the pre-war period, and a still further increase in the occupation period. On the other hand, figures from Regierungsbezirk Coblence indicate a steady decrease from the pre-war period through the war period and the occupation period. In view of the fact that the diagnosis of diphtheria was frequently not verified by cultures, too much reliance cannot be placed on the figures. The death rate from diphtheria in Regierungsbezirk Treves shows an increase in this rate of .05 per 1000 in the war period over the pre-war period, and a still further increase of .02 in the occupation period over the war period. It is of course probable that these figures are more accurate than those regarding incidence of the disease.

The death rate from diphtheria in Regierungsbezirk Coblence shows an increase of .06 in the rate per 1000 of the war period over the pre-war period, and a decrease of .02 in the rate per 1000 in the period of occupation over the war period. For the whole occupied territory, the average increase of the war period over the pre-war period is .067 and exactly the same rate for the occupation period as for the war period.

Typhoid Fever.—The figures for the incidence of typhoid and para-typhoid fever in Regierungsbezirk Treves, show an increase per 1000 of .07 for the war period over the pre-war period. Similar figures in Regierungsbezirk Coblence show an increase in the rate per 1000 of the war period over the pre-war period of .018. A summary of the incidence of the whole American occupied territory shows an increase in the rate per 1000 of .0628 for the war period over the pre-war period, and an increase of .590 of the rate per 1000 of the period of occupation over the war period. It is believed that the apparent increase is due to the more careful investigation and supervision of the civil population by American medical officers than during the war period. This is especially indicated in the outbreaks of typhoid in Kreise Adenau and Ahrweiler, where case after case was brought to light by American medical officers which had been diagnosed as influenza.

Dysentery.—At no time during the American occupation, up to January 1st, 1920, did dysentery in the civil population cause serious uneasiness to the medical department. In the pre-war period, the records were badly kept and furnish little real knowledge of the existence or non-existence of dysentery. It is perfectly evident, however, that in 1917 and 1918, there were severe outbreaks of dysentery in several Kreise of Regierungsbezirk Treves and Coblence. In Regierungsbezirk Treves, these outbreaks occurred notably in 1917 and 1918 in Stadt- and Landkreis Treves. In 1917, in Regierungsbezirk Coblence, there were severe outbreaks in Kreise Mayen and Ahrweiler, also in Mayen and Stadtkreis Coblence in 1916.

There can be little question that these outbreaks were due to military conditions. In 1915, the records show a total of 76 cases in Kreis Mayen, due to an outbreak in the insane asylum at Andernach, which at that time was also used as an emergency military hospital. All these outbreaks had practically come to an end at the beginning of the American occupation, and did not recur until August and September, 1919, practically disappearing in October. At the time of the occurrence of these cases in 1919, general instructions were issued to the civil population through the Regierungsund Geheimer Medizinalrat, regarding the nature of dysentery, and the usual means of its transmission and prevention. All cases of 1919, as far as is known, were of the "Y" type.

Scarlet Fever.—The disease is not regularly reported, and no conclusions can be made from official figures regarding the effects of war on the disease. The same statement may be made regarding measles.

Influenza.—Until the time of occupation, influenza was not a reportable disease. From 1909 until the armistice, no figures were available from the official records. By order of the military authorities, reports were required of attending physicians of cases of influenza under their care until June, 1919, when they were discontinued, owing to the fact that influenza had greatly decreased. In November, 1919, owing to an outbreak of influenza in Westerburg, the order was reissued. These reports, however, do not indicate serious conditions in any part of the American occupied territory, and are therefore not included in the tables.

Street railways.—The street railways in Regierungsbezirk Coblence are generally well managed. The cars are kept in a clean condition and there is little expectoration. The tendency towards a lack of ventilation, however, is very noticeable. It is the custom to close the ventilators on or about October 1st, and to keep them closed until the following Spring. On account of an apprehended recurrence of influenza in Coblence in November, 1919, an order was issued requiring all ventilators in tram-cars to be kept open at all times, as well as the enforcement of the regulation against smoking in closed cars. Prior to issuance of this order, in the course of a week of close observation, travelling in the tram-cars on an average of twice daily, eight violations of the anti-smoking ordinance were observed. These eight violations were distributed as follows: German 1; French 1; American soldiers 6. In these six were included two officers. There can be no question that the German, prior to the occupation, observed the anti-smoking ordinance. Observations made during the two weeks following issuance of the order concerning ventilation in the street railways of Coblence, Ehrenbreitstein, Bendorf, Vallendar and Niederlahnstein, revealed no evidence of compliance. The order was worded as follows:

HEADQUARTERS
AMERICAN FORCES IN GERMANY

OFFICE OF CIVIL AFFAIRS

5 December, 1919.

To: The Regierungspräsident,
Coblence.

1. On October 18th, the following order was transmitted to you:

"1. In view of the recurrence of influenza in Regierungsbezirk Coblence and the Unterwesterwald, it is desirable that all side ventilators of street-cars be kept open at all times, and that the prohibition against smoking in closed cars be strictly enforced.

2. Please issue necessary orders to carry out these precautions without delay."

2. Many inspections have been made of street-cars since that, and in not one single instance has the order been found to

be complied with in full. Frequently it has been found that no attention whatever has been paid to it.

3. The Commanding General, American Forces in Germany, therefore directs that you publish the following notice at once:

"NOTICE."

"In view of the recurrence of influenza in the Regierungsbezirk Coblence and Unterwesterwald, the Commanding General, American Forces in Germany, has heretofore directed, that the side ventilators of all street-cars be kept open at all times and that the prohibition against smoking in closed cars be strictly enforced.

Inspections show that there has been a flagrant disregard of this order by those concerned. He, therefore, directs that this notice be printed in both English and German and conspicuously posted in every street-car in the American zone of occupation.

The posting of this notice will be considered a compliance with its terms on the part of the responsible street-car officials; thereafter its execution rests with the conductor of each street-car while on duty. Failure on the part of the officials properly to post this notice, or failure on the part of the street car conductors properly to enforce it, will render those concerned liable to trial by American military tribunals.

Likewise passengers on street-cars resisting the enforcement of this order will, upon complaint properly filed by street-car conductors, be brought to trial for violation thereof.

This order will take effect at 12:00 o'clock noon, December 15th, 1919.

HERR VON GRÖNING,
Regierungspräsident."

Respectfully,
I. L. HUNT,
Colonel, Infantry,
Officer in Charge, Civil Affairs.

This episode is mentioned at length for the reason that it shows the usual method of enacting ordinances or issuing orders. What may have been the result prior to the revolution is uncertain, but the present result is, as a rule, a lack of compliance, unless necessary force is given to the matter by the occupying military authorities. In this case, further orders of a drastic nature produced adequate ventilation.

Sewage disposal.—In smaller communities, wash-water is generally disposed of by being thrown on the ground, and urine and faeces received into primitive privy-vaults, usually badly constructed, practically never protected from flies and often allowed to become filled to overflowing. Wells are sunk without any regard whatsoever for these possible sources of pollution. The contents of the privy-vaults, when emptied, are generally thrown into the manure-pit, which is in front of every house or premises where horses or cows are kept. Frequently there is no privy at all near the premises, but the faeces and urine find they way into the manure-pit in some manner. Still more frequently, a badly constructed privy is placed immediately over the manure-pit in front of the house, or near the manure-pit and connected therewith by badly constructed drains. The contents of these manure-pits, mixed animal and human dejecta, are used for fertilizer. That serious local outbreaks of typhoid fever, paratyphoid fever and dysentery occur under conditions of this kind, is not remarkable. Local health officials were interviewed concerning the danger of this method of disposal of faeces, owing to the possibility of carriage of disease by flies. The information was always received from these German officials that there were no flies. Personal observation showed, however, that at least during the hot months flies were numerous.

Prevention of nuisances.—In the prevention of nuisances, the Kreisarzt depends entirely upon complaints received, making no personal investigations. In other words, public nuisance is not a public nuisance unless it annoys someone sufficiently to produce a complaint. While the Kreisarzt has authority to enter houses for the purpose of correcting insanitary conditions, it is apparently the general impression that he cannot do so until an infectious disease has broken out, so that the real correction of nuisances or conditions inimical to health or life depends very largely on the citizen. The sanitary conditions of private premises, in both the rural communities and the large towns or cities, is really more of an indication of the individual householder's cleanliness, or lack of it, than the index of the efficiency of the public health administration.

Street cleaning.—As far as small communities, villages and towns are concerned, this amounts to little. In large towns and cities, there is a force of street-cleaners under the Bürgermeister's office, whose work is most inefficiently performed by reason of lack of necessary implements. Horse-drawn sweepers are not used and the brooms are made of bundles of twigs, and are extremely ineffective. In the city of Coblence, the main thoroughfares are kept reasonably clean, but the less frequented and less important ones are generally neglected. Receptacles for paper and refuse of various kinds were placed at various points on order of the Officer in Charge of Civil Affairs. In this connection, however, it is to be remembered that the average German citizen is much more careful regarding the littering of streets with paper and other refuse than are American soldiers.

The public thoroughfares of Coblence are divided into two classes: one, for which the Bürgermeister is responsible and which includes by far the greater number of streets, and another, for which the Regierung is responsible and which consist of the streets bordering on buildings used by the Regierung. In addition to these, there are many alleys and passages for which private persons are responsible and which are generally not maintained in clean condition.

Removal of waste.—In smaller communities of American occupied territory, ashes are disposed of in any convenient locality, while garbage is regularly fed to animals or thrown on the manure pile. Refuse of various combustible kinds is burned. In larger cities, such as Coblence, ashes and garbage are collected, usually together, and used as filling of suitable places where dumping grounds are established. Refuse of various kinds is carefully sorted and everything of any value removed and sold. The Coblence City Dumping Ground is carefully managed and causes little or none of the nuisance that usually prevails in these places.

Dwellings.—In general, both in small communities and in large cities, dwellings are maintained in clean condition. The notable thing about them, however, is the general lack of adequate ventilation. The almost invariable practice is to close all windows and outside shutters at nightfall and to leave them closed all through the night. In the colder months, it is not even customary to open them during the daytime. The method of heating, in the larger communities at least, is generally by means of steam or hot water, but the bulk of German houses are still heated by large porcelain stoves, usually set in the walls so as to serve adjoining rooms.

Milk supply.—The milk supply is chiefly dealt with in the chapter on food. In regard to the sanitary conditions under which milk is produced and handled, however, it may be said that very few precautions are taken to prevent infection. Milk cows are commonly badly housed in dark and ill-ventilated stables, and are usually filthy or cleaned only at rare intervals. Milk containers are seldom washed with hot water and are never sterilized. There are practically no facilities for cooling milk sufficiently immediately after it is drawn. It may be said, however, that for the prevention of milk against infection and the protection of the civil population, dependence is placed upon the sterilization of milk by boiling before it is used.

The rules which govern the sale and handling of milk, with few exceptions, are so general in character as to be of little value in the protection of milk supply. The interesting points of these regulations are as follows:

1. The addition of milk-ice to milk is permitted, which, it would seem, removes possibility of determining the chemical composition of the original milk;

2. The fat standard is 2.7%;

3. The milk rules are, according to police regulations, to be enforced with discrimination;

4. Pasteurised milk is understood to be milk heated to at least 70 degrees centigrade. Apparently the length of time it is kept at this temperature does not matter.

5. There is apparently no regulation that provides for a definite amount of total solids. The only note of chemical constitution made is with regard to the fat content.

There is apparently no effort to use a "permit in writing" system. Milk from cows whose general health is bad, may be used if the veterinarian states that the milk is fit for sale. Milk taken from cows afflicted with tuberculosis is not permitted to be used, but as there is no provision for application of the tuberculin test, it would appear that the rule is not of much value. Milk containing less than 2.7 fat,—the minimum required,—may be labelled "unskimmed milk, second quality."

Water supply.—The water supply of communities in American occupied territory is divided into three classes:

1. That of small villages or Gemeinden in which some, usually not all, households, have an individual driven or dug well, placed with regard to convenience but with disregard of sanitary considerations. Frequently in such communities several houses use one well. Occasionally, even in small communities, this individual supply is supplemented by a spring or water-course in the neighborhood. The water from these wells is usually polluted and its use therefore forbidden to troops.

2. That of villages or towns of larger size, which have a water supply usually taken from a well or wells, on or near the banks of some water-course, the level of the water in these wells fluctuating coincidentally with that of the river, i. e., the water of these wells is merely river water, more or less purified by filtration.

3. This class is not materially different from Class 2, except in extent of supply and use of storage reservoirs. An example of this supply is that of the city of Coblence obtained from six wells on the banks of the Rhine at Oberwert, above the city. The water of these wells is undoubtedly merely Rhine water, purified to a considerable extent by filtration, except when the Rhine level is high and filtration becomes too rapid to be effective. There are three pumping stations near these wells, by which the water is pumped into a small conduit, which, dividing into two branches, carries the water by one branch directly into the service pipes of the city, and by the other, to a small reservoir on the hill above Coblence. The capacity of this reservoir is such that in event of a breakdown, it could supply the town for a few hours only. When everything is working properly at the pumping plant and water is supplied in normal quantity to the city, the amount supplied per head per day is approximately 50 liters. It is an interesting fact when

the pumping plant was built, the pumps were placed on a level that would prevent their being worked effectively in case the Rhine rose to an unusual height. When, in December, 1919, the Rhine rose to a level unknown for many years, namely 8.13 above its mean level, this caused a flooding of all the pits containing pumps, put several of them entirely out of condition and interfered materially with the working of the others. Serious damage occurred from the impossibility of oiling the pumps as required, on account of the water in the pits. That a big city like Coblence, the capital of the Rhine Province, and a city of undoubted wealth, should not have a storage reservoir of sufficient size, placed on the hills above the city, the purity of which, under all circumstances, would be beyond question, is most astonishing.

Throughout the period of occupation, the water supply of the city of Coblence has been plentifully treated with chlorine by order of the American authorities. In passing, it is interesting to note that the German sanitary authorities in American occupied territory do not believe in chlorination for purifying doubtful or contaminated water.

In times of doubt or contamination, sole reliance has been placed on the advice published to the community in the newspapers, to boil water before drinking it.

Prostitution.—At the beginning of the occupation, prostitution in American occupied territory was regulated by the Germans under the Prussian law. The essential provisions in this law are as follows:

Professional prostitutes are required to appear for examination at certain places and on certain days. If found to be affected with syphilis, chancroid or gonorrhoea, they are compelled to undergo medical treatment. In order to facilitate this medical treatment, it is the custom to designate certain public consultation hours. If the persons affected with any of the above diseases do not visit the physician during these consultation hours with sufficient frequency, or if suspicion exists that they carry on their trade in spite of their disease, they are immediately removed to hospital, where they are confined until a cure has been effected. There are, of course, local rules and customs, for the purpose of carrying out the provisions of this law, all of which are unsatisfactory in obtaining any diminution in prostitution and its results. Under this law, it is the practice of the Germans to arrest a woman on the bare suspicion that she is a prostitute. She is often deprived of her liberty for many days, while an exhaustive laboratory investigation is made to determine her state of health. During this period, she is also subjected to questioning by the police, for the purpose of obtaining if possible an admission that money has been received from men with whom she has consorted. If during the questioning, the answer is "Yes", she is then brought to trial and practically invariably convicted of carrying on the trade of prostitution without a license. If she says "No", even under severe police pressure, she is then *ipso facto* not a professional prostitute. In either case, she is examined medically before being placed before a judicial tribunal for trial. In either case, also, if found diseased by the medical examiner, she is sent to hospital for treatment and kept confined until cured. If a pecuniary transaction is admitted and if she is sent to hospital, she may be kept there for weeks or even months and, when released, brought before a civil court, which always inflicts severe punishment. No consideration is given to the fact that the woman may have been confined in hospital for a long period of time. In case of a woman who denies the taking of money and who is found to be diseased, she is treated in hospital until cured and then released unconditionally. If found not diseased, she is released in a similar manner, after the few days necessary for the examination. It may be readily understood that only the very young and unsophisticated women made admission of their financial dealings and are punished, and that the older, better informed and more dangerous ones, make persistent denials and so escape punishment. After serving sentence of confinement for practising prostitution without a license, the woman comes under what is called "Control", i. e., she is required to report regularly for examination to the Kreisarzt or his representative. By this system, very young girls, frequently fourteen or fifteen years of age, who might possibly by other methods be saved from this kind of life, are practically forced into it. Old, hardened, sophisticated women, often suffering from dangerous chronic diseases, are not under any surveillance or control except in the case of positive laboratory findings. This dependency upon a laboratory diagnosis has a two fold danger. First, any married woman infected by her husband, innocent herself of any wrongdoing, may be arrested upon suspicion or false accusation of a spiteful neighbor and deprived of her liberty for months at a time in a prison hospital, and this without ever having been before a judicial tribunal of any kind. Second, a professional prostitute, affected with a chronic, incurable gonorrhoea, if the laboratory diagnosis, is negative, is set free to resume her trade unmolested for the time being. It is a well known fact that laboratory findings in chronic gonorrhoea will be positive at times only. From the beginning of the occupation until October 18, 1919, the army made use of this German civil procedure, merely adding the expedient of deporting without trial such persons as seemed undesirable, and bringing them before a military court if they returned. This order of deportation, however, was not effective, for the reason that it consisted merely of the words "expelled from American territory", written on some part of a Personal Ausweis (pass). It was expected that the German police on duty at the railway station, instructed to that effect, would promptly turn back any persons coming into American occupied territory with such an Ausweis. The practical fact, however, was that the German police merely glanced at the Ausweis and seldom read, or perhaps could not read, the English words thereon. Late in 1919 the military authorities adopted the procedure of issuing this order in writing to each person deported. This was in effect, however, for only a short time before the regulations of October 18th, 1919, which regulations changed the whole course of procedure. In spite of tremendous activity, however, under the German civil methods of procedure, large numbers of arrests and large numbers of persons sent to hospital for treatment, the venereal rate of the army slowly but steadily increased from the week of January 17th, 1919, when it was at the remarkably low figure of 9 per thousand per annum, to the extremely high figure of 422.65 per 1000 per annum on October 21st, 1919. The figures shown in a chart which forms a part of Appendix No. 56 (Vital Statistics [not printed], indicate the unsatisfactory conditions which obtained. Although the proceedings already outlined as those of the German civil procedure were carried out vigorously, numerous prophylactic stations were provided and the use of prophylaxis was urged. The German civil hospital in Coblence became so overcrowded with women suffering from venereal disease, that it was unable to perform its functions in regard to other diseases. Hospital accommodations therefore were set up by the Landräte of the various Kreise, either at the chief town of the Kreis or in a neighboring town, under orders from the Office of Civil Affairs, A. F. G. Orders were also issued, requiring that all women arrested in any Kreis and found to be suffering from venereal disease, should be confined in hospital and treated in the Kreis in which arrested. Constant vigilence was required, however, to prevent both the military authorities and the civil officials from sending cases of this kind to the Coblence hospital. On account of the overcrowding of the Bürger Hospital of Coblence, it became necessary to establish quarters for these women in some other place, and to relieve the burden which was interfering with the proper work of the hospital. In the latter part of July, 1919, after considerable pressure exerted by the Office of Civil Affairs, the Bürgermeister selected the building of the Staatsarchiv, and removed a sufficient number of the records therein to provide dormitories and proper treatment-rooms for sixty-one patients. The hospital was established against the desires and wishes of the civil authorities and in spite of all obstacles they could place in the way of getting the necessary premises, the apparent desire being to induce the military authorities to provide hospital accommodation for this purpose. Place after place was suggested by them and rejected by other civil authorities or by the Army, as unsuitable or undesirable. It is quite possible that it was thought that the military authorities would in time become tired of seeking a civil hospital and would construct or use one of its own military buildings. Pressure was continued, however, until the Staatsarchiv was selected and fitted up for the purpose. The place was prepared and formally opened on August 14th, 1919. During the period in which the hospital has been working, it has received and cared for 347 cases of venereal disease, distributed as follows:

Disease	August	September	October	November	Total
Gonorrhoea	140	70	53	27	290
Syphilis	11	2	5	6	24
Chancroid				1	1
Syphilis and gonorrhoea	8	14	7	2	31
Chancroid and gonorrhoea			1		1
TOTAL	159	86	66	36	347

There remained on November 30th, 1919, the following cases:

Disease	
Gonorrhea	22
Syphilis	7
Chancroid	4
Syphilis and chancroid	10
TOTAL	43

The following points with regard to these figures are to be noted:

1. The marked and steady diminution in the total number of cases;
2. The marked and steady diminution in the cases of gonorrhea;
3. The increase in the proportion of the cases of syphilis and the remarkably small number of chancroidal infections.

During the operation of this hospital the great difficulty has been to keep the women properly employed. The large majority of the inmates were perfectly able to do any kind of manual work. There was not, however, enough work in the hospital to keep them all busy, and it was difficult, in the presence of this idleness, to maintain discipline and order. Some occupation was provided by turning one of the rooms into a reading room and providing books and reading matter for the patients. Many attempts were made to obtain sewing and other employment, but without result; all who had any such work to give out refused to give it to the hospital, on account of the supposedly dangerous maladies with which the patients were affected. Attempts were also made to give lessons in English to such of the inmates as desired them. This also had to be discontinued, on account of the small number who were desirous of doing this work. A series of readings was arranged for, but was discontinued for the reason that only a few of the inmates cared to attend. These read-

521572—43——11

ings were not religious nor instructive, but intended merely to be interesting or amusing. Religious services, Catholic through the German authorities, and Protestant through the Salvation Army, were held later.

Realising that the methods in use were not producing any effect on the venereal rate, nor reasonably protecting the Army from venereal infection, the following regulations were put into effect on October 18th, 1919:

HEADQUARTERS

AMERICAN FORCES IN GERMANY

COBLENZ, GERMANY.

CIVIL AFFAIRS BULLETIN No. 34. — 15 October, 1919.

VAGRANTS AND JUVENILES

1. There is hereby created a Court for Vagrants and Juveniles. The Court shall consist of one officer and shall be appointed by the Officer in Charge of Civil Affairs, Coblence. The territorial jurisdiction of the Court shall be co-extensive with the zone of American occupation.

2. A vagrant within the meaning of this order is:

(a) Any person whose usual place of abode is outside of the zone of American occupied territory, present therein without visible means of support; or,

(b) Any woman who solicits or has illicit sexual intercourse with any person serving the United States or any associated government.

3. A juvenile offender within the meaning of this order, is any person under the age of sixteen years who violates a law of war or any order of the American military authorities.

4. Any male person found guilty as a vagrant shall be punished by imprisonment for not less than two and not more than six months.

5. Any woman found guilty as a vagrant:

(a) Shall be punished by indeterminate imprisonment for not less than two nor more than six months.

(b) She shall, immediately following conviction, be examined by a physician designated by the Chief Sanitary Officer, Office of Civil Affairs, A. F. in G., and if found infected with venereal disease, shall be confined in the hospital for treatment until released from such hospital by authority of said Chief Sanitary Officer. Time of confinement shall be counted as a part of the sentence of imprisonment.

6. Vagrants under such terms and conditions as the Court may prescribe, may be released from imprisonment by order of the Court, provided no woman in the hospital under treatment for venereal disease shall be so released. A vagrant whose usual place

of abode is outside the zone of American occupation, may be ordered deported and the sentence of imprisonment suspended during his or her absence therefrom.

7. A juvenile offender on conviction may be sentenced to pay a fine of not exceeding five thousand marks and to imprisonment not exceeding six months, or to both such fine and imprisonment. On conviction, such an offender, according to his age and circumstances and the nature of the case, may be delivered by order of the Court into the custody of civil authorities, for such confinement as the Court may direct, in a reformatory or industrial school, or into the custody of his parents or guardian. Periodical reports will be required by the Court at such times as the Court may order, as to the conduct of any juvenile offender delivered into the custody of civil authorities, parents or guardians.

8. No sentence or order of the Court shall be put into effect until approved by the authority appointing the Court.

By Command of Major General ALLEN:

WM. W. HARTS,
Chief of Staff.

Official:

S. G. TALBOTT,
Adjutant General.

Recapitulation.

1. Reliance was placed at first on:

(a) Medical examination of women;
(b) Hospitalization of women suffering from venereal disease;
(c) Prophylaxis for men;
(d) Anti-fraternization regulations.

Later was added:

(e) Deportation of undesirable persons.

On October 18th, 1919, were added:

(f) Prostitution made a crime; and a
(g) Vagrancy Court established and operated for purpose of making proof of actual prostitution unnecessary.

2. *Deductions.*

1. (a) and (b) would not protect against venereal disease;

2. (c) did not add sufficient protection, as statistics showed that half the men contracting venereal disease had taken prophylaxis, (probably improperly, since properly taken it insures almost complete immunity);

3. (d) prevented only the better class of women from associating with soldiers, and did not prevent the worst class of dissolute women from so doing;

4. (e) helped but little, because deported persons returned with slight interference, as was shown by investigation of German police methods at railway stations.

The activities of the Vagrancy Court undoubtedly diminished the number of exposures to venereal disease, and it is a fact that the venereal rate slowly but steadily decreased after its establishment. This may possibly be said to be *post hoc* and not necessarily *propter hoc*, but it is fair to assume that the large number of women in the jail and hospitals, and the large number deported, must necessarily mean a lessened number of such women in the streets and in cafes, and many observations of reliable persons indicate that this is the case. On September 27th, 1919, the anti-fraternization order was rescinded, with the hope that this would induce the soldiers to associate with the better classes of the civil population. Thereafter no woman was arrested and subjected to the humiliation of a physical examination except after trial and conviction in a court presided over by an officer specially qualified and selected for this duty.

Population.—The official census under German law is taken every five years, the last being taken in 1910. In intermediate periods, or whenever, in an emergency, officials desire to know the population figures, they are obtained by estimates compiled from police records of removals from the area, and the surplus of births over deaths. Since 1910, especially during the war years, a number of estimates of the population were made for purposes of food and fuel distribution, and for war purposes in general. These estimates, experience shows, are extremely inaccurate, and figures obtained from the same Kreis at different times, but for the same period, show a variation often as high as twenty per cent. In view of the discrepancy in the figures obtained from the German authorities, the populations of the different Kreise in American occupied territory have been determined solely from the birth and death relationship, using the 1910 official census figures as a basis. The figures given in this report therefore represent the population production of the Kreise and of the whole American occupied territory, not the figures which might be given by German estimates. No corrections can be made with the latter method for removals of persons from one Kreis to another, but the results obtained are so close to the averages from German estimates, that the figures herein given represent a probable, close approximation to the actual state on June 30th, 1919. Deaths from the war are throughout included in the reports from Kreise, since such deaths were reported, when possible, to the home Kreis of the death, and included in the year's total mortality of the Kreis. Unfortunately, no attempt was made to obtain data of these war deaths from the entire area, but from a few Kreise, these figures have been obtained, and will be referred to later.

The population of the area from 1909 to 1914 increased rapidly, with a constantly diminishing outgo of immigrants, with the result that the curve of the population the pre-war period shows a dense population in the area, increasing from 836,931 on January 1, 1909, to 884,676 on January 1, 1914, an average yearly surplus of births over deaths of 9,509. This rapid increase of the population is in no wise limited to American occupied territory, but was characteristic of the State of Prussia as a whole, as well as of the remainder of the German Empire, and produced an intense population pressure; a population more numerous than the country was able to support, and productive of conditions that must lead either to extensive emigration, or to the acquisition of added territory. With the onset of the war in 1914, and through the year 1914, the increase in the population was maintained at about its normal pre-war yearly average, but in 1915 there was a sharp change, such that the surplus of births over deaths was only 1.53 per 1000. From the close of 1915, onward, deaths exceeded births in most Kreise, whereas, in the area as a whole, the excess of deaths over births produced a relatively rapid diminution in the population. The net decrease in 1916 was slight, increased in 1917 and became alarmingly heavy in 1918, so that at the close of the war, the population had decreased to about the figures of 1913, which shrinkage continued during the first six months of 1919.

On June 30th, 1919, the estimated population of the area was 872,649, as compared with its maximum figure on January 1st, 1916, of 894,478. This is shown graphically in Figure 1 [not printed], which illustrates the sharp rise through the pre-war years and during the first war year, and the sharp decline in the later war years. The psychological effect of a rapidly diminishing population upon the German, with his well known desire for rapid increase, could not have been otherwise than profound; and, as essentially the same condition prevailed throughout Germany and was generally known, may be accepted as one of the many contributing causes of the total collapse of the German *morale* in 1918.

In one hundred years of Prussian history, this is the worst population condition encountered, and is the only period in which there was a markedly sharp incidence of deaths over births, with a diminution of population continuing year by year. At various other periods in war-time, the population has slackened its increase until the surplus of births over deaths was only 1 or 2 per 1000; but at no period since the wars of Napoleon has there been anything at all comparable to the results found in the last three years of the recent war.

Births and Deaths.—The births in the Kreise of American occupied territory during pre-war years, in rate per 1000 population, show a general downward trend of the yearly birth rate. This was common to the entire German Empire, but most pronounced in large urban areas of cosmopolitan character, less in smaller towns and least marked in rural districts. There are divergencies in the birth rate of different Kreise, which are indicative of the purely local effects of religious, social and economic conditions. In American occupied territory, urban Kreise show a much lower birth rate than do rural ones; but these are pre-war trends, and not the result of the war.

About a year after the war began, there was a sharp diminution in the number of births, and this continued progressively, as is shown in Appendix No. 55, Table 1 [not printed], until at the end of the war in 1918, the birth rate per 1000 was only half what it had been at the beginning, and was steadily decreasing, although at a less rapid rate than in the years 1915, 1916, 1917 and 1918. Deaths, on the other hand, were in pre-war years about normal, but attained variability during the war years, due primarily to casualties of war and to disease and other causes amongst the civil population. The curve of the mortality per thousand of individual Kreise and for the whole American occupied territory, shows deaths exceeding births throughout the last three years of the war, so that the population constantly decreased and this during the last year, 1918, with great rapidity.

Illegitimate Births.—Much was written during the war concerning the number of illegitimates in Germany and of the wilful production of illegitimate children to replace the rapidly diminishing population. Attempts to gather information upon this subject have not met with conspicuous success. The German officials, through the inadequacy of their official records or from other causes, either could not or would not supply data upon the subject. Information which may be accepted at face value was obtained in six Kreise.

If the figures from these six Kreise are accurate, the increase in illegitimacy was slight, being but .039 per 1000. One, however, is inclined to be not only critical but highly suspicious of the returns made by the German authorities on the question of illegitimacy, for the reason that even in the pre-war years the rate per 1000 of illegitimates is distinctly lower than the rate per 1000 of illegitimates given in the official Prussian reports for the corresponding years and is very much below the rate for illegitimacy in Germany as a whole, which is, as is well known, high. Respecting the other aspect of illegitimacy,—the wilful production of illegitimate children for purpose of population replacement,—little concrete information is available, and only of individual instances and not of wide-spread practice of this nature. There is no indication that this practice of the wilful production of illegitimate children for the purpose of population replacement was at all wide-spread or commonly practised in the portion of the Rhine Provinces occupied by the American Army.

Still births.—With the decreasing number of births characteristic of the war period, still-births show an actual decrease of .109 in the 1000 as compared with the pre-war period. To what agency this was due, it is difficult to say. It may be only one of those accidental phenomena common in vital statistics, and hence without significance.

In general, the effect of the war was from its beginning immediate and acute, causing a diminution of the civil population and an increased death rate, reaching a climax in the last two years of the war which no people could long sustain and survive. The total result of the war on the population, as far as numbers are concerned, is to turn it backwards approximately seven years, and when to this are added the number of people who have suffered injuries, and whose effectiveness is impaired through various agencies of war, the actual decrease in efficiency of the populace as a whole is great indeed.

CHAPTER No. 8

FOOD SUPPLY OF CIVIL POPULATION IN AMERICAN OCCUPIED TERRITORY

Conditions on Arrival of Third Army; German System of Distribution of Food Supplies; German Estimates of Food Situation; Salient Facts on Food Situation; Summary of Conditions; Causes of Food Shortage; Military Action Regarding German Food Supply; Circulation of Food; American Foodstuffs Sold to Civil Population; Conditions on January 1st, 1920.

Conditions on arrival of Third Army.—When the Third Army entered its area of occupation, it found the principal foodstuffs rationed, as had been the case for several years. In brief, the situation may be outlined thus, prior to the war, the average food consumption for the German population, expressed in calories, was about 3500 calories per person per day. According to German figures, this had shrunk to 3000 calories in 1914, 2000 in 1915, 1500 in 1916, and to 1200 in the winter of 1917–1918.

All the principal foodstuffs had been rationed during the war, and, on paper at least, every resource of the Empire in the way of food was entirely under control and carefully distributed.

The ration at the beginning of the occupation was essentially as follows:

Bread	260 grams per head per day
Potatoes	500 grams per head per day

The main reliance for sustenance was placed on the above two foods and, except in the large cities, where the supply was subject to much fluctuation, the amounts indicated, or more, were fairly consistently provided during the whole of the year 1919.

In addition, the following substances constituted a part of the ration in the amounts indicated:

Meat	200 grams per head per week. Frequently reduced in amount, and often not issued at all.
Fat	150–200 grams per head per week. Later became very scarce.
Butter	20 grams per head per week. Practically never issued in the ration.
Sugar	600–750 grams per head per month.

Marmalade-------- 200 grams per head per week. Often unavailable.

Milk------Not issued at all to the population in general, on account of its scarcity. Issued only to children under 6 years of age and, on physician's certificates, to the sick, nursing mothers, pregnant women and the aged. One half to one litre per day.

Fresh vegetables, in general, were not rationed and were fairly plentiful. Additional substances, such as rice, oats, grits, margarine, sausage, "Ersatz" (substitute) coffee, eggs, and additional flour, were added to the ration from time to time when available.

German system of distribution of food supplies.—The supply of grain and potatoes, as the staple foods, was supposedly under strict control of the central government in Berlin. The estimates of production, carefully compiled from statistics of previous years and a complete knowledge of the areas under cultivation, were probably very nearly correct. Deliveries by the producers, however, were very much below the estimates and became increasingly so during the occupation. This was due to two causes; first, the low prices, fixed by the government, to be paid the farmers for their produce and the consequent temptation to the farmers to sell privately and secretly at higher prices whenever possible; second, the tendency of the producers to keep more for their own use than that to which they were entitled under the law. At the beginning of the period of occupation, the supply of grain on hand was the lowest of any time during the war, being only 350,000 tons. As 250,000 tons were required for monthly distribution, it is quite evident that there was only a small margin of safety.

In order to offset the difficulty of price, the government had attempted to obtain larger and more prompt deliveries of grain by offering premiums. In 1915, twenty marks per ton, over and above the maximum price, were offered, with considerable success. In 1916, fifty marks per ton were offered as a premium with less result, and in 1917, 100 marks with almost no result. In 1918, as high as 150 marks per ton as premium were offered, but without increasing the amounts delivered. During that year, according to statistics, five million tons of grain should have been delivered to the Grain Office, but up to November, only two million tons had been received.

In regard to potatoes the situation was not so bad. Except in the large cities, the full ration of seven pounds per week per head was consistently maintained. In the cities, however, shortages often caused a temporary reduction of two or even three pounds per head.

The situation in regard to meat was stated to be very bad, due to the fact that large numbers of cattle and swine had been killed during the early years of the war, on account of the difficulty of procuring food for them. This had caused a tremendous and immediate loss in the supply of milk and milk products, and a plethora of meat at first was soon followed by a scarcity. This situation, however, was found to be in fact not less bad than had been stated. It is not, however, at all surprising that the Germans viewed the situation with much alarm, and that the newspapers published daily articles showing the great scarcity of foods, and giving the inference that the whole population was on the verge of starvation. It is an interesting fact that prior to the armistice, the newspapers described food conditions as "fairly satisfactory", considering the four years of war and the blockade, and that after the armistice and revolution, they immediately became pessimistic and gloomy on the subject. This was due, of course, to the fact that prior to the armistice and the revolution, the newspapers presented the views which the Imperial Government wished to have presented, and that after the armistice and revolution, the views of any editor, author, politician, or individual, were allowed. It was natural, therefore, that prior to the revolution the expressed views should be encouraging and consistent and that thereafter they should be widely divergent, varying and pessimistic.

German estimates of food situation.—The German Armistice Commission stated that "Germany is on the verge of starvation. The harsh armistice terms of the Allies merely precipitate this tragedy. Famine leads to anarchy and Bolshevism, which now menace Germany." The newspapers reflected the opinion of the Armistice Commission, appealing to America to provide food for 70 million starving people and thus prevent the spread of Bolshevism.

The following statement was made by the Imperial Food Office:

"Continuance of the potato ration seems to-day entirely out of the question. At most only half the present ration can continue to be given.

"In our early estimates, we counted on grain from the East. This failing us, we can now look for grain sufficient only until February 7, 1919, and this, only with the help of last year's ration. The present transportation situation, however, makes even this seem entirely impossible. From February 7th on we shall be able to give a ration of only eighty grams of flour—a third of the former one. By reducing the temporarily higher ration on December 1st, instead of on January 1st, as normally would have been the case, not much is gained.

"The fat and meat questions are serious ones. We cannot import oleaginous seeds from Russia, and the fat supply is threatened by the anticipated break-down of the milk-supply on April 1st. The ration between now and then in butter and in margarine will average 3.3 grams a day per person. The meat ration will be 100 grams per week. The substitutes, which partly replace this, will largely have been consumed by February 1st.

"The result is evident when it is considered that the present is but a third of the normal ration, and that it will drop to a half of what it is at present. It signifies slow but sure starvation."

In the *Vossiche Zeitung* of December 5th, 1918, was published the following article on food conditions:

"The German food supply is on the brink of a catastrophe. The situation has been too much influenced by internal changes to fix the responsibility on any group of individuals. From the beginning of the revolution, the hope has been that Germany would be saved from disaster by other countries. The decision remains with our enemies whether they will pay the price to save Germany from hunger and anarchy. The Allies have already expressed their intention of supplying Germany with foodstuffs, if order is maintained within the country.

"As regards present food supplies, the German Armistice Commission is unable to bring appreciable help; their efforts thus far have not had the least success. One asks with increasing astonishment how it is that in the first report of the armistice, food supplies to Germany were mentioned, and in subsequent communications, the question was passed over entirely. By a series of wireless messages and notes of protest, it has again become an actuality. Does anyone really think that the Allies cannot be informed of the situation in Germany, without all these beggings and entreaties? Do they really expect to win the sympathy of enemy lands through steps which compromise our national dignity?

"It depends more upon the Allies than upon France to get us out of our difficulty. The Allies know already what measures we want them to take, and they are ready to help Germany when their conditions are fulfilled. Their first demand is that order be maintained in Germany. They ask that Germany not pass from one revolution to another, that its economic life be held under control so that the outbreak of anarchy will become impossible.

"Our future hangs on our decision. There is still time for the German people to win back everything they have lost and the essentials they cannot do without—unity, the respect of the world, and daily bread. But the twelfth hour is here; it demands prompt action."

On November 26th, 1918, in the *Vorwaerts*, appeared the following article:

"There are still apparently some people who do not realize the frightful gravity of our economic position. We are wandering along the edge of a precipice and our only safety is in fulfilling our highest duty—self-sacrifice, greatest efforts in work, and iron discipline—these alone can save our people. If we do not succeed in giving food, light, heat, shelter, and clothing to the people, despite the frightful difficulty in maintaining communications, then we are lost, because first comes hunger, then anarchy, civil warfare, the fall of the State, and on the heels of this, the intervention of a merciless enemy.

"Each one must realize that on him depends the existence or the destruction of our people, and, in the first place, his own life. Above everything else, everyone must work to keep us from going on the rocks. Despite fatigue, everyone must continue to do his duty.

When, for example, a workman does not help in preparing materials or foodstuffs, it means that somewhere else there will be comrades without bread. Every strike brings national catastrophe nearer."

A careful observation of food conditions and food distribution in American occupied territory during the whole of 1919, leads inevitably to the conclusion that the official, semi-official and press estimates of the food situation were grossly exaggerated. That there was a considerable shortage of foods, particularly those to which the German was accustomed, and of food luxuries, cannot be denied, but that seventy million people were on the verge of starvation is untrue. It is perfectly true that the comparative scarcity of accustomed articles of food, and the probability of the conditions becoming more pronounced, was producing an ever-increasing social unrest. This is particularly so, as the condition was depicted in more or less exaggerated form and constantly held before the people in public print.

Salient facts on food situation.—A few salient facts regarding the statements and figures submitted by the Germans to support their contentions, will be of interest.

It was claimed that, before the war, the German workman had, and required, about 3500 calories per day; that this was gradually diminished during the war until in 1918 the ration contained not more than 1400 calories.

While German food authorities do not agree, nevertheless American food experimenters have definitely proved that the average adult workman, cannot only live, but also thrive and do the most effective work, on a ration of from 2,300 to 2,800 calories per day. This is for workmen; for any community of people containing a large proportion of children, as is the case with Germany, the caloric value of the ration could be considerably reduced without danger.

The caloric value of the official individual German ration in 1918 was stated to be about 1,400, and the inference was, that the population was getting that amount of food only. As a matter of fact, however, it became increasingly apparent during the occupation, that a very considerable proportion of the foodstuffs produced never went out of the possession of, and were consumed by, the producers, or were sold at high prices in illicit sales to the wealthy, and were never considered in the food statistics. In other words, the rural population was in general well provided for in American occupied territory.

In every large city in the world, there is always a certain proportion of the population which is destitute and unable to obtain necessary food, because of lack of money, though there may be no scarcity of food. No cognizance of this class was taken by the Germans in their various allegations made to show the results of food shortage caused by the "inhuman" blockade of the Allies.

During the last years of the war, every man had received his bread ration on time, and yet, immediately after the armistice, a prominent official publicly stated that "within a few weeks", proper nourishment of the people with bread, in the principal districts where bread was needed, could no longer be assured. In spite of this statement, the bread ration was increased in December, 1918. No general shortage of bread occurred, although no additional nor unexpected supplies of flour were obtained until the following April, when some purchases were made from American army stocks. It is quite true, however, that the people did not like the quality of the official flour, as there was little refining in its preparation, practically all the bran being ground with the grain (94% instead of 70%, to which they had been accustomed.)

Summary of conditions.—Conditions on arrival of the Third Army may then be summarized as follows:

There was a shortage of all foodstuffs, compared to the abundance which prevailed before the war, though there was no danger of actual starvation.

There were no large reserves in view, which caused a feeling of panic among the population.

Misleading statistics and statements were published in order to give the worst possible aspect to the existing condition.

The most serious shortages were of meat, fat, milk and milk products, due in some part, at least, to the inability or inefficiency of the food officials in enforcing the food regulations.

As far as grain production is concerned, the whole population of Germany is divided as follows:

Self-providers, i. e., tillers of the soil—14 millions.

Those living in self-providing districts, but not themselves engaged as farmers—24 millions.

Those who must be entirely supplied, including the army—32 millions.

Total—70 millions.

More than one half of the population, therefore, are provided for by their own production, or by the local production of the districts in which they live; in other words, are provided with grain without any action on the part of the Central Grain Office. The figures are essentially the same for the potato supply, and show how far from actual starvation was the population as a whole at the beginning of the occupation.

Causes of food shortage.—The rapid retreat of the German army in the last weeks of the war, resulted in a tremendous loss of foodstuffs, particularly grain, through theft and destruction.

Owing to rapid demobilization of the army after the armistice, the number of self-providers was greatly increased (Class 1 above). The effect of this, at first, was to lessen the amount of food, especially grain and potatoes, available to send to the Central Food Office for distribution. Demobilization also increased the number of non-producers living in surplus production districts (Class 2 above), and of the non-producers who had to be supplied by the Central Food Office. At first sight, it would seem that the demand of the army would be proportionately lessened, but this was not the case, on account of the tremendous losses from army supplies through destruction and theft, as already mentioned.

During the war, crops had progressively diminished, particularly in 1917 and 1918. Many causes operated to effect this diminution, the principal ones being:

(a) Lack of seeds, due to the blockade, the poor quality of crops and the necessity of using a larger proportion than usual of the crops for food;
(b) Scarcity of labor, particularly efficient labor, due to the large number of men absent with the army, and the ineffectiveness of labor of prisoners of war. Immediately after the armistice most of the prisoners of war were liberated but as they did not work, simply contributed to the class that had to be supported;
(c) Lack of fertilizers due to the blockade; lack of transportation, which shut off part of the supply of nitrates, phosphates and potash; rapid diminution of live stock, causing a decrease in manure production.

Other factors contributing to the food shortage were:

Loss of control of the Baltic meant less fish from the sea and from Scandinavia.

The revolution and consequent strikes, disorders and unemployment, together with the large number of returning soldiers and prisoners of war, little disposed to work, and filled with the idea that their privations entitled them to a life of ease and plenty.

Little or nothing harvested from the enemy countries, particularly wheat from the Ukraine and Roumania and potatoes from Poland.

Lack of coal for threshing purposes.

Difficulty in transporting and circulating the diminished crops, due to the scarcity and deterioration of rolling stock and deterioration of road beds, maintenance of which was dependent upon inexperienced female labor during the war; lack of engines and coal for the same. Social unrest and strikes or threatened strikes also interfered with transportation.

Smuggling and illicit trading in foodstuffs. During the whole of the period during which food had been rationed, this had been one of the great difficulties of the Food Administration, due very largely to the absurdly low wholesale prices paid by the Central Government. As long as there were people with money willing to give more than the government price for any foodstuff, they were producers anxious to obtain what they considered a fair or good profit by illicit sales. As soon as the revolution occurred, all respect for officials and fear of punishment disappeared as if by magic, and smuggling and illicit trading went on practically without restraint.

Military action regarding Germany food supply.

Regulations.—All regulations for conservation of German food supply have been quoted in the Appendix on Food-Supplies (No. 20) [not printed], and, for the most part, explain themselves. The following comments, however, may be of interest:

Early in the occupation, regulations were promulgated by the armies prohibiting officers and soldiers from eating in German restaurants. When it was later found that the food-scarcity was somewhat less pronounced than had been believed, and when it was realized that in any case the amount consumed in this way was neglible, the prohibition was removed. Later, all army messes were forbidden to buy anything except fresh vegetables (not including potatoes), from German sources. Late in 1919, all American military persons and employes and American personnel of the Inter-Allied Rhineland Commission were prohibited from buying any article of the German ration from German sources. This was done only when the larger cities were feeling a potato shortage and had been obliged to seriously curtail the potato ration.

The Lebensmitteleinfuhr, G. m. b. H.—In order to allay the social unrest due to scarcity or impending scarcity of food, and to the non-distribution of allotted supplies, the United States Army, early in April, 1919, began the sale of flour from army stocks, at cost price, to the civil population.

On account of the unstable condition of the German government, no credit could be given, and payment was required,—at first, when the request for certain quantities of supplies was submitted, and, later, an advance payment at the time of the request, followed by payment in full upon delivery. As each one of the sixteen Kreise in American occupied territory had to estimate its own needs and to finance its own request for supplies, it became immediately evident that some central purchasing agency must be established. At the instance of the Oberpräsident of the Rhine Province, a company was formed by public-spirited citizens, and incorporated early in April, 1919, known as the Lebensmitteleinfuhr, G. m. b. H. or, in English, "Food Distribution Company, Limited."

This Company was an undertaking in the public interest. Its employes were paid a fixed salary without any share in profits that might be obtained. According to Paragraph 2 of the Articles of Incorporation, the object of the Company was "the purchase and distribution, in the public interest, of foodstuffs and other commodities of daily use."

Article 9 ran as follows:

"The Company pursues a public aim exclusively, and its endeavours will tend to improve the food situation in general. Its capital stock bears no interest. The yearly net profits, after writing off an appropriate sum, will be used for the public interest, according to a plan to be agreed upon by the shareholders, with consent of the President of the Board. The shareholders expressly waive repayment of their shares. No part of the profits is turned over to any private interest, of any nature whatever."

The President of the Board, referred to above, was the Oberpräsident of the Rhine Province, who by virtue of his official status was the head of the Food Administration of the Province.

The first methods of the Lebensmitteleinfuhr were slow, laborious and ineffective. On being informed that flour was available from army stocks for sale to the civil population, and being requested to state how much it desired to buy, the Company attempted, before submitting its request, to obtain by letter or telegram from each Landrat and the Oberbürgermeisterei of Treves and of Coblence, a definite statement of the amounts required. Each one of the chief officials named, felt that he must consult his sub-officials, and each sub-official "felt the pulse" of his constituents before making an answer. When the answers finally came to the Lebensmitteleinfuhr, they were generally qualified by stipulations as to prices to be paid. As these prices had been previously fixed at the cost of the supplies to the American army, and by the official rate of exchange of dollars and marks, there was, of course, no possibility of bargaining. On the other hand, there was frequently so great a lapse of time between the offer of the supplies and its acceptance, either in part or in whole, that either the supplies were no longer available for sale, or the rate of exchange had droped so materially as to make the Company undecided what to do. Although this indisposition on the part of the Company to assume responsibility without wasting valuable time interfered greatly with the purchase and distribution of food, nevertheless during the whole of 1919 the difficulty remained. As late as November, 1919, the Company stated officially that it had made many requests for supplies which had not been granted, for reasons unknown. As a matter of fact, supplies had time and again been offered, which the Company had failed to purchase within a reasonable time, and which were then disposed of elsewhere.

Practically speaking, the duty of the Lebensmitteleinführ was to finance the transfer of foodstuffs from various sources to the civil population. It received requests from food officials, transferred these requests to the army or other source of supply, sometimes making an advance payment, received the deliveries and made payment for them, and transferred the deliveries, charging to, and collecting from the various Kreise, the cost of the supplies, according to the distribution of the food administration. On the whole, the system worked very well. The inability to grasp opportunities quickly, however, accompanied by complaints that the Company was often unable to obtain the supplies it desired, caused much irritation on both sides.

Circulation of food.—The method of distribution was that the Oberpräsidium, as the central food administration of the Rhine Provinces, forwarded all supplies to a central or convenient point in each Kreis, putting them in custody of the Landräte or the Oberbürgermeister of the cities of Coblence and Treves.

From that point distribution was left entirely to those officials. The policy and method of distribution, therefore, often materially differed in different localities. In Adenau, for example, the channels of distribution were well laid out, the Landrat displayed initiative in obtaining the aid of American motor lorries for transportation, and the food was on sale within a few hours after delivery. In Ahrweiler, on the other hand, some of the first supplies of American food were not distributed to the shops until three weeks after delivery, because there was not enough delivered to redeem all the food tickets which had been issued. The delay might have been still greater, had not the matter, quite by accident, come to the attention of the Office of Civil Affairs, and orders been issued to proceed at once with distribution. In only one Kreis was the plan of distribution itself bad, viz. in Landkreis Treves. In all other cases, difficulties were in the main temporary, and due to two causes,—first, the disposition to hoard supplies; second, inability of the food officials to alter quickly the plan of food distribution or of food allotments, when a change had been made in the quantity or kind of food.

Sale of American foodstuffs was by the ticket system. The following is a quotation from a letter of the Oberpräsident, which indicates how some of the difficulties were met:

"The distribution of American foodstuffs has been operated on the ticket system, which is in general use here, and no difficulties worth speaking of have been encountered. Unfortunately, the prices of these commodities are very high, owing to the unfavorable rate of exchange on German currency, and the poorer civil population experiences great hardship in providing money for their purchase. To quote an instance of these unusually high prices, a four-pound loaf of bread of German flour costs between M. 1.05 and M. 1.15, whereas a one-pound loaf of bread of American flour costs M. 1.25.

"A similar difference of price obtains in various other commodities. It should not be forgotten in this connection that it was due only to the very welcome complaisance of the American authorities that the prices were maintained at the April rate of exchange. If we had been obliged to pay the June rate,—equivalent to an additional 17%,—there is no doubt that a great majority of the population would have been unable to pay the price. A proposal is now on foot for the Kreise and cities to grant facilities, according to their financial capacities, to the poorer classes when making their purchases. The city of Treves was the first to distribute special bonus tickets to a limited circle of poorer people to be used in part payment for foodstuffs; thus, the bonus tickets for bacon represent a value of M. 2.80, so that the purchaser gets it at five marks a pound. In the case of flour, the bonus tickets are worth 55 pfennigs, reducing the cost of this commodity to one mark per pound; in the case of rice, they represent 60 pfennigs, reducing the price to M. 1.22.

"The whole system is somewhat intricate; whether it will be of permanent use, experience must prove. As it has been introduced only in Treves, no conclusive opinion can be given. Of course, the number of people enjoying such advantages must be limited, as otherwise the expenditure would be too great, considering the sums that are turned over."

The method of using the food tickets, was, in brief, as follows. Each person or head of family was issued the number of coupons to which he was entitled, each coupon authorizing him to buy a certain quantity of a certain food, within a certain period of time. There were also coupons, not bearing the name of any particular food, but numbered, to be used for foods not regularly issued. The coupons entitled the holder to register with some merchant of his own choice, who in turn registered his total requirements with the local Food Office. When the food arrived, each merchant was allotted the amount necessary to provide for the persons registered with him. Announcements of distribution were made in the daily papers. The merchant kept the coupons to indicate the amounts and regularity of his sales. Any allotment not entirely sold, might be returned to the food authorities or debited against the next allotment. The system seemed simple enough, but there was considerable traffic in tickets. Well-to-do people seldom suffered for lack of food, though they may often not have been able to get the variety they wanted.

American foodstuffs sold to civil population.—The amounts, in round figures, of the more important foodstuffs of American origin distributed and sold in American occupied territory during the year 1919 are as follows:

Flour	17,000	tons
Bacon	930	"
Sugar	1,660	"
Rice	790	"
Lard	200	"
Canned Beef	550	"
Salmon	795,000	cans
Milk	652,140	"

It is an interesting fact that although less than 22,000 cans of milk had been sold to the population for the whole American occupied territory since the first of August, 1919, and although complaints had been made of the insufficiency of the supply, there were still on hand in the City of Coblence alone, in the middle of December, 1919, 162,000 undistributed cans of milk. A protest from the Office of Civil Affairs prevented a wasteful, sentimental, and extravagant distribution of a large part of this supply on Christmas Day without regard to age or state of health.

Conditions on January 1st, 1920.—In the present area of American occupation, the food conditions on January 1st, 1920, were not alarming.

Meat.—This continued to be scarce. According to the figures submitted from the German food authorities, the amount of meat available for supply to the civil population was about one-third that supplied before the war.

Potatoes.—In the area of American occupation, with the exception of the city of Coblence, the supply of potatoes was possibly sufficient to ensure the ration, or nearly so, of seven pounds per week until the next harvest.

There was a moderate deficiency of supply in Kreise Neuwied and Ahrweiler, but this deficiency could be easily made up from surplus production areas, such as Mayen and Cochem.

A large part of the surplus production of Kreis Mayen was destined however for the cities of Cologne and Bonn.

The city of Coblence was on this date receiving only five pounds per week, instead of seven pounds, as normally expected. This, however, was calculated on the basis of the available supply, divided by the total population, but as a matter of fact a considerable part of the well-to-do of the population were already supplied until the next harvest.

Milk.—The milk situation remained about the same as in 1919. There was a considerable supply of undistributed canned milk on hand, more than 100,000 cans in the city of Coblence alone. The supply of fresh milk was about 16% of what it was before the war. Of the amount required for distribution to women, children, the sick and the aged, only about 45% was available.

Flour.—The flour situation changed materially for the better, in that the full bread ration was maintained practically everywhere throughout the area, and the flour distributed was of a much better quality than previously, being an extraction of 82% instead of 94% which it had been early in the year 1919.

In spite of the deficiency of potatoes and meat, there was little danger of actual starvation, as substitutes were found from time to time to take the place of these commodities.

Butter and Eggs.—These were practically unobtainable on the market at retail prices, corresponding reasonably with the wholesale prices fixed by the government. Both of these commodities, however, could be obtained by any purchaser who was well known to dealers, though at prices far in advance of the legal standard. The price established by the Government for fresh butter purchased from the producer varies from 5-½ to 6-½ marks per pound. It is regularly sold at the present time in secret sale only, at a price varying from 25 to 28 marks per pound.

Margarine and Dutch Cheese were on sale at most shops at reasonable prices. The population in general, however, showed no liking for margarine.

Sugar.—There was no scarcity of this, and large quantities have been available to be sold to wine growers.

The ration as officially provided for the population in the latter part of December, 1919, amounted to about 2,200 calories, and seemed sufficient to maintain the average working-man in a proper condition of health. The question in American occupied territory is not therefore so much a question of food shortage as it is of food cost, since the cost of food has increased in much greater ratio than the increase in wages.

CHAPTER No. 9

PUBLIC WORKS AND UTILITIES

Formation and Scope of Public Utilities Department of American Army of Occupation; General Policy of Supervision and Administration; Ownership, Control and Direction of Public Utilities; Gas Works; Water Works; Power and Light; Coal; Sub-Coal Commission and Inter-Allied Rhineland High Commission; Labor Troubles.

Formation and scope of public utilities department of American Army of occupation.—When the advance from Luxemburg into Germany was begun, the Officer in Charge of Civil Affairs was directed by G. H. Q. to assume charge of all public utilities such as gas works, electric plants, etc., in order that the proper operation of such plants should be coordinated and the interests of the American army, as well as those of the civilian population of the occupied area, protected. Up to the time that Army Headquarters were established in Coblence, the actual operation of the Department was of necessity more or less restricted; but once this establishment was effected, immediate steps were taken to exercise supervision over all public utilities in the area. As supervision of the various public utilities called for a certain amount of technical knowledge, it was agreed between the Officer in Charge of Civil Affairs and the Chief Engineer, that the latter should take over their actual operation and supervision, referring matters of general policy to the Officer in Charge of Civil Affairs, to be decided in accordance with the spirit of the broad policy adopted toward the inhabitants of occupied Germany.

The most important duties of the department were confined to the city of Coblence and its environs, as the largest and most important city in the American area, as well as the seat of Army Headquarters. Supervision in the Headquarters area of the army was exercised directly by the Public Utilities of the Third Army. A Divisional Public Utilities Officer administered utilities in each Divisional area, referring to the Army Public Utilities in matters of policy and for any necessary liaison with other Divisions or points outside the occupied area. Advance G. H. Q. being at Treves, all public utilities there were administered by an officer attached to that organization, and no control was exercised by the Army.

The utilities in which the army chiefly interested itself were gas plants, electric plants, and water works. While several others, such as public baths, slaughter houses, etc., received some attention, they were of so little importance that they may be neglected in this report.

As has been said before, the utilities of Coblence were the most important in the whole army area, therefore, this report is based on the operations at that point, but it may be assumed that the outlying towns experienced similar difficulties which were met by local action in most cases. Wherever any feature of especial interest occurred in outlying districts, it is mentioned in the report.

General policy of supervision and administration.—It was at once apparent in establishing the policy of the department, that two distinct methods of administration were possible, either direct control of all public utilities, or administration thereof through the German civil authorities already in charge. The latter method was adopted for several reasons, and proved to be an entire success. Its advantages were:

(a) Saving of military personnel;
(b) Facilities and benefits derived from established organization;
(c) Placing of operating responsibility on officials accountable to their own government;

In addition to the above, there were other small advantages of position in regard to the civil population.

Directors of public utilities were informed that they would be held responsible for the efficient operation of their plants, but that they would be given reasonable aid by the army in procuring coal, oil, etc., also in adjusting difficulties which might arise from the extra burden placed on their plants. A fair idea of this extra burden may be obtained from the fact that Coblence, with a normal population of 60,000, was at one time housing over 80,000 troops and civilians.

The proof of the wisdom of the adopted policy is shown by the fact that every effort was made by the directors of public utilities to please the military authorities, and that no insurmountable difficulty was ever encountered.

Ownership, control and direction of public utilities.—In general, title to public utilities in Germany is vested in the city or town in which they operate, but in several cases, notably of electric light and power, large private corporations serve several towns or cities. This is the case with the power company supplying Coblence, a description of whose activities is given in a later section.

In larger cities, the directors of gas and water works are appointed by the municipal authorities under certain fixed rules and requirements. The same is true of light and power plants if they are publicly owned. In smaller towns and communes, the Bürgermeister is director of all public services.

A survey was made of all public utilities plants in the American area, soon after the Office of Public Utilities was placed in operation. They were found to be in a deplorable condition. All repairs that could possibly be avoided had been put off during the war. Plants which were not even in a condition to properly meet their ordinary deliveries, were now called upon to supply not only these demands, but those of the Army of Occupation as well. It was at once decided that a system would be established whereby American troops and all industries on which they were dependent should be favored. It became necessary at times to use the regulations issued to that end, as a means of spurring on directors to further efforts to meet all demands.

A description of the public utilities in Coblence, their plants, their difficulties of operation and of the steps taken to meet these difficulties, will be found in the following pages. Technical details are purposely omitted.

Gas works.—The gas supply of the city of Coblence is derived from two sources,—three benches of eighteen retorts each and further equipment described later in this chapter. These are owned by the city and are situated on the Moselle river near its junction with the Rhine. A supplementary supply is obtained from the Concordia-Hütte, or Smelting Works, at Bendorf, a town not far below Coblence on the right bank of the Rhine. This plant also supplies the city of Neuwied and the towns of Engers, Bendorf, Vallendar and Ehrenbreitstein, all on the right bank. There are forty-one retort groups at the Concordia works; but a large amount of the gas generated is consumed industrially at the plant.

There is very little commercial consumption of gas in Coblence, practically one-quarter of the total amount being utilized in the operation of the city water works, a description of which will be found later in this chapter. The system of distribution has the great weakness that the water works are connected with the mains used for household supply in such a way that in case of a shortage, the domestic supply can be rationed or cut off only by shutting off individual districts or houses, as well. If any attempt is made to close the mains, the pumps at the water works are also deprived of fuel, there being no container there.

History of Operation During Occupation.—This history actually begins with the year 1919 and finds the city of Coblence with two of its three benches in operation, one being held in reserve to meet emergencies. About two-thirds of the 30,000 cubic meters of gas consumed daily was furnishd by them, while the Concordia Hütte of Bendorf furnished about one-third, under a contract already in force when the Third Army reached the Rhine. This contract calls for a delivery of two-thirds of the gas used in Coblence, and had it been supplied, would have permitted the city to make some badly needed repairs in its own retorts. It appeared on investigation that the exact terms of this contract had never been carried out. It was therefore determined to force the Concordia Works to deliver more gas, and the directors were informed of this determination. The directors proving that their retorts were in such a condition that repairs must be immediately made, their schedule of repairs was approved, with the understanding that they might reduce the quantity furnished the city until April 15th, when the full amount called for in the contract was to be delivered. It was necessary to aid the Works in procuring the necessary material for these repairs.

From the first survey made of public utilities, it was at once apparent that the coal supply was a matter of difficulty, as the production of German mines had been greatly reduced. In the case of gas plants, the deterioration in the quality of the coal and the fact that a mixture of the gases generated was employed with a view to economizing coal, reduced the thermal value from 5200 normal to an actual 4000.

In order to enable the Concordia Works to make payments for coal and other supplies, it was decided, after a survey of their financial condition, that it would be necessary to allow them to export part of the finished product on hand to unoccupied Germany. This plan was recommended to, and approved by, the Economic Section of Advance G. H. Q. at Treves, and the Works were thus relieved of a situation that threatened to become serious.

During the month of April, 1919, the Hütte again took over the furnishing of a major portion of the gas used by the city, and the latter immediately retired two of its three benches of retorts from use for repairs, and reconstruction and excavations for the foundation of a new and fourth bench, were commenced. From this time on the city furnished approximately one-third of the gas consumed, the rest being furnished by the Hütte. This arrangement was on the whole satisfactory, except during a period covering the last half of May and the first half of June, when lack of coal, combined with the poor quality obtained, caused a falling off in the supply furnished by the Hütte. It became necessary to put the other towns supplied by this company on a ration. The ration system was not feasible in Coblence, as has been explained earlier. After the plan of rationing the smaller towns had been placed in effect, there was still a deficiency in the quantity delivered to Coblence. As it was felt that this was partly due to lack of interest and cooperation on the part of the directors of the Hütte, it was decided to apply a stimulus to their efforts, and their press works, where car-wheels are manufactured, was therefore closed for several days. The directors at once made every move to produce more gas, and as soon as they reached the required quantity, the press works were reopened. A further difficulty was experienced with the Hütte during the month of June, owing to friction between the directors and the Wirtschaftsstelle, the latter being a board for coal distribution whose function is explained in another section of this chapter. The result of this friction was that the reserve coal supply at the Hütte was reduced to the danger point. The directors were summoned to Coblence and informed that if the city suffered through their negligence, they would be prosecuted before the Military Commission. They took immediate steps to obtain a supply of coal, and the threatened gas crisis was averted. Numerous petty difficulties, which naturally arose from time to time, were disposed of, usually without action, by the Public Utilities Officer. It is worthy of remark that every effort to please was made throughout by the officials of the gas works, and that their service was remarkably efficient when the extra load placed on their plant is considered.

In order to facilitate supervision of the gas plant, daily reports were required of coal consumed, coal received, coal on hand, gas generated, gas delivered from the Concordia Hütte and other minor details. These reports were represented graphically in the office.

Water works.—The water works of the city of Coblence are on an island in the Rhine just above the main part of the city and derive their supply from wells. It might be supposed that the water in these wells would come from the river, but such is not the case, as the river-bed has an impervious layer not far below the ground-line. The water for the wells comes from the hills southwest of the city.

The pumping plant, in December, 1918, consisted of the following units:

Pump House No. 1; built in 1886.—Three gas engines of an old design, where ignition is caused by a flame similar to a Bunsen burner, and three plunger-pumps, of 120 cubic meters per hour capacity each. It has been the practice to hold one unit in reserve.

Pump House No. 2; built in 1904.—Three gas engines of a modern type and three plunger-pumps of 150 cubic meters per hour capacity. All units in this plant are operated continuously.

Pump House No. 3.—Two Diesel oil engines and two centrifugal pumps, of 250 cubic meters per hour capacity, each. One of these units is held in reserve.

A survey made of the water works revealed the fact that as the weather grew warmer, the existing plant would prove inadequate to meet the additional demand caused by the Army of Occupation. An electrical pumping unit was therefore ordered by the army and this was installed as Pump House No. 4. This consisted of a motor and a centrifugal pump of 250 cubic meters per hour capacity. This unit was held in reserve and was operated only when demand for water justified.

There are no facilities for water storage, aside from a tank on a hill overlooking the city, which contains only enough to supply the needs of a barracks on the same hill. All the large units at the plant pump direct to the mains and are assisted by two electrical pumping units near the Moselle, which are used as boosters, and by two gas units which pump to the tank at the barracks.

The town of Matternich, near Coblence and on the left bank of the Moselle, is also connected with the mains of the Coblence system. In case of a shortage of its own water supply, Metternich, can call on Coblence for assistance, but this procedure cannot be reversed.

History of Operation During Occupation.—As the original survey of the water works disclosed the fact that the system was inadequate to meet the increased demands caused by the occupation and that new machinery was necessary, it was decided to do all that was possible to decrease consumption until such time as the new pumping unit, by which the capacity of the plant was increased thirty percent, should be ready. An investigation was made of the requirements of the principal consumers and certain restrictions were imposed. An interesting fact in connection with the consumption of American troops is, that there was an increase at the several barracks occupied by them, ranging from 100% to 1000% over the amount consumed by German garrisons in the same buildings. During the month of June a continued drought made drastic measures necessary. The civil population was warned through the Bürgermeister to reduce consumption, and similar orders were issued to troops by the Army Commander, forbidding the washing of motor vehicles at garages, and horses at stables, ordering that such operations be carried on at the river; also that special care be exercised in regard to running taps, etc. By this means consumption, which at one time had reached 48,000 cubic meters daily was reduced to an average of 36,000 cubic meters.

The town of Metternich suffered from an inadequate water supply during the month of June, and aid was extended by means of the connection with the Coblence mains, already mentioned. Considerable anxiety was felt in this matter because of lack of sufficient head for proper fire protection at the Ordnance Depot, but the reduction of consumption relieved this situation. This deficiency in supply also extended to the fortress at Ehrenbreitstein, which has a semi-private source of supply. Steps were therefore taken to connect another spring to the system, fed entirely from natural springs.

The gas engines described in previous paragraphs of this chapter use illuminating gas for fuel and are connected directly with the city mains which supply gas for household consumption. The difficulties experienced in maintaining a requisite gas supply were all reflected at the water works and, in fact, most of the steps taken at the gas works were rendered imperative by their connection with the water supply.

It was difficult to obtain fuel for the oil engines, during the months of January and February, but no great trouble was experienced in later months.

When it is taken into consideration that the machinery at the pumping plant had undergone, during the war, only such repairs as were absolutely essential to operation, it is surprising that few breakdowns occurred during the period from January to July, 1919. The greatest source of trouble was the motors of the booster plant on the Moselle. When these had been repaired in Cologne, they worked perfectly.

The extreme high water in December, 1919, and January, 1920, was a partial cause of the interruption of the water service. In each instance the army sent engineer troops to help salvage electrical equipment and maintain some of the pumps in operation during the floods. The Office of the Chief Engineer is at the present time looking into the question of extending the present system in order to provide for the increased population of Coblence, due to the various civil commissions now residing there.

The water works were required to furnish daily reports to the Office of Public Utilities, showing amount of gas and oil consumed, amount of oil on hand and amount of water pumped. A graphical record of these reports was kept.

Power and light.—Power and light for the city of Coblence are supplied by the Coblenzer Strassenbahn Gesellschaft, or Coblence Street Railway Company, a privately owned corporation which not only furnishes light, power and transportation in Coblence but in the nearby towns of Bendorf, Sayn, Ehrenbreitstein, Arenberg and Lahnbrücke. The larger part of the power is supplied by a modern up-to-date generating plant at Höhn, a town in the neutral area not far from Marienburg, in which the company holds a half interest.

Power at Höhn is generated by steam, which in turn is produced by boilers with an automatic overhead feed of lignite. This is obtained from a nearby mine owned by the company, and is delivered direct to the fires by a conveyor. The power is delivered to a substation at Horchheim, across the river from Coblence, at 50,000 volts, and there reduced to 10,000 volts for local distribution.

The company also operates a steam power plant in Coblence, whose installation is described further on in this chapter.

History of Operation During Occupation.—More trouble was experienced in the administration of power and light than in any other branch of public utilities. This arose from two principal causes; first, the fact that the Street Railway Company is privately owned and the owners objected to American intervention; secondly, the fact that the power plant at Höhn lies in the neutral zone, and direct intervention at that point was impossible. This fact will be alluded to later.

When the Third Army entered Coblence in December, 1918, a survey of the electric plant seemed to indicate that the machinery was in good condition. This later proved to be a mistake, and, while a superficial inspection did not disclose the fact, the boilers and electrical machinery were nearly all on the point of breaking. These breakdowns did, in fact, commence at once, the first trouble being with the boilers. Several of the boilers failing at the same time, early in January, a system of power and light rationing was at once placed in force, not only in Coblence, but also in other towns supplied by the company. An effort to make necessary repairs to the boilers by using the acetylene welding plants at the American Motor Repair Shops met with no success. Workmen were subsequently obtained from Düsseldorf and by the end of February, the boilers had been repaired to such an extent that the system of rationing was relaxed. While numerous boiler breakdowns occurred from time to time after February, no serious difficulties were met with.

Much the same condition prevailed with electrical equipment, not only at Coblence but also at Höhn. It was constantly necessary to submit machinery to extensive repairs, one 4200 k. w. generator from Höhn having been sent to Berlin for that purpose. It at first appeared that there was no material for effecting minor repairs in the occupied territory, but a search disclosed the fact that a sufficient supply of such material existed, although in some cases the owners objected to parting from it.

The occupation of Coblence by troops necessitated a large number of extra installations for lights, power, moving picture machines, etc. In order to expedite matters, these installations were usually made with American material by American troops. The Office of Public Utilities insisted that authority for such installations be obtained there, in order that transformers serving certain districts of the city should not be overloaded. In spite of an order issued to this effect, numerous unauthorized installations were made, until rigorous punishment of the offenders stopped the practice. This order applied not only to troops but also to civilians, and several of the latter were prosecuted for unwarranted consumption of power.

As the Street Railway Company had no extra meters to care for the increase of installations, an arrangement was made for flat rates until 250 meters were obtained from Strassburg, and every effort expended to limit consumption in order to satisfy the just claims of the company.

While there was no actual shortage of coal at Höhn, the mine being at the plant, the question of supply for the Coblence plant was a grave one, as it was equipped to burn only coal, or, in case of necessity, briquettes mixed with steam coal, this mixture, however, being injurious to the plant. Serious trouble was threatened during the month of March, when the Inter-Allied Coal Commission cut the allowance for the Coblence plant from 1000 to 400 tons. Following a complaint from the Office of Public Utilities, an allowance of 900 tons was agreed upon.

Administration at Höhn, as has been stated before, was a point calling for considerable diplomacy. This plant was handicapped by two technical difficulties. The first was that one generator, the one sent to Berlin, was already bad and another was in such condition that no continued load could be placed on it. The second was that the copper wire of the transmission line to Horchheim had been taken out and replaced with iron wire, with a subsequent loss of power in transmission. An economic difficulty further lay in the unrest of the miners, partly due to lack of food and clothing, partly to the social upheaval in Germany. The company appealed to the army for help and officers were sent to Höhn to interview representatives of the miners. The employes were told that if they remained at work, the question of wages and hours would be adjusted between them and their employers and food would be sold to them from the supplies of the army at cost, but that if they persisted in striking, it might mean the extension of the American bridgehead to embrace Höhn, in which case they would be forced to work. The former alternative was accepted and an arrangement made, whereby the miners received a 90% increase in wages and a seven hour day in the pit; food was also sold them as promised. The demands of the laborers being considered just, the directors of the company were told that they would be held personally responsible if the increase in wages and the change in hours were not accorded, and if, as a result thereof, the miners struck. The directors were also informed that in case of a strike and subsequent lack of power in the army area, tramway service would be suspended and only those houses supplied with light in which Americans were billeted.

The Street Railway Company was required to furnish such daily reports as were necessary to keep the Office of Public Utilities fully informed as to its activities. These reports were kept in a graphical form.

Coal.—Although coal supply as a whole was not a function of the Office of Public Utilities, yet a brief sketch will be given of it here. Responsibility for coal, as far as the army was concerned, lay with Advance G. H. Q., a representative also being attached to the Inter-Allied Coal Commission at Cologne. This Commission met once a month and passed on the coal requirements of the various industries and public utilities in Alsace-Lorraine, the Palatinate and the Rhine Province. With this organization the Public Utilities Officer had little or no direct contact, as the requirements of his area were presented through the American member.

The supply of coal for all uses in American occupied territory was looked after by the Wirtschaftstelle, or Industrial Board at Coblence. All large consumers and wholesalers were required to obtain their coal through this board, which operated very efficiently, having extensive powers of confiscation, etc. It often became necessary to take up the matter of lack of coal for some special utility with the Wirtschaftstelle, and in every case utmost efforts were made to supply the need. In some cases the necessary coal was taken from other utilities in which a surplus existed, and in other cases from private industries. The utility to which the coal was delivered was in all cases called upon to make regular payment to the party from whom it was taken.

The strike which occurred in the Ruhr district during March and April, 1919, was not felt in American occupied territory until May and June; but it then became very difficult to obtain sufficient coal to meet demands.

Almost the whole question of administration and supervision of public utilities in American occupied territory between August 10th, 1919 and January 10th, 1920, on which date peace was declared and the Inter-Allied Rhineland High Commission assumed control, resolved itself into one of procuring sufficient coal to operate them. It was apparent as early as February, 1919, that it would be difficult to obtain a sufficient supply, and with this fact in mind, the Public Utilities Department suggested that coal be sent from America to cover the requirements of the utilities in the American area. This proposition was not complied with owing to the cost of transportation and to the depreciation of the mark.

The first serious coal shortage occurred at the Coblence Power Station about August 15th, 1919. On September 1st the Plaidt Station, whose transmission system includes Kreise Mayen and Ahrweiler, was extremely short of coal. The allotment of the Plaidt plant was 900 tons per month, but in view of the fact that the Silesian forces were to occupy Kreis Mayen, thereby increasing the population by approximately 10,000, it was necessary to attempt to procure an increase in the allotment. This, however, proved very difficult.

Several gas works were shut down during the winter months for varying lengths of time, ranging from one to five days as a maximum. In cases in which the water supply was dependent upon proper operation of the gas plant, these utilities were supplied with coal from army stocks, thereby preventing water scarcity. In order to make certain the delivery of the allotted coal to the Coblence Power Plant and the Plaidt Power Plant, a system of shuttle-trains was organized. A shuttle-train of eleven cars was operated and convoyed from Coblence to Homberg, (Ruhr Basin, unoccupied Germany), to haul nut coal. The capacity of this train was about equal to the daily consumption. In like manner, a shuttle-train was convoyed to Wiesling for raw brown coal, also for the benefit of the Coblence Power Station. A shuttle-train of fifteen cars was operated and convoyed to Frechen for the Plaidt plant. Operation of these shuttle-trains guaranteed continual supply to the important electrical utilities in the area, but did not meet the immediate demands which these plants faced. For that reason, the Public Utilities Department loaded the barge *Karl Wilhelm* with 585 tons of raw brown coal and a similar quantity of generating briquettes, the idea being to tide over the critical conditions and build up a small reserve.

An attempt was made to have the local branch of the Reichskohlen Kommissar increase the allotment to the utilities in the occupied area in order to provide for the increased consumption attendant upon the increased population. The Reichskohlen Kommissar, however, absolutely refused to increase these allotments so that it was necessary for the army to requisition the coal needed, which was delivered by the shuttle-trains mentioned in the previous paragraph.

The average allotment to utilities was 66⅔% of that of the year 1918. Had this quantity been delivered, it would not have been necessary for the army authorities to step in; but the deliveries by the German Distribution Board was actually in most cases about 30% of those of 1918.

The next serious coal shortage occurred during the first flood, during which time it was necessary to resort to the requisitioning of barges on the Rhine river. Barges were requisitioned at three separate times,—once during the first flood (December, 1919), again during the second flood (January, 1920), and finally during the Sparticist uprising in the Ruhr coal-fields (March–April, 1920). In all, about three month's coal supply was commandeered for public utilities by means of requisitioning barges. In every case, the original consignee was in unoccupied Germany. The question of the grade, quality and suitability of the coal was determined by an analysis made at the laboratory of the municipal gas works.

Sub-Coal Commission and Inter-Allied Rhineland High Commission.—The Coal Commission of the Inter-Allied Rhineland High Commission caused a special Amtliche-Verteilungs-Stelle to be created in Cologne, whose duties with respect to the occupied area were the same as those of the Reichskohlen Kommissar for unoccupied Germany. The functions of the Amtliche-Verteilungs-Stelle were conditionally based upon its success during the month of January, 1920. Close account was kept of its activities during that month, but it failed

521572—43——13

to live up to the program outlined, due primarily to the closed traffic on the river Rhine because of the floods. The receipts of coal in February very nearly equalled the consumption. A decided improvement was shown in March as well as a possibility of building up a working reserve for the utilities of the American area, which is, therefore, assured of continual supply of coal. As a spur and aid to the efficiency of the Amtliche-Verteilungs-Stelle, the requisitioning of barges on the river Rhine is however still occasionally resorted to.

From the following copy of a report to Colonel Miller, Chief Engineer, American Forces in Germany, of a coal conference at Cologne with the new Amtliche-Verteilungs-Stelle, some idea of the problem will be gained:

HEADQUARTERS

AMERICAN FORCES IN GERMANY

OFFICE OF CHIEF ENGINEER

January 5th, 1920.

MEMORANDUM: TO COLONEL MILLER.

SUBJECT: Report on Coal Conference held at Cologne on January 3rd, 1920, under Director Kruse, Herren Jordan and the Inter-Allied Rhineland High Commission.

I. The report is discussed under three headings, as follows:

1. The technical requirements of all occupied areas;
2. Production within the occupied areas and that which is to be supplied from outside sources;
3. General remarks and information.

1. Requirements.

(a) The entire occupied area will require for the month of January 1,280,000 tons. This figure was determined by taking the averages delivered during August, September and October of 1918 under the war direction program and distribution system then in existence. It is thought that this is the most favorable figure. However, in addition to the above figures must be added:

TABLE A

For	*Amount*
Palatinate (Bayerische Pfalz)	155,000 tons
Special Area near Saarburg	3,800 tons
TOTAL	158,800 tons

making a total of 1,438,800 tons per month.

(b) The technical estimated deliveries for the month of January are as follows:

TABLE B

For	Amount		% Delivery
Railroads	110, 000 tons		100
Armies	30, 000 tons		100
Industries, Private Railroads and Water Works	16, 200 tons		100
Domestic	160, 000 tons	(225, 000)	75
Essential Industries	38, 000 tons	(44, 300)	70
Electrical and Gas Works	69, 200 tons	(144, 300)	66⅔
Industries: Glass, Cement, Iron, etc	198, 900 tons	(663, 000)	30
TOTAL	622, 300 tons		

to which total must be added the requirements of the Palatinate and the special area near Saarburg, making a grand total of 781,100 tons. Therefore, the total approximate requirements will be 800,000 tons per month. This is the figure which the new distribution will try to approach for the month of January.

2. Production.

(a) *Raw brown coal and briquette production for Cologne area:*

	Tons
The total average monthly production is	425, 000
with a deduction of	235, 000
which leaves a net for distribution of	190, 000

Summary of Deduction:

TABLE A

For	Amount (tons)
France	75, 000
Holland, Belgium and Switzerland	15, 000
Unoccupied Germany	70, 000
Local Sales	75, 000
TOTAL	235, 000

Of the net for distribution, the following approximate allotment is for the occupied area:

TABLE B

For	Amount		% Delivery
Railroads	34, 000 tons		100
Armies	64, 000 tons		100
Utilities	30, 400 tons	(38, 000)	100
Domestic	22, 400 tons	(28, 000)	100
TOTAL	150, 800 tons		

(b) *Aix-la-Chapelle.*

1. The unit per man shift production is 1/3 less than it was before or during the war.

2. The average monthly production for October and November, 1919 was 175,000 tons with a deduction of 127,000 tons which leaves a net for distribution of 48,000 tons.

Summary of Deduction.

TABLE C

For	*Amount (tons)*
Officers and workmen	32,000
France	68,000
Holland and Luxemburg	7,000
Local Sales	10,000
Belgian Army	*10,000
TOTAL	127,000

The conference claimed that there would be approximately 51,000 tons for distribution in the occupied territory.

(c) *Remarks.*

(a) The Aix-la-Chapelle and Cologne areas, therefore, can provide 241,000 tons monthly, which leaves a deficit of 559,000 tons to be supplied from the right and left banks of the Rhine, through the Amtliche-Verteilungs-Stelle, Amt Essen, or an estimated daily delivery of 19,000 tons. It might be well to note here the daily delivery from the Essen-Ruhr district as reported in last weeks Bulletin, G–2. It will be seen from this what the output is for this group of mines.

3. General remarks.

(a) The program was drawn up with the understanding that no coal would come from the Saar Basin. Formerly, 10,000 tons monthly were supplied to gas and water works in the occupied area from the Saar Basin. The average daily delivery to the occupied area from the Saar Basin was 2,200 tons per day. The question of the method of shipment was not discussed in detail.

(b) The Herren Kruse made the assertion that, river and rail transport being favorable, there was no reason why the program should not be carried out. They promised to furnish a railway-table showing the distribution of the coal. The fact was brought out that it required 12 locomotives to haul one special, on the assumption that each locomotive travels 100 kilometers per day and will not operate over 8 hours. It is of interest to note that at present, approximately 1,030 cars of coal go through the American occupied area per day. Our demands, including army and public utilities, hardly average more than 80 cars. I attempted, or rather requested, that with the transaction of the meeting, we be furnished a table showing the kind of coal, the amount and the purpose for which it was to be used, which would be furnished to each area, American, British, French and Belgian. This request was overruled. After one month's operation, I can see no reason why the new Distribution Board should not be compelled to supply us with this information.

*The conference stated that the 10,000 tons alloted to the Belgian army are distributed as the Belgians see fit.

At the beginning of each month, a new delivery program will be delivered, showing the amounts requested and the probable percentage deliveries in totals for armies, water works, gas and electrical works, essential and ordinary industries, and kind of consumption.

(Signed) F. R. LYONS,
Captain, Corps of Engineers,
Officer in Charge of Public Utilities and Roads.

It may be well to digress a moment at this point and very briefly outline the relations between the Reichskohlen Kommissar, the Amtliche-Verteilungs-Stelle, the Kohlen-Wirtschaft-Stellen and the Kohlen-Handels-Gesellschaften.

The Reichskohlen Kommissar corresponds to the American Fuel Administration Board. He supervises the production and distribution of all combustibles in Germany. In addition, he supervises the distribution of gas, electricity and all component by-products.

The Amtliche-Verteilungs-Stellen are the Official Distribution Offices created for each mining district. On the fifth of each month, by the aid of a coal card, and in accordance with the general program of the Reichskohlen Kommissar, they state the probable delivery of coal to each consumer. The clerical work involved and the accuracy of the figures are attended to by the Kohlen-Wirtschaft-Stellen, or Coal Managing Offices.

A Kohlen-Wirtschaft-Stelle is established in a certain number of Kreise of a given state or province. The Coal Managing Office states the amount of fuel required by each consumer and superintends the distribution according to the lists made by the Official Distribution Office (Amtliche-Verteilungs-Stelle).

The Kohlen-Handels-Gesellschaften are simply the middlemen who handle the coal from the mines to the consumers and are subject to the orders of the Amtliche-Verteilungs-Stellen. The Kohlen-Wirtschaft-Stelle assures itself that the Kohlen-Handels-Gesellschaften execute the orders of the Amtliche-Verteilungs-Stelle.

The Economic Section of Advance General Headquarters, A. E. F., retained the records of coal consumption for every industry and utility in American occupied territory. These records were apparently taken to the United States, following the lifting of the blockade on Germany, about June 15th, 1919.

Labor troubles.—The Public Utilities Department has in several instances acted in the capacity of an arbitration board between employers and the employees. This is particularly true in the case of the Höhn Generating Station and the Gas Works of the Concordia-Hütte Steel Mills. This policy was departed from at the Höhn Generating Station when the excess army food stocks were disposed of and the question of arbitration was passed over to the regular German channels. There was a general strike at Höhn about one week after American food ceased to arrive there. The army then felt that it had done its share to maintain an amicable relation between employers and employes. As a result of the Höhn strike, the tramway system of the city of Coblence was suspended for about three days. The pressure brought to bear by the motormen's and conductors' union of the Coblence Street Railway system had the desired effect upon the miners at Höhn, who went back to work as soon as they discovered that the American authorities had stopped the tramway service in Coblence.

A special non-strike order was published at the Concordia-Hütte which provided for cases in which the workmen refused to accept the decision of their own arbitration boards, and an officer from the Office of Civil Affairs was to be the final judge. This order had not been resorted to up to January 10th, 1920.

CHAPTER NO. 10

INDUSTRY AND LABOR

Industrial Conditions During Occupation; History and Effect of Restrictions Imposed on German Industry by Allies; Labor Conditions During Occupation; American Policy Towards Strikes; Requisition of Labor; Insurance of Laborers in Service of United States Army.

Industrial conditions during occupation.—When the American army entered the Rhineland, in December, 1918, industrial conditions were in a very chaotic state. The signing of the armistice had of course destroyed the market for war-supplies which, for four years, had consumed the larger part of the production of Germany. The demobilization of the army after the armistice had had an even more disastrous effect. Hundreds of thousands of soldiers released from military duties, thronged into the cities, creating a vast army of unemployed which could not be absorbed by industry. These difficulties were serious enough, but the surrender of 5,000 engines and 150,000 railway cars to the Allies, by paralyzing transportation, placed almost insurmountable obstacles in the way of the resumption of industry. The very existence of the German people depended on their being able to keep their industries going and on their being able to change from a war to a peace basis despite the difficulties under which capital was laboring. How great these difficulties were, may be judged by a consideration of the restrictions imposed by the Allied governments on exports and imports.

History and effect of restrictions imposed on German industry by Allies.—The reasons for imposing a commercial blockade on occupied and unoccupied Germany even after the armistice, as well as the reasons for not making that blockade absolute, have been set forth in Chapter No. 3 of this report. The duties and operations of the Central Economic Commission, the Interallied Economic Committee of Occupied Territories and the Interallied Railways Commission, have been described in that chapter as well as the procedure whereby authorizations for imports and exports were requested and granted.

It should be clearly understood that a double commerical blockade of Germany had been established. There was first a dead-line drawn along the 1914 western boundary of Germany, from the borders of Holland to those of Switzerland. A second dead-line was established along the line of the front of the Allied and associated armies. Neither of these lines could be crossed without special authorization from the Allied High Command. There never was a blockade on goods passing from one of the army zones to another, although there were temporary restrictions on the circulation of individuals between the zones. Goods could always be forwarded between any two points in occupied Germany, but persons were required to have a special pass to travel from one army zone to another. Another point that must be constantly borne in mind is the fact that there was never a commercial blockade of Germany by the neutral powers whose borders touched hers, for the very simple reason that the Allies had no means of enforcing such a blockade except indirectly. There was, however, a complete blockade of the German coastline, the British fleet, after the armistice, having entered the Baltic from the high seas.

It will be remembered that as the German armies withdrew from the western front, they were required to pass certain imaginary, numbered lines on definite dates, these lines being fixed in a protocol to the armistice. During the forward movement of the Allied armies, no trains were permitted to pass between the hostile armies except such as carried workmen to and from work in the industrial and mining centers, and a few freight trains which carried supplies to the Allied and associated armies. In certain other cases, complete trains of ore, coal, coke, etc., and trains of supplies for civilian populations were permitted to pass, subject to the rules of the Interallied Railways Commission.

An order was published on December 2nd by the High Command, which was to become effective on the day on which the Allied armies reached the Rhine. This order provided that no passenger, freight or empty train should thereafter cross the Rhine unless it fell within one of the following classes: (1) military trains of the Allied armies (troop, military material, or supply trains) destined for the territory forming the bridgeheads of Cologne, Coblence and Mayence; (2) trains of returning empties, which had been used for transporting the above mentioned supplies, material or troops; (3) trains bringing raw material from the right bank of the Rhine, destined for Alsace-Lorraine and the Rhine Provinces; (4) similar trains returning empty; (5) trains of rolling stock which were being delivered by Germany to the Allies under the armistice. This order was the first attempt to favor Alsace-Lorraine commercially and to prevent the closing of factories and consequent industrial disaster in the occupied territories. An explanatory order was shortly issued to the effect that the left bank of the Rhine should be interpreted to include the three bridgeheads of Cologne, Coblence and Mayence.

The Interallied Railways Commission issued Orders No. 16 on December 16th, in which five Rhenish bridges to be used for railway crossing were named. The order provided that the following trains might cross the Rhine freely: (1) Allied military trains and returning trains of empties; (2) rolling stock from Germany to be turned over to the Allies; (3) all freight trains, provided they carried only the articles listed below, to wit: raw materials necessary for the operation of factories; food supplies, fuel of all kinds (coal, coke, gasoline, crude oil, etc.); tools for shops or factories; spare parts; transmission belts; motors; light supplies for shops and factories; automobile spare parts and rolling stock parts; machinery necessary for operation of factories, powerhouses, etc. All these articles required special authorization from the nearest representative of the Interallied Railways Commission. These authorizations were obtained by the prospective consignee, on application to one of the offices of the Interallied Railways Commission.

It will be noticed that no supplies of any kind were permitted to pass from occupied into unoccupied Germany. The Interallied Railways Commission issued Orders No. 27 on December 28th, which regulated railway transportation with greater elaboration of detail. It will be found complete in Appendix No. 47 of this report. [Not printed.] In brief, it established the following regulations in respect to the blockade at the bridgeheads: freight trains originating in unoccupied territories were permitted to pass into occupied Germany by certain fixed routes, provided they contained the following raw materials: coal, coke, lime, limestone, iron ore, zinc ore, lead ore, phosphates, saltpeter, mine props, gravel, plaster, cement, leather, hides, lubricating oils, heavy oils, newsprint paper, tobacco, sand, calamite, copper ore, crushed stone, basalt, lead scraps, salt and rock salt, kerosene, gasoline, clay, grain, seeds, all kinds of fuel and necessary transportation for supply of the civil population, foodstuffs, necessary spare parts for factories and electric plants, transmission belts, motors, etc., spare parts for automobiles and railway equipment. All of the above articles were permitted to enter without special authority. Certain others, necessary for the operation of factories in the ocupied areas, were allowed to enter only on special authority. Goods trains which carried goods listed above were also allowed to pass between the bridgeheads through occupied territory, provided the cars were sealed beforehand. Empty trains which had brought in articles authorized by the Commission, were at the same time permitted to return to unoccupied territories. Commercial traffic originating in occupied Germany and consigned to unoccupied Germany was still forbidden, but special permission for the transportation of certain raw materials and manufactured products might be obtained under conditions laid down by the Interallied Annulment Committee.

The blockade was still further relaxed on January 18th. Imports from unoccupied Germany were permitted without special authorization or formality when they consisted of (a) necessary civilian supplies; namely, food, including beverages, hay or grain for fodder, agricultural seeds; (b) necessary raw materials for industries, including fuel of all kinds; (c) manufactured articles necessary to industries. This last applied to industries only, and did not authorize dealers to sell manufactured products, either wholesale or retail, in occupied territories. In addition to this list, the following commodities might be imported if special permission were obtained from the occupying army: (a) products not mentioned in paragraph 1 and which were considered necessary to the economic well-being of occupied territories. Permission for import of such articles could be granted by the Economic Section of the army, acting under instructions of the Luxemburg Commission.

Later transportation from unoccupied to unoccupied Germany through the bridgeheads was permitted in the case of food, beverages, raw materials and empty railway equipment, provided the transportation was accomplished in complete trains, unmixed with cars destined for the occupied zone.

Commercial traffic originating in occupied Germany and consigned to unoccupied Germany was, in principle, forbidden, but as this prohibition in practice destroyed the principal market of factories in occupied territories, the armies came to realize that exceptions would have to be made. Exportation of certain manufactured products was authorized on condition that each consignment be strictly limited to the base production of the factories, and that the stock on hand on January 15th was thereby not diminished. Exportation of raw materials was forbidden, in principle, in the case of the following commodities, but was permitted under special circumstances: coal and fuel, sheet iron, girders, rails, mouldings, commercial iron, railway sleepers, wires, special steel, zinc, lead, aluminum, copper, motors, locomotives, cars, machine tools, electrical apparatuses, looms, cement, plate and roof glass, tiles, sawed timber, mine stulls, leather, hides, sugar, chemicals of all kinds, dyes, pharmaceutical products, paper.

This order of January 18th remained in effect until July 1st, when order No. 76 of the Interallied Railways Commission was published. The new order took notice of the fact that Germany had assented to the peace terms, and provided that imports from Allied countries into occupied Germany should be unhindered, excepting only arms, munitions and articles specially manufactured for war; that imports from neutral countries should be subject to previous authorization of the Interallied Economic Committee. Imports from unoccupied Germany were to be free in the case of grain and grain products for fodder, and seeds and fertilizers. For all other articles imported into the Rhineland from unoccupied Germany, authorization had to be obtained from the Economic Section concerned.

Exports from occupied Germany to Allied and neutral countries were permitted upon permission from the Economic Section. Economic Sections were further authorized to grant permits for exportation from occupied to unoccupied territory of all goods, provided they were not entirely manufactured from material imported from Allied or neutral countries or contained 25% of such material.

The following list of articles required special authorization from the Central Economic Committee for their exportation into unoccupied Germany: Foodstuffs, living animals, window panes, roofing glass, bottles, empty phials, gold, platinum, silver, copper, locomotives, cars, electrical machines, material for spinning or weaving, agricultural machines, leather, skins, boots, shoes, transmission belts, manufactured fur, chemical products, coloring matters, except pharmaceutical products. In the case of these articles, requests were forwarded by the Economic Section concerned, with its recommendation, to the Interallied Economic Committee. Fuel, minerals, coal, coke and bricketts were permitted to pass from unoccupied Germany either into occupied Germany or into Allied countries.

The blockade of Germany was formally lifted on July 12th, except for import and export of firearms and ammunition and other war material. The exportation of coal and coke was permitted to continue, although the Allies reserved the right to make further regulations for these articles. Export of dyes, chemical drugs, platinum, gold, silver and foreign securities, was prohibited unless authorized by the Interallied Rhineland Commission at Coblence, which, among its other duties, had taken over those of the Interallied Economic Commission of Luxemburg. All other goods might circulate freely. Restrictions on imports and exports were cancelled as far as Luxemburg was concerned. None of these army orders of course attempted in any way to alter the customs-regulations of Allied and neutral countries.

The removal of the blockade was the result of a promise made by the Allies to Germany towards the end of June, to induce her to sign the peace treaty as speedily as possible.

It will be observed that during the period of occupation, the severity of the blockade was constantly lessened. This was because the Allies realized that it had become too absolute and severe, and dreaded lest its severity might drive the German people into Bolshevism and their industries into bankruptcy, thereby causing the Allies to lose not only the restitutive indemnities which they hoped to collect under the peace terms, but also prevent the economic restoration of Europe for an indefinite period.

Severe as were the restrictions on exports and imports, they were not the only handicaps under which German industry labored. The railways were in very poor condition owing to the shortage of rolling stock and primarily by its prolonged intensive use during the war with inadequate repairs and the surrender of locomotives and wagons to the Allies after the armistice. A strict censorship had been established on all postal, telephonic and telegraphic communications. The severe regulations on circulation also tended to incommode the free movement of buyers and salesmen. The post, too, functioned with great difficulty, due not only to poor traffic conditions, but also to the censorhip restrictions. The shortage of coal was so great that it kept every industry in constant dread of being forced to close. In the face of such tremendous impediments to successful business, it is immensely to the credit of industrial Germany that she was able to overcome the obstacles which confronted her.

Labor conditions during occupation.—The situation confronting German labor during the occupation was, if possible, even more serious than that confronting German capital. The interests of capital and labor being vitally related and interdependent, labor was confronted with the danger that all, or at least the bulk of industrial plants would be forced to close down indefinitely and that laboring men would thus be thrown out of work. In addition to this danger, prices were exorbitantly high and kept rising as the value of the mark decreased. High prices were also partly due to diminution in the supply of the necessities of life. The demobilized army flooded the country with unemployed men, and the fact that industries needed time to change from a war basis to one of peace, prevented them from absorbing the unemployed. The armistice had also shattered all hope that ex-soldiers might get employment in the country. To meet this condition in occupied territories, thousands of men were set to work repairing roads.

In addition to these very material dangers which confronted German laborers at the beginning of what threatened to be a long and hungry winter, they were also faced with certain psychological dangers that were no less real for being immaterial. The former Imperial German Government, which they had been trained to believe was the strongest and finest of human institutions, had crumbled in a week. The "invincible" German army had, for all practical purposes, capitulated, and, exhausted and defeated, was vanishing like the snows of spring. The Kaiser, whom they had been taught to honor and revere almost as a god, had shamefully deserted them, animated by most ignoble human fear. Their country was invaded, the Rhine, for generations their traditional bulwark, was controlled by the enemy. The world they knew and understood was gone, incomprehensibly swept away. They were hungry, unhappy, dazed, and for many months did not know which way to turn. Russian Bolshevism with its false promises of wealth and a social millennium was urging them to throw off all restraint and indulge in an orgy of destruction.

When one considers the tremendous physical and psychological pressure bearing on them, it is to the credit of the stable character of German workingmen that they did not really rally to the red lure of disorder, even in unoccupied territory. In occupied Germany, where the strong hand of the Allies held them to the paths of duty, the matter was of course entirely different.

American policy towards strikes.—When the industrial condition of Germany in 1918–1919 is considered, it must be admitted that a certain number of strikes were inevitable and that only great tact and, on occasions, a judicious display of force by the military, combined with a forbearance on the part of employers and laborers, prevented a complete standstill of industries in the occupied territories. The military authorities, for the purpose of better dealing with them, divided strikes into two classes: those purely industrial in .character and not involving the army; those involving the army, either (a) directly, when laborers were employed by the army, or (b) indirectly, as in the case of strikes in public utility plants which were serving the army. These two classes of strikes were dealt with in different manners.

Strikes and labor disputes were due to a variety of causes. These have been mentioned previously in a general way. Specifically, the chief causes of strikes were: decreased purchasing power of the mark with relation to all the necessities of life, and especially to foodstuffs; unsettled trade conditions, which made employers unwilling to grant increases in wages until they had assurance that they could market their product; transportation difficulties; shortage of raw materials; radical tax laws; distrust of the new government; mental unrest and unwillingness to work on the part of many laborers as a result of the revolution; radical agitation, often emanating from unoccupied territory.

It was obviously to the interest of the military authorities to prevent any sort of strike, whether or not they were directly interested, for any strike is a fertile source of disorder, and disorder cannot be tolerated in a military government. No effort was, however, ever made to compel strikers to work, nor employers to come to terms with workers, if the strike involved purely industrial concerns. Only one statement of policy was ever issued in the matter of purely industrial strikes. This was in the form of an indorsement to a communication of the Commanding General of the III Corps. It is quoted below:

> "1. The policy of the army in the event of strikes, as per instructions received by this office, is as follows:
>
> (a) There is no objection to strikes when not in conflict with Memo. Third Army, under date of April 6th, dealing with the subject of strikes (i. e., those directly involving the army.)

(b) There shall be permitted no picketing; in the event of such procedure, pickets shall be arrested and fined.

(c) There shall be no labor agitation in conjunction with any strike. In the event of such procedure, the leaders shall be arrested and tried.

(d) There shall be no protracted idleness on the part of strikers. Strikers must within a few days, if no agreement is reached between the employer and employe, secure some supporting employment elsewhere.

(e) Strikers who refuse to secure work in the event of disagreement between employer and employe shall be deported without the area occupied by the Allied forces."

The first three clauses are practically the same as the American laws dealing with strikes at home. The only difference was, that in Germany the army, not the civil authorities, enforce them. The last, which under other circumstances might well be construed as unfair to labor, was necessary for the preservation of order.

The difficulties confronting both employers and employes in these disputes had to be carefully investigated before any such dispute could be properly dealt with. Employers were harassed by many difficulties, which the laborers were either unable or unwilling to understand, and the latter were driven by the ever-increasing cost of living and the agitation of labor leaders. Because it was a well-known fact that the cost of living was beyond the rate of pay obtained by the average worker, the American authorities adopted the policy of allowing it to be known that they were interested in every strike, but would remain neutral except when their own interests were directly concerned. It was early seen that any other attitude would have placed the American army in the position of being forced to arbitrate strikes, and the difficulties and complexities of the many questions involved forbade this. The army had to be very careful, for if strikes were forbidden, unscrupulous and short-sighted employers might easily have utilized the decision of the American authorities to hold the wages of their laborers down to the minimum, and discontent would have grown on all sides. On the other hand, if strikes had been allowed *ad libitum*, the fires of agitation would have been fanned and workers encouraged to make unreasonable demands, which, if not complied with, would have resulted in disorder and violence. It was the primary concern of the American authorities to prevent this.

The policy was early adopted of confining all labor troubles to the status of disputes and of preventing strikes and lockouts if possible. After a body of workmen had once laid down their tools and gone on strike, it became doubly difficult to effect an adjustment, as the labor leaders would not permit the workers to go back without gaining some advantage, fearing a loss of prestige for themselves, and the semblance of a defeat for their cause. The employers once having said they would not accept the demands of the workers, were not thereafter inclined to submit easily. It therefore became necessary that local Officers in Charge of Civil Affairs should be notified of labor disputes as soon as they occurred, and before the dispute could become a strike. Orders were accordingly issued to all Landräte and Burgomasters that they were to notify their respective Officer in Charge of Civil Affairs whenever a strike appeared imminent. As soon as the Officer in Charge of Civil Affairs of a Kreis was thus notified, he was directed to summon both parties to the controversy to appear at the Office of Civil Affairs. These summons were printed forms bearing the official stamp of the office, and thus threw the moral weight of the occupying forces into the dispute. The effect produced on employer and employe was equal, and pleased the laborers in particular, who felt that their side of the dispute was considered by the army just as seriously as was the employer's. The time set in the summons for the conference was usually two or three days in advance of the anticipated strike, depending on the nature and seriousness of the controversy.

When both sides came to the conference, each was allowed to state his case, the workmen generally having precedence. The calm statement of both sides, which was required always, had a good effect. It was intended primarily to show the Officer in Charge of Civil Affairs the extent of the dispute, and whether it was possible to settle before it came to an actual strike. After each side had stated its case, the Officer in Charge of Civil Affairs explained the position which the United States would adopt in the matter: that in all disputes between employers and employes, the military authorities remained neutral and would not interfere unless its own interests were threatened. He further explained what the interests of the American army as an occupying force were, and what their probable action would be in case disorder followed the strike. He then stated that it was practically impossible for the Officer in Charge of Civil Affairs or any other military man to act as an arbitrator for the dispute, inasmuch as he was relatively ignorant of such fundamental matters as transportation, normal cost of living, trade union precedents and regulations involved; and that the wiser course was for both parties to settle the dispute amicably among themselves, rather than trust to a representative of a foreign army, particularly since he was untrained for that duty and would likely enough settle it to the dissatisfaction of both sides. Finally, the Officer in Charge of Civil Affairs asked employers and employes whether they believed that they could not settle the matter themselves without further participation on the part of the occupying forces.

One conference of this sort was generally sufficient to induce both sides to reach an agreement. In the great majority of cases, both employers and workers promised that they would effect an adjustment. In a large number of the remaining cases, the workers returned to work pending a settlement between the employers and their own representatives. In a few cases, a second conference was necessary.

After such a conference as is described above, the matter was in most cases dropped officially for several days, but the Officer in Charge of Civil Affairs made it his business to call personally at the plant, in order to see how the matter was progressing. He also, incidentally, listened to the statements of both parties to the dispute. If any agitators were believed to be busy, their names were carefully noted, and if a later investigation disclosed the fact that they were strangers in the locality, a hint was dropped that their presence was not desired. By such a system, both employers and workers were impressed with the fact that the military authorities were taking a very active interest in seeing that the dispute was settled without a strike. Parenthetically, it may be stated that the Vagrancy Court established in October, 1919, and described at length elsewhere in this report, was intended in part to be used to punish labor agitators who were without visible means of support.

When strikes involved the army directly, as, for instance, when the army was employer, or, indirectly, when the army was dependent on the product of the striker's labor, they were handled in a different manner. In December, 1918, when the railways were requisitioned, or rather taken over from the German government, the personnel was also requisitioned. Paragraph 3, Orders No. 1, Interallied Railways Commission of the Rhine Provinces, November 24th, 1918, which dealt with this matter, reads as follows:

"All the personnel mentioned above will be considered as requisitioned and will come under the military regulations of the Allied armies in the zone of operation. This personnel may be sure, if it performs faithfully its duties, that it will be treated well, and paid at the present German rate, and will have its food supply assured. On the other hand, any one who abandons his post, will be subject to court-martial. Any negligence in the execution of the service will be severely punished. It is everyone's interest to remain at his post. It is also in the interests of the population, for their supplies and the maintenance of the economic life in occupied territory could not be assured if transportation did not work normally."

As a matter of fact, railway employes never actually struck in the American zone. They threatened to do so at the beginning of June in protest at the establishment of the "Rhine Republic" in the French area. The President of the Interallied Railways Commission of the Rhine Provinces at once issued an order, stating that due to external political conditions, the German railway personnel had threatened to strike, and that they must not participate in political demonstrations of any sort and must remain peacefully at work. It further stated that no digression from this policy would be tolerated, no matter under what pretence. The Officer in Charge of Civil Affairs, Third Army, simultaneously informed Herr von Groote, Oberpräsident of the Rhine Province, that in view of the threats of strikes by government officials and by employes of public service corporations, such as railways and trolley systems, he wished to call attention to the fact that the American authorities could not permit strikes by employes engaged in such vital matters to the army as transportation, inasmuch as they would materially impede the administration of government of the occupied area. As a result of this pressure, the strike was averted and thereafter none was even attempted.

In regard to another class of strikes, in which German laborers were engaged in work on which the United States army was directly dependent, the restrictions were likewise very severe. On March 28, 1919, Major General Dickman issued a proclamation on the subject, which read as follows:

HEADQUARTERS THIRD ARMY

AMERICAN EXPEDITIONARY FORCES

LETTER OF INSTRUCTIONS No. 17. — 28 March, 1919.

* * * * * * *

PROCLAMATION

All German laborers who have voluntarily entered the service of employers engaged in work for the United States Army of Occupation, or who are working for the said army, under requisition or otherwise, are forbidden jointly or in conspiracy to plan, attempt or carry out any scheme to strike or abandon work. Any breach of this order will be punished by the American Military Tribunals.

By COMMAND OF MAJOR GENERAL DICKMAN:

MALIN CRAIG,
Brig. Gen., G. S., Chief of Staff.

OFFICIAL:
W. A. HAVERFIELD,
Adjutant General.

A memorandum was issued on April 6th to Corps Commanders and Chiefs of Sections in support of this proclamation. The memorandum required that:

(a) Immediate steps should be taken to ascertain the name, occupation, residence, etc., of the leaders of the various labor unions in the

American occupied territory and of any other persons living in the American occupied territory who might be considered as leaders of organizations of any kind that might be concerned in a possible strike.

(b) Should it become fairly certain that a strike is about to be called and that such strike would involve directly or indirectly, interference of any kind with the supply or the arrangement of the American Army of Occupation, the leaders mentioned will be immediately arrested, with a view to their trial for attempting to interfere with the supply or other arrangements of the American Army of Occupation.

(c) If not already arrested, immediately on the calling of the strike, should such strike be thought to involve directly or indirectly, interference of any kind with the supply or other arrangement of the American Army of Occupation, the leaders mentioned in paragraph (a) above will be immediately arrested with a view to their trial for interfering with the supply or other arrangements of the American Army of Occupation.

In spite of the severity of these regulations, strikes of this nature did occur in two instances, when the stevedores engaged in unloading American supplies at the ports of Andernach and Bendorf left their work simultaneously. The men were working directly for the United States government. The strike was severely dealt with at Andernach, all the prominent strikers being arrested, tried and sentenced. It is rather humorous to note in this regard that they served their sentences doing gratuitously the very work for which they had previously been hired and paid. The officer in charge at Bendorf did not act so promptly as his colleague at Andernach, and before he had taken any action, the strikers at Bendorf, hearing of the method adopted by the American authorities at Andernach, returned to work after an absence of only a few hours. No disciplinary action was taken in this strike. As far as can be ascertained, these were the only two strikes in the area in which the United States figured directly as employer.

In the case of strikes of laborers in such public utilities as electric, gas and water works, quarries furnishing stone for the United States engineers, and other concerns engaged in work directly affecting the army, matters were treated more drastically than was the case in purely industrial strikes, but less so than when they were aimed directly at the American army. At first, even such strikes had been prohibited, as will be noted by an examination of the preceding order; but it was soon realized that such drastic action was unfair to the laborers, as well as impracticable. Restriction by the army of the use of the strike by the workingmen in public utilities, soon caused wages in these establishments to remain fixed; wages in commercial establishments rose with the cost of living. The result was a natural flow of machinists, electricians, and skilled workingmen, to the purely industrial establishments. This was gradual, and there was no strike, but at the end of a short time, public utilities found themselves without skilled help. Employers were then faced with the problem of either voluntarily raising their wages or losing a considerable number of their employes.

If positions outside public utilities were not available, skilled workers often found that they were paid less wages than unskilled laborers. Requests that they be allowed to strike were repeatedly made to the army by laborers in public utilities. These came to the attention of the Officer in Charge of Civil Affairs and efforts were made by labor to induce him to use his influence to obtain satisfaction for laborers from employers. The military authorities realized that when employers were unable or unwilling to meet the demands of the workers, steps would have to be taken to adjust the matter amicably, as, otherwise, a continual agitation and unrest would be going on under the surface and would furnish a fertile field for radical propaganda. A policy was finally adopted for labor troubles in public utilities, which was almost identical to that used in the case of purely commercial strikes, the only difference being that in the case of public utilities, the pressure exerted by the military authorities was greater, since their interests were more involved than in others.

Requisition of labor.—The right of an occupying army to requisition laborers from the inhabitants of the occupied territory is well recognized and is especially provided for in Article 52 of the Hague Rules. The Hague Rules and the customs of civilized warfare require, however, that the laborers shall receive cash payment for their services, and that the occupying army shall not employ them in work directly furthering military operations.

During the march from Luxemburg into Germany in 1918, the roads were found to be in bad repair. Weather and the heavy traffic incident to the retreating and advancing armies had necessarily considerably damaged them. The problem of repairing them arose therefore, even before the Rhine was reached. Requisition of German labor to accomplish this task was one of the first solutions suggested. As has been mentioned, the entire German railway personnel in occupied territories had been requisitioned by Marshal Foch shortly after the armistice. It therefore seemed an unwise extension of policy to requisition labor for use on roads. There was no doubt of the legal power resident in the army to make such requisition, but when the matter was submitted to the Commanding General, it was decided not to do so, inasmuch as the Americans would later have to pay the laborers direct or, at least, furnish them with rations, which was at that time impossible.

On December 24th, 1918, a few days after the occupying forces had reached their permanent stations, Memorandum No. 36 was issued by the Commanding General, Third Army, under the title of "Instructions for the Procurement of Supplies, Services and Billets by Requisition on the German Authorities, and Reports to be made in Connection therewith." One paragraph of this memorandum stated that local commanders were not authorized to requisition labor without the approval of higher authority, except in cases in which the repair and the maintenance of existing buildings were such as to require immediate action. The "higher authority" was described as the first sections of the staffs of Divisions and Corps and of Army Headquarters.

It was also announced to be contrary to the policy of the army to requisition direct labor except in emergencies. In the event of such emergencies, labor might be requisitioned for such government buildings, barracks, depots, hospitals, waterworks, electric light plants, etc., as had been taken over by the Americans and which already had a working force of janitors, charwomen, mechanics, etc. The requisition for the building was permitted to include the necessary personnel. The local civil authorities were required to pay these employes out of their public funds at rates similar to those in force prior to the occupation. American officers in charge of these buildings were directed to see that the employes were paid promptly by the officials concerned. When private buildings and utilities, such as hotels, hospitals, factories, etc., which already had a working force were taken over by requisition and it was necessary to retain such force, the following procedure was observed. Individuals included in the working force were paid either from United States funds or from funds turned over by the German government, at the same rate that they received prior to the occupation. It was made incumbent on all United States officers and agents in charge of private buildings or utilities to take such steps as would insure prompt payments.

Road work was carried on for a long time by voluntary German labor, but towards the end of February, 1919, this labor being no longer forthcoming, the Commanding General authorized Corps Commanders to serve requisitions upon local officials for such as was necessary to repair the roads, if sufficient voluntary labor could not be obtained at current prices. Requisitioning of personal services was not authorized except under these conditions.

Conditions had undergone a change by March, 1919, and a great deal of German labor was then required by the army, partly because the coming of Spring had further damaged the roads and partly because the change of base of the army from France to Rotterdam and Antwerp required the employment of German stevedores to unload the Rhine boats. The Commanding General therefore issued a Letter of Instructions to Corps and Division Commanders on March

6th, revising the previous regulations regarding the employment of German labor. The instructions read as follows:

> "As a general policy, American soldiers serving within the limits of the American Army of Occupation will not be employed on duties which will tend to belittle their dignity or lower their status in the opinion of the German population. This will apply in particular to such work as the construction and maintenance of roads, which, while necessary to the Army of Occupation, will benefit directly or indirectly the German population. Under such circumstances, whenever practicable, German labor will be employed under American supervision, but this American supervision will not go to the extent of supplanting German road-overseers or others who are in a position to more efficiently conduct the labor under consideration."

Payment for these services was made on payrolls in German marks supplied by the German government, as required by the terms of the armistice. Because of the shortage of food in Germany and the necessity of properly feeding the laborers in order that they might perform the heavy work, a policy was adopted of giving them a part of their pay in food. It was optional with the laborer whether or not he took his pay entirely in money or partly in money and partly in food. He was not allowed to draw more food for a day's work than that authorized in the German ration, this to prevent his profiteering in American food supplies. An elaborate table was worked out, giving the German ration in the component equivalents of the American ration. The cost of each component was stated in dollars and marks. It was fairly easy in this way to deduct the amount of food taken by each laborer from his daily wage. The balance was paid in cash. The tables figuring the cost of supplies in marks had to be frequently changed, as the rate of exchange on marks became more and more unfavorable.

It was the policy of the American army to procure voluntary labor whenever possible, but a sufficient number of volunteers was frequently not obtainable. In such cases, a requisition for additional labor was served on German civil officials. The labor for the port of Bendorf was obtained through a requisition served upon the Regierungspräsident of Coblence. The American forces paid the laborers so requisitioned with money furnished by the German government to defray the cost of the occupation. The requisition calling for 300 laborers at Bendorf was the only one over which any trouble arose, and this occurred through a bureaucratic row between the Acting Regierungspräsident and the Oberbürgermeister as to who should furnish the labor. As the labor was not forthcoming at the appointed hour, both officials were arrested and given twenty-four hours additional in which to comply. It was then furnished as directed. No further action was taken by the American authorities, other than a severe reprimand; but the Regierungspräsident—a superannuated government official—was soon after retired by the German government on its own initiative.

Insurance of laborers in service of United States Army.—It has been seen that throughout the period of the occupation, the army authorities employed a large number of civilian laborers, recruited either voluntarily or by requisition upon the civil authorities. These laborers were paid at the rate current for such work. When, toward the end of March, owing to the change of base from Brest to Antwerp and Rotterdam, a greater number of laborers was required than theretofore, the question of government insurance of these laborers was brought up, and had finally to be settled by a military order.

All workmen whose incomes are deemed insufficient to keep them from want in case of disability through illness or injury, or to enable them to save a competency for old age, are required by German law to insure themselves in the government insurance bureau. They must pay weekly premiums to this bureau, fixed according to the amount of their earnings. These payments are duly recorded in a book commonly called the "stamp book", which is kept by the beneficiaries. One "stamp book" is for old age pension and another for disability insurance. Not only must the German workingman pay the premiums on both kinds of insurance, under penalty of a fine, but the employer must also, under a like penalty, pay a prescribed premium on the insurance of each of his employees.

The military authorities took no notice of this law during the early part of the occupation, with the result that the insurance of many of its German employes lapsed. This was felt to be very impolitic, and the Officer in Charge of Civil Affairs made a recommendation on April 21st to G-4, that the American authorities should pay the insurance premiums of its employes according to German law and charge the cost to the German government. This recommendation was approved, and an order issued on May 16th, providing that the employe should pay his part of the insurance premium and that the employer's share be paid by the Disbursing Quartermaster upon presentation by the German civil authorities of a certified statement from the government insurance bureau of the amount due in each case under German law.

The Regierungspräsident of Treves objected to this order, on the ground that German laborers and employers had in practice changed the mode of operation of the law, so that the employer, as a matter of fact, paid the entire premium, deducting the employe's share from his wages, and that the laborers, being required to pay the premium themselves, would not understand the procedure and would regard it as an imposition. This objection was considered but was finally disapproved, partly because of the additional work that its adoption would have thrown on the Department of the Disbursing Quartermaster, and partly because it was found that the practice by which the employer paid the entire premium was by no means universal. It was therefore arranged that the Bürgermeister should regularly send a representative on the days when wages were paid, and that this representative should collect the employes' share of the premiums when they received their money from the Disbursing Quartermaster. These amounts, together with the share of the premiums paid by the American authorities at the same time, were sent to the local government insurance bureau. Inasmuch as this arrangement threw a very considerable burden of administrative work on German officials, the American authorities lent every possible facility to them in order to help them to calculate the amounts of insurance. These officials were always to be notified before payments were made, and were to be given access to the civilian pay-roll in order to determine the proper premiums.

CHAPTER NO. 11

RELATIONS BETWEEN ARMY AND CIVIL POPULATION

Orders Regulating Personal Relations between Army and Civil Population; Marriages; Saluting; Crimes Committed by Members of American Forces; Lack of Education and its Relation to Crime—Statistics of the Army of Occupation; Offenses Committed by German Civilians against Members of American Army.

Orders regulating personal relations between army and civil population.—The attitude which it was desired should be adopted by the American Army in regard to relations with the civil population of occupied Germany, was announced by General Pershing in an order issued three days before the Allied armies entered Germany. This order, in part, is given below:

G. H. Q.

AMERICAN EXPEDITIONARY FORCES

GENERAL ORDERS No. 218. } FRANCE, November 28, 1918.

In view of the extraordinary conditions under which that part of the American Expeditionary Forces which constitutes the Army of Occupation of German territory is serving, the Commander in Chief desires to acquaint the officers and men composing it with the expectations which he entertains as to their conduct. You have come not as despoilers or oppressors, but simply as the instruments of a strong, free government whose purposes towards the people of Germany are beneficent. During our occupation the civil population is under the special safeguard of the faith and honor of the American Army.

It is, therefore, the intention of this order to appeal directly to your pride in your position as representatives of a powerful but righteous nation, with the firm conviction that you will so conduct yourself in your relations with the inhabitants of Germany as will cause them to respect you and the country you have the honor to represent. While you appear among them as a conquering army, you will exhibit no ill-will towards the inhabitants.

On the other hand, you are warned against conduct unbecoming your position as instruments of military rule. So long as a state of war continues, Germany remains enemy territory, and there must be no intimate personal associations with its inhabitants. A dignified and reserved attitude will be maintained on your part at all times.

It is not believed that any acts of pillage or violence will be committed by members of the American forces, but, should any persons prove themselves unworthy of this confidence, their acts will be considered not only as crimes against the sufferers, but as dishonoring the American Army and as a direct insult to the flag of the United States. Such transgressions, should they occur, will be punished with the severest penalties known to our military law.

G. O. 218, G. H. Q., was shortly followed by an order from the Third Army, which prescribed that members of the forces of occupation should confine themselves to official relations with the Germans and should limit their personal relations to an attitude of "courteous tolerance." This order popularly known as the "anti-fraternization order", became probably the most discussed regulation ever issued by the Army of Occupation, and affected the private life of practically every member of the army.

At the time it was issued, however, its full effect was not appreciated by the army in general, as they had not yet recovered from the emotions produced on the field of battle, which tended to make them distrust and hate all things German. The idea that perhaps as time went on, other feelings might come to the fore, was still undreamt of, and the anti-fraternization order awoke neither resentment nor surprise. During the first days of the occupation, feeling on both sides was in fact so intense, so distrustful and so bitter, that it is surprising that no serious consequences occurred. A vast majority of Americans undoubtedly believed at the time, that the Germans respected no agreement or moral law, that they took pleasure in evil for evil's sake and sought, above all, revenge for their defeat. The inhabitants had also received their full share of propaganda, by which the American Army had been conjured up as an undisciplined mob of semi-savages. Knowing themselves in its power, a not unnatural sense of fear became prevalent. In consequence, the whole attitude of the civil population became so extreme conciliatory that both officers and men of the occupying army looked upon their obsequiousness with contempt.

The American soldiers were, of course, beginning with their first night on German soil, billeted in the houses of the civil population, who, inspired partly by fear and partly by orders they had received from the Bürgermeister and other officials, did everything in their power to placate them. The soldiers, for the first time in months, slept in beds. The peasant women prepared a supper on the family stoves, probably believing that they were required to furnish food to the invading soldiers, as they had been to their own. After supper, the soldiers probably sat in the warm kitchen surrounded by the family, which is a luxury to be appreciated only by those who have spent long, cold nights in the field. Soldiers and children almost always strike up a strong friendship, and the American soldier and the German boy or girl were no exception to the rule. Soldiers of German extraction who could understand the language, of course got promptly on friendly terms with the civilians with whom they were billeted, despite the rigors of the anti-fraternization order. By the time the troops reached their permanent stations, their mental attitude towards the inhabitants had entirely changed. Though by no means friendly to them, they no longer regarded them with the nervous hatred engendered by fear of sudden treachery. Furthermore, they saw that a great deal of what they had read about the innate viciousness of all Germans was not true, and their war memories in consequence began to fade. A soldier was even occasionally heard to admit grudgingly that the Germans were "pretty good fighters", which remark foreshadowed the eventual revolution of sentiment.

The arrival of the various organizations at their permanent stations produced an even greater change in the relations of the troops towards the inhabitants. No soldier had until this time spent more than one or two nights in the same house and, therefore, the men had gained no personal acquantaince with individuals. In making his billet in a German home, the soldier found that he could not avoid such acquaintance with his host, and both parties made offers of friendship. The shortage of fuel and the fact that the kitchen was as a rule the only heated room in the house, practically forced the soldier to spend his evenings with the family, and he became friendly or hostile according to how their temperaments coincided with his own.

The military authorities soon realized that it was impossible to enforce the anti-fraternization order in billets, particularly in the small villages, where the people were poor and the houses had a single living-room, which, by force of circumstances, had to be used by soldier and civilian in common. Moreover, it was impracticable to prevent the soldier from visiting his company comrades who were billeted in other houses of the town. Consequently, in a relatively short time, the average soldier personally knew most of the inhabitants of his village. The result was inevitable. At the end of a few months, the only effect of the anti-fraternization order on the average soldier was, that it prevented him from appearing in public with his German acquantainces.

The theory on which the order had been promulgated was entirely in accord with the fundamental principles of military government. If the Americans succeeded in maintaining an attitude of dignified aloofness from the Germans and did not mingle with them at all on a basis of equality, the military authorities would be better able to enforce their orders to the civil population and, incidentally, to maintain discipline. It was based on the identical theory which prevents an officer from mingling on terms of equality with soldiers,—the fear that familiarity will breed contempt, to the detriment of discipline. The anti-fraternization order was sound in theory and undoubtedly succeeded in achieving its purpose as far as the officers of the army were concerned; but its failure to achieve practical results among the soldiers soon become notorious. The officers, being billeted in the better class of houses, had rooms to themselves, were furnished with coal to heat their rooms and were in fact isolated from the population in general. Soldiers, on the other hand, were thrown into the most intimate contact with the inhabitants, shared their lamp and fireside and enjoyed the same comforts and the same amusements as they during the long winter evenings. The average soldier found that he could no more maintain an attitude of dignified aloofness from the people than could a lieutenant bivouacking for a long time with a detachment of three or four men in the field.

The order, as might naturally be supposed, irritated the German population considerably, largely because they misinterpreted its purpose, and believed it due to a feeling of personal superiority on the part of the occupying army. The imagined contempt therefore wounded their pride. However, at least as far as it affected officers, the order probably had an excellent effect, inasmuch as by requiring them to remain aloof from the population, it thereby lent them a halo of superiority similar to that which had formerly been accorded to German officers. On the other hand, by preventing open association with the civil population, it precluded their acquirement of the broadening influences which usually are obtained from residence in a foreign land. The officers, finding it practically impossible to learn the language, also found it difficult to study the institutions and customs, the history, laws, methods of manufacture and the commerce of the country. But the Government took the viewpoint that the forces had been sent into Germany to insure the fruits of a hard-earned victory, not for their personal pleasure or profit.

The anti-fraternization order continued in force during the entire existence of the Third Army. With the return of the combat divisions to the United States, during the summer of 1919, leaving only a small garrison for permanent duty after the signing of peace, a situation arose which necessitated its rescission. The troops were in barracks instead of in billets. Under these conditions, they no longer had an opportunity to mingle with German families and thereby see a softer side of life than their routine duties permitted. German women of better families would not associate with soldiers on account of the army order, and their feminine companionship, if they sought such, had to be found among such types of prostitutes,—usually the lowest ones—as would risk arrest for the chance of a few marks. The venereal rate among the troops in consequence increased with great rapidity. As a preventive measure and in order to allow the soldiers to associate with the better class of women, the anti-fraternization order was, on September 27, 1919, more than nine months after its publication, at last rescinded.

Curiously enough, there was almost immediately a striking increase in the venereal rate among the troops. This, however, was believed to be due less to the revocation of the order than to a large influx of prostitutes from France, who had come to Germany upon the departure of the American troops from the former country. The establishment of a vagrancy court, which is described in Chapter No. 7, and other drastic measures, soon rid the American zone of these unwelcome visitors. The venereal rate thereafter began to decline rapidly.

During the year 1919, it must be admitted that the relations between the American soldiers and the civil population were, on the whole, all that could be expected. The good feeling of friendship on both sides increased, slowly but steadily, from the day on which the troops settled in their permanent billets. There were of course certain minor lapses from this general tendency. About April 15th, rumors of the severity of the coming peace terms began to be noised through the American zone. Coupled with the critical food and fuel situation and the shadow of Bolshevism hovering over unoccupied Germany, these rumors brought home to the people, as nought else had, the fulness of the defeat they had suffered. A general spirit of discontent was not long in making itself felt, and the people in the American zone no longer tried to hide their dislike of the Allies. The soldiers' attitude towards the Germans had never been in a true sense one of friendship. Like most Americans towards a foreign race, they got along with them, tolerated them, but kept their feeling of superiority the whole time. The moment, however, that the Germans showed an insubordinate spirit, the troops reacted violently and their once forgotten dislike flared up anew. Numerous fights occurred between disorderly elements of the army and similar elements among the young demobilized German soldiers, most of which, however, originated in drunken brawls or jealousy over German girls. This era of ill-feeling reached its height about the middle of May, when the belief became prevalent that the Germans would reject the peace terms. Though it is believed that the ultimate source of ill-feeling was the peace terms, there were other apparent causes, which probably grew out of this primary discontent. Among these caucses were discrimination in prices against Americans, increase of drunkenness among Germans, incivility toward officers and soldiers, insulting remarks to woman welfare workers. Incidents such as these were quick to inflame the anger of the American soldiers, and their sleeping resentment consequently awoke. In a number of instances, jealousy also arose among the young men because the soldiers had captured the hearts of their former sweethearts. Several priests took occasion to use their pulpits to warn their congregation against permitting German girls to become friendly with American soldiers, and in some cases went so far as to post lists of girls known to have associated with soldiers, on the doors of the churches. Some of the girls who had been on friendly terms with Americans were even subjected to insults from German men. Friction under these circumstances was inevitable.

The German civil authorities, quick to realize the fact that things could not go on as they were, issued a proclamation cautioning their citizens against the practices which had irritated the occupying army, and urging them to keep relations with the soldiers on as friendly a basis as possible. The American Army on its part, issued orders enjoining the soldiers to be more orderly and to maintain the attitude insisted on from the beginning, of treating the defeated enemy both justly and firmly. A Rhine steamer which had been used by the army during the Spring as an excursion boat for soldiers, was loaned to the German Red Cross, in order that that organization might take maimed soldiers in the hospitals of Coblence on an excursion. The result of these conciliatory measures on both sides, together with the signing of peace, was to reestablish peaceful if not friendly relations by the beginning of July.

Marriages.—It was to be expected that some of the 250,000 members of the Third Army would, in the six months they lived in the occupied area, fall in love with and desire to marry German girls. The order against fraternization of course tended to check such marriages, but due to its practical non-enforcement outside of Coblence, there were many cases in which soldiers fell in love with German girls and in the course of time became engaged to them. It is the custom among lower class Germans to consider an engagement as the equivalent of a marriage. In the spring of 1919, a number of soldiers came forward, announced that they were engaged to German girls, and admitted that they were responsible for the fact that these girls were about to become mothers. They requested permission from the army to marry, stating that they were perfectly willing to stand trial for disobedience of the anti-fraternization order, but were bound in honor to marry the girls. This was a situation which had not been foreseen when the order was promulgated. The object of the anti-fraternization order had been primarily to increase the honor and respect in which the German should hold the United States and its soldiers. If the soldiers were now not permitted to marry and were forced by the army to leave their prospective brides to bear illegitimate children, the dignity of the United States would certainly not be raised. On the other hand, if the soldiers were not punished for violating the anti-fraternization order, discipline would suffer. If they were merely punished for violation of the order against fraternization and still not permitted to marry, the government would be placed in the position of forcing the soldiers to commit the crime of seduction.

Again, if they were punished and permitted to marry, the government would be in the position of encouraging the soldiers not to marry the girls whom they had seduced, inasmuch as it had placed a premium on their non-acknowledgment of their guilt.

After considerable hesitation, the Army Commander decided to permit soldiers who had seduced girls on promise of marriage, to marry, if their victims had through them become pregnant. The soldier was required to submit a written statement acknowledging his responsibility and declaring that he wished to marry the girl, while the girl was required to submit a certificate of an American army doctor or a reliable German physician to the effect that she was pregnant. The soldier was, under these circumstances, not punished.

Approximately one hundred marriages of American soldiers with German girls were performed under the provisions of this regulation. This is a surprisingly small number when one considers the fact that the army of occupation consisted of more than 250,000 men, most of whom had just spent months of hardship at the front, without any relation with women.

Unfortunately, many soldiers desired to marry who did not come under the new regulation. Some of these were transferred from the divisions going home to the organizations which were to remain as the permanent garrison, while it is to be feared that others deliberately attempted to place the girls to whom they were engaged in a condition which would permit them to marry. At any rate, so many soldiers of the permanent American forces in Germany applied for permission to marry, alleging the pregnancy of their betrothed, that the Commanding General very naturally became suspicious. Married soldiers in any military unit reduce its efficiency and are a constant source of trouble, inasmuch as they sleep out of barracks. Furthermore, an organization with a large proportion of married men is difficult to transfer from one station to another.

Consequently, on October 9th, the Commanding General published a circular letter to organization commanders, directing that all soldiers desiring to marry, even those whose prospective brides were of Allied nationality, must apply through military channels for permission. A supplementary letter shortly thereafter directed commanding officers to refer all such applications to the first section of the General Staff at Headquarters for final approval. These measures effectually checked the growing number of marriages, although they undoubtedly worked a great hardship on many German women who, with a view to complying with the previous regulation, had permitted themselves to become pregnant in order that they might be allowed to marry. Whenever a case appeared particularly distressing, the Commanding General of course authorized the marriage, provided an investigation proved the former good character of the girl.

Saluting.—The American authorities at no time insisted on any outward manifestation of respect or friendliness from the civil population, nor did they require German men to uncover in the presence of American officers or of the national colors. The only courtesy on which the American authorities insisted was that demanded by the common usages of society.

All uniformed officials were, however, required to salute all Allied officers. There are a great number of German officials who wear uniform when on duty—police, gendarmes, railway guards, trainmen, tram employes, postmen, customs officials, state and city foresters, etc. Under the former *régime* these officials had always been required to salute German officers, and there was little trouble in forcing them to recognize our own. Such failures to salute as happened were generally due more to carelessness than to defiance, though a few cases were unquestionably due to the latter. These were severely punished.

Crimes committed by members of American forces.—Crimes have been committed to a greater or less extent by every military force occupying a hostile country in time of war. The reader of history knows that crimes by the military are an inseparable, though regrettable, accompaniment of such an occupation. Always appreciating that fact, it became the earnest desire of the American authorities to decrease the number of crimes committed by members of our forces to the lowest possible minimum, and strict orders were issued to soldiers to uphold the dignity of their country. Crimes nevertheless did occur and, during April and May, became a matter of considerable concern to the Commanding General.

It was learned about this time that German officials were keeping a record of all crimes and offenses alleged to have been committed by the Allied forces or members thereof. It was probably their purpose to publish later the statistics thus amassed, as a proof to the world that the Allies had also committed iniquities, and that the German misdeeds in Belgium were by no means unique. While it is quite probable that they did this to exculpate themselves from the odium felt by the world because of their conduct in Belgium, it is also probable that they hoped by its publication to prevent, or at least mitigate, the punishment of those Germans who were charged by the Allies with responsibility for the Belgium atrocities. The average German to this day disbelieves the stories of the Belgian atrocities which have been so widely spread through Allied countries, and feels that they were simply a means of propaganda for the Allied press. He bitterly resents their imputation of national savagery and is anxious to rid his country of their stigma by narrating tales of similar Allied brutalities at every opportunity.

The Officer in Charge of Civil Affairs, Third Army, brought this matter to the attention of the Army Commander in a memorandum, dated June 10th, 1919, and made a recommendation that, in order to protect the army against false accusations, every complaint against soldiers, however trivial, should be investigated carefully and recorded. This recommendation was approved and, shortly thereafter, the Officer in Charge of Civil Affairs directed all German officials in American occupied territory to submit all accusations of misconduct on the part of troops, however trivial, in order that the American authorities might take suitable action. This order was published in Paragraphs 33 and 36, Civil Affairs Bulletins, June 9th and 10th, 1919, respectively:

33. **Reports of alleged offenses against civilians by the military.**—All Officers in Charge of Civil Affairs will immediately call upon their respective Landräte and Oberbürgermeister for a complete report of all cases on file in their respective offices of crimes or offenses alleged to have been committed by American soldiers against civilians. This report will be submitted in English and forwarded to the Officer in Charge of Civil Affairs, Third Army, so as to be received not later than June 15. In case there are no such records on hand, a statement to that effect by the proper official will be forwarded. (See G. O. No. 8, these Headquarters, 29 May, 1919.)

36. **Action on complaints and petitions.**—All former complaints made by civilians against the military and all applications for clemency by civilians convicted by military courts will, upon being brought to the attention of the Officer in Charge of Civil Affairs, be reduced to written form and will be promptly acted upon by the Officer in Charge of Civil Affairs, or by him presented to such authority as is authorized to act on the matter, and in every case, whether the matter complained of is or is not remedied, or whether the application for clemency is or is not granted, the complaint or application, together with a full report upon the facts and circumstances involved, will be forwarded to the Officer in Charge of Civil Affairs, Third Army.

Following the memorandum of the Officer in Charge of Civil Affairs, the Commanding General published an order requiring that every complaint submitted by a German should be investigated by a commissioned officer. The report submitted was to include the name of the complainant, the name of the accused, if known, the nature and circumstances of the offense charged, and a statement of any disciplinary action taken or, in default of this, the officer's recommendation in the matter. This report was to be submitted through channels to the Officer in Charge of Civil Affairs, Third Army.

Claims for damages, when placed with the military, were forwarded to the Rents, Requisitions and Claims Service. If a claim for damages involved an accusation of misconduct on the part of American troops, a copy was forwarded to the Office of Civil Af-

521572—43——15

fairs. Records were also kept of these complaints, which included the name of the complainant, the name of the accused, if known, the name of the investigating officer, the findings of facts, and the action taken or recommended to be taken. The mass of information thus acquired was kept carefully indexed and tabulated. At the same time, the Judge Advocate was required to furnish the Office of Civil Affairs with copies of all general court martial orders, so that a complete record might be obtained of all cases, including evidence of punishments awarded. The result of these various methods proved entirely satisfactory and, within two months, a record was on file of almost all crimes alleged to have been committed by American troops against German civilians, which included a report of the facts, a statement of the action taken, the punishment inflicted, or the reason why the latter was impossible. Disciplinary action was not taken in some cases because the charge made was found to be either frivolous or untrue, while in many others it was impossible because there were no means of identifying the malefactor.

Altogether, some 800 complaints were received up to October 1, 1919, ranging from murder to accidental destruction of property. Two hundred and fifty-six of these complaints were accusations of felonies, divided into the following classes: homicide, 25; rape, 17; attempted rape, 17; misconduct toward German women (not legally felonies but, because of their seriousness, tried by general court martial), 14; inducing children to commit sodomy, 2; burglary, 13; robbery, 79; larceny, 53; assault with deadly weapons, 36.

It is believed that, in nearly all cases, the officers in command of troops endeavoured to prosecute their soldiers for crimes against German citizens with as much energy as they would have shown had the crimes been committed against American citizens, and that the cases in which officers tried to shield offenses committed by soldiers (if there were any such at all), were negligible.

Records of trials by general courts martial up to October 1, 1919, of American soldiers charged with committing offenses against civilians, disclose the following facts:

Of twenty-five homicides laid by the Germans at the door of the American army, eleven soldiers were tried and six convicted. Three of the acquittals were based on the fact that the manslaughters were accidental and two on lack of proof. The average of the sentences of the six convicted American murderers of Germans was greater than that of the two Germans convicted of murdering Americans. There were seventeen complaints of rape, resulting in eight trials. As a result of these trials, five soldiers were convicted and three acquitted. There were seventeen complaints of attempted rape, with five convictions and four acquittals. In both the foregoing classes of offenses, the acquittals were the result of failure to prove the crime and not of failure to identify, and probably most of the complaints which did not result in trial were dismissed for similar reasons. There were fourteen alleged cases of misconduct toward German women, i. e., cases of non-sexual assault, resulting in nine trials and six convictions. There were eight convictions and one acquittal for burglary, out of a total of thirteen complaints. There were seventy-nine complaints of robbery, sixteen soldiers being tried, and eight convicted. The reason for the small number of trials and convictions for this offense is probably due to the fact that the soldiers who were said to have committed the robberies were generally masked and therefore could not be identified. Fifty-three complaints of larceny by soldiers are of record, forty-two soldiers having been brought to trial. Twenty were convicted and twenty-two acquitted. From thirty-five complaints of assaults with deadly weapons, twenty-three convictions and one acquittal resulted. There were therefore altogether 118 trials of felonies as against 256 complaints. There were, in addition, seven complaints of mistreatment of German prisoners, two soldiers having been convicted for such offenses by general court martial. The reason why there were no trials in the other five cases or in three somewhat similar cases in which Germans complained that they had been robbed in prison, was because the complaints were not submitted to the army until after the organizations to which the accused belonged, had returned to the United States, identification being therefore impossible.

In the above figures, only the general courts martial records of the Army of Occupation have been considered. A great many American soldiers were tried by special and summary courts as a result of complaints by German civilians. Since there was no central office of record for these cases, it is now impossible to compile statistics based on them. However, as practically all the cases in which felonies were charged against soldiers were tried by general courts martial, these statistics offer ample evidence of the army's desire to deal justly with the civil population.

Of the 250,000 men who formed part of the Third Army for six months, only 256 were accused of felonies—a rate of two per thousand per annum. It is of course by no means certain that all these felonies attributed by the Germans to the American forces were indeed committed by them, but even assuming that the complaints were in every case true, the total is very small when one considers that the bulk of all crime—in fact almost all crime—is committed by youths and men between fifteen and thirty years of age, and that the army was composed almost entirely of men of those ages. Most of these men had been away from the restraints of good women and home life for nearly a year and, during the four months preceding the occupation, had been subjected to the demoralizing influence of continual campaigning in the presence of the enemy, where neither the sacredness of human life nor the rights of property was given a second thought. Furthermore, as investigation later disclosed, nearly 4% of the American soldiers were illiterate, and illiterates are exceptionally dangerous during the occupation of a foreign country, inasmuch as they are incapable of amusing themselves in a better way during their leisure-hours and are apt to give way to self indulgence and dissipation. The majority of American soldiers were not used to alcoholic stimulants, and the result of their stay in the Rhineland, the heart of the German wine country, in no sense taught them temperance. Almost all the acts of violence committeed by American soldiers can be definitely traced to over indulgence in alcohol.

In addition to the alleged felonies which have been discussed above, there were 367 complaints of common assault on civilians by soldiers. There is no record of trial of soldiers for this, by far the most common offense, since soldiers who may have been brought to trial were turned over to inferior military tribunals and to special and summary courts.

It must be admitted that the number of trials of Americans for alleged offenses against German civilians appear few in comparison with the number of complaints lodged. This was probably due to the fact that the civilian could seldom bring any more substantial clue to identify the malefactor than that he was an American soldier. It was therefore extremely difficult for the American authorities to apprehend the offenders, most especially since a good many of the crimes were undoubtedly perpetrated by soldiers absent without leave from their organizations whom the American authorities were themselves seeking to apprehend. The difficulties confronting the military authorities in apprehending soldier criminals were further increased by the fact that the Germans generally seemed to be afraid to make a complaint direct to American officers, but delayed their complaint until they could transmit it through their Bürgermeister. The result of this timidity was very unfortunate, inasmuch as the American authorities usually did not learn of the misdeed until several days after it had been committed. It was by that time, of course, difficult to find the offender. The bare figures by themselves appear discreditable but it was in effect only by the most vigilant efforts that the army could bring to trial the proportion which it actually succeeded in doing.

Lack of education and its relation to crime—statistics of the Army of Occupation.—With regard to the lack of education in the Army of Occupation, it is advisable to call attention here to the report of the Director of Education of the Third Army, Dr. Benton. His criterion for judging literacy was the ability to read and understand a newspaper, and to write a letter which could be censored by an ordinary literate man. As a result of his investigation, 9,929 illiterates were discovered in the American Army of Occupation. This investigation, it should be added, was in no sense complete.. Even this number, however, is discreditable to the army of a nation which boasts that it has the finest system of public education in the world, and has long prided itself on the fact that its citizens are more intelligent and better educated than those of other nations. Post schools were started by orders of the Army Commander in every little hamlet in the area, and placed in charge of regimental chaplains or enlisted men who had been teachers in civil life. As a result of these efforts, the number of illiterates was gradually reduced to 1311 in June 1919.

Offenses committed by German civilians against members of American Army.—A careful record was kept of the trials of civilians before provost courts of the American army. Up to January 10, 1920, the offenses and number of convictions were as follows:

Sale of prohibited alcoholic drinks	517
Unauthorized sale of light wine and beer	335
Sale or unlawful possession of U. S. property	1580
False statements to U. S. officers	68
Unauthorized assembly	314
Larceny of U. S. property	781
Violation of circulation orders	2317
Failure to obey lawful orders of military authorities	556
Drunkenness and disorderliness	177
Unlawful possession of deadly weapons	302
Insulting language concerning U. S. Army	143
Violation of sanitary regulations	918
Disrespect to American officers	81
Practising prostitution while diseased	91
Discrimination in prices	16
Interfering with U. S. troops (assault on American soldiers)	357
Damage to, or destruction of, U. S. property	36
Vagrancy	401
Miscellaneous	712
	9702

In addition to the above, three Germans were tried for murder by military commissions.

In order to understand correctly the apparently large number, one should take into consideration the list of similar offenses by civilians against United States troops, which offenses number 357. Practically all assaults, whether committed by Germans on Americans or by Americans on Germans, were the result of drunken brawls. In most cases, both parties were about equally to blame, and the figures are consequently as discreditable to one side as to the other.

Certain facts which may be of value in event of a future occupation may be deduced from the figures of classified offenses kept by provost courts. The most numerous class of offenses was violation of circulation orders, that is, infractions of the Allied orders regarding identification cards and passes for travel. Of this class, 2,317 offenders were tried by American provost courts, that is, about one quarter of the entire number of trials. The circulation orders referred to required every German to provide himself with an identification card, and to obtain permission from the military authorities for all trips. These requirements were a constant burden on German civilians, who often were forced to stand in line for hours, in all kinds of weather, in order to get a pass for a journey of a few miles. It must also be confessed that the regulation was of little value to the Allied authorities. Any German who had a mission so important as to induce him to take the risk, could travel wherever he wished by forging a pass or borrowing that of a friend. The only result obtained by the circulation orders was unnecessary irritation of the civil population. On one occasion, certain German criminals actually had duplicate passes printed and identical rubber stamps made, and for several weeks carried on a thriving business by selling forged passes to such civilians as preferred to pay for a duplicate pass, rather than wait in line for one regularly issued. Such Germans as were tried were usually more careless than defiant, and practically no one was caught trying to elude observation.

The next most numerous class of offenses on record was of sale or unlawful possession of U. S. property, for which 1580 civilians were convicted up to January 10, 1920. Most of these were cases either of larceny of American goods or of receiving stolen American property. The civilians implicated were however not tried for larceny, inasmuch as that offense is a difficult matter to prove, but their cases should be considered in connection with the 181 cases of actual larceny tried.

The third most numerous class of offenses consisted of the 918 violations of sanitary regulations. Most of these cases arose directly from the policy of the army to inculcate American ideas of sanitation among the Germans. In the small country village of the Rhineland, the majority of the houses are built around a central court, which is apt to be unsanitary from the American viewpoint. When the American Army arrived in Germany, an order, aimed at reducing the danger of influenza and insisting that the inhabitants keep their windows open at night, was quick to arouse a great deal of resentment, so utterly opposed were the German ideas of sanitation to American ones. Many a soldier had his outraged landlord say to him during the first days of American occupation, "What is the use of my heating your room for you all day if you let all the heat out at night?" Summer had come before the people had got used to the idea of opening windows at night, and the American authorities had by that time issued new orders to the civil population.

The manure pile is one of the German peasant's most prized possessions, usually being kept in close proximity to the family kitchen. Inasmuch as manure piles are breeding places for flies, the Commanding General directed that they be scattered over the fields, and that none in future be permitted to accumulate. This order caused no end of annoyance, and really worked a distinct hardship on the people, inasmuch as it compelled them to alter their carefully arranged system for fertilizing their fields. Most of the 918 cases of violation of sanitary regulations arose from infractions of this order. Enforcement of a change of customs on any people, no matter how insignificant the change may be, is more provocative of dissension than is anything else.

There were 335 trials of civilians for selling wine outside of legal hours, and 517 cases in which distilled liquor and champagne were sold in violation of orders. These 850 cases should be considered as a unit, and they make together the next most numerous class of offenses committed. Here again it will be seen that it is extremely difficult to change long established customs by a sudden order. The German had always taken alcohol where and when he pleased, naturally could not understand the purpose of the order, and resented what he considered as an interference with his proper habits of life. No other class of offenses compares numerically with those already discussed.

Disregarding larceny and unlawful possession of U. S. property, we find that the three most numerous classes of offenses arose over matters in which the American orders directly interfered with national customs and traditions.

The orders published by the army were on the whole faithfully complied with, if we except those dealing with alcohol, circulation and sanitation. Probably in the case of the army, at least two of these three were absolutely necessary for the well-being of the army. Yet it is instructive to note that these three rules, inasmuch as they upset the people's scheme of life, caused us more trouble and resulted in more irritation, than did the enforcement of all other American orders put together. This is evidence that it is wise to study the national habits and customs of an occupied country before issuing orders, and that, when issued, those orders should be so framed as to avoid interference with national habits and customs as far as possible.

Sentences of the American provost courts, it must be confessed, were severe, particularly in the assessment of fines. The commanding generals naturally felt that fines under the courts of a military government should be more severe than in civilian courts. However, as most provost court officers computed the fines in dollars and mentally converted the amount into marks, and as the value of the mark had shrunk at one period to a twenty-fifth of its former value, it will be understood readily why the fines were often unreasonably high.

A military occupation which accomplishes its purpose with the least possible friction with the inhabitants, is, in the end, the most successful. The Rhinelander today feels that, although we were severe, we were pre-eminently fair in our dealings. He also realizes that we did not indulge in petty spite and meanness, that we told the truth and pursued our objectives openly and directly, without chicanery or double dealing.

CHAPTER NO. 12

FINANCE AND REQUISITIONS

Payment for Maintenance of Armies of Occupation; The Comptroller's Decision; Method by which Partial Payment was Obtained; Requisition of Billets and Supplies; Payment for Billets and Supplies; Claims; Rate of Exchange; Allied and Neutral Securities in Enemy Hands; Banking Institutions; Taxation and Budgets; German Wine and Luxury Taxes.

Payment for maintenance of armies of occupation.

Article IX of the Armistice states:

> The upkeep of the troops of occupation of the countries of the Rhine (not including Alsace-Lorraine) will be charged to the German government.

From this it is clear that the German government was under obligations to pay for the maintenance of the American Army of Occupation, except as to that part of the army which was stationed in the Grand Duchy of Luxemburg.

No means of payment being laid down in the Armistice, it was necessary that the intricate question as to how Germany should fulfill this obligation be taken up by the Armistice Commission. The cost of the occupation included far more than the pay of the troops, for the cost of food, clothing, supplies, billets and requisitions all had to be settled. It was clear from the first that Germany would be unable to provide food and other supplies, since she had none to spare. It was equally clear that as far as the troops were paid in cash, Germany would have to fulfill this obligation at once, since, inasmuch as they occupied Germany, it was essential that they be paid in marks, and Germany alone had the necessary currency. Even, however, in the matter of pay, Germany of necessity could furnish only about half the required sum to the United States, for approximately half the pay of the Army of Occupation was not given to the men direct, but was disbursed for them in the United States in the form of allotments, war risk insurance, etc. Nevertheless, the Army of Occupation was certain to need many millions of marks, not only for the troops, but also for the payment of requisitions, claims, etc. It was soon realized that the Commanding General of the Third Army would have large amounts of funds received from Germany to disburse and account for, and this brought up the important administrative question as to exactly how these funds should be handled.

The Comptroller's decision.—It was possible to consider the monies paid by the Germans to the Third Army as paid to the United States Government, in which case, in accordance with Section 3617 of the Revised Statutes, 1904, they would have to be turned over to the U. S. Treasury. This section reads:

> The gross amount of all moneys received from whatever source for the use of the United States, except as otherwise provided in the next section, shall be paid by the officer or agent receiving the same into the Treasury, at as early a day as practicable, without any abatement or deduction on account of salary, fees, costs, charges, expenses, or claim of any description whatever.

Once in the Treasury, money could of course be disbursed only in accordance with Clause 7, Section 9, Article 1, of the United States Constitution, which provides that:

> No Money shall be drawn from the Treasury, but in Consequence of Appropriations made by Law; and a regular Statement and Account of Receipts and Expenditures of all public Money shall be published from time to time.

On the other hand, it was equally possible to consider the money to be turned over by Germany, as a military contribution levied on the German government by the American army. If this latter view prevailed, the Commanding General could disburse the funds received, whenever and wherever he deemed necessary, since the money could then be considered as a war levy. Obviously, it was greatly to the advantage of the army that the latter view prevail, since, if the funds acquired from Germany were turned into the Treasury, they could be disbursed only in accordance with the laws laid down by Congress. It was believed certain that many situations, never provided for by Congress, would arise during the occupation, since such an occupation had of course never been foreseen.

Shortly after army headquarters had been established in Coblence the Army Commander submitted a request to the Commander in Chief, asking that he be permitted to treat the funds collected from the German government as a war levy, and disburse them as a part of the internal economic administration of the army. The Commander in Chief concurred entirely in this request and recommended its approval, but since the matter involved points of law which were not directly within the province of the military, he submitted the request, together with his recommendation and a supporting brief prepared by the Judge Advocate, A. E. F., to the Assistant Comptroller of the Treasury, who was at that time representing the Treasury Department in Paris.

In the accompanying brief, which appears in full in Appendix No. 23, [not printed], the Judge Advocate stated that he was perfectly aware that under the Constitution, Congress had the power "to raise and support armies" and to "make rules for the government and regulation of the land and naval forces," and that, under Section 3617 of the Revised Statutes of 1904 (already quoted), any money received by any officer or agent of the United States from any source whatsoever should be turned into the Treasury as miscellaneous receipts. He admitted that this section was wide enough in its language to cover funds about to be received from the German Government under clause IX of the armistice. He argued, nevertheless, that these funds were not so covered. In the first place, when Congress passed these statutes in 1904, it never intended to control funds received in such a way as these; for the occupation of a foreign country could not possibly have been foreseen in 1904. Since Congress did not intend to apply the law to an occupation, he did not believe this argument to be valid. In the second place, he contended that even if it were argued that Section 3617 actually covered the receipt of the monies in question, so much of that statute as applied to war levy was later repealed in 1907 by the treaty ratifying The Hague Convention, since it is a well settled principle that when a treaty is ratified, it automatically becomes law and repeals all prior legislation in conflict therewith. Article 49 of The Hague Convention, which was ratified in 1907, and therefore assumed the status of a treaty, provides:

> If, in addition to the taxes mentioned in the above article (48), the occupant levies other money or contributions in the occupied territory, this shall only be for the needs of the army or for the administration of the territory in question.

The Judge Advocate argued that Article 49 directly repealed Section 3617 of the Revised Statutes as far as that section applied to war levies, if indeed it applied to them at all. As further support of his contention, he cited the precedent of General Scott in Mexico. The latter had made levies and disbursed them for the support of his army, without ever turning them into the United States. (The same had been done during our whole administration in Porto Rico, the Philippines, Cuba, and at Vera Cruz. These of course were, in their nature, trust funds held by the United States for the ultimate use and benefit of those countries.)

In order to support his contention further, the Judge Advocate pointed out the innumerable difficulties which would arise if the monies paid over by the Germans had to be disbursed under the statutes of the United States. Experience has shown throughout history that unless requisitions are paid for either by cash or on short term credit, they soon cease to be a source of supply. This is due to the fact that the inhabitants of a hostile land feel that the requisition is in fact a confiscation and not a purchase, and therefore hide their property rather than bring it forward on the order of the occupying military authorities. Furthermore, requisitions without payment are a certain cause of friction between the troops and the inhabitants, and friction should be avoided above all else in a military occupation. The Congress of the United States had provided for the payment to Allied nationals of claims and requisitions by the army, but had entirely ignored the identical question in occupied territory. Therefore, if the funds were turned over to the Treasury, there would be no legal means of paying Germans in the Rhineland for property which was either damaged or requisitioned. The Judge Advocate inferred that it would probably be necessary to make still other payments which could not then be foreseen, and which, if the money were turned in direct to the Treasury, would be difficult ever to settle.

If the United States, by means of vouchers payable by the German government, paid the individual Germans for such property as it requisitioned, it thereby placed its commercial honor in the hands of a government which the experience of the last few years had shown was without honor in its international engagements. History has proven that no nation has ever honored the vouchers of a hostile occupying force. It was felt that Germany might fail to honor these vouchers, and that the requisitions of the army in the Rhineland would therefore not be requisitions but mere confiscations. If this occurred, it would place an almost insupportable burden on the inhabitants of occupied territories. The American Army did not desire thus to discriminate against the civil population of the territory under its control. Besides, at this period, the policy of the Allies in regard to the Rhineland had not been definitely decided, and many thought that it eventually might be expedient to separate the province from the rest of Germany. If this policy were adopted, it would be unwise to entangle the interests of the inhabitants of the Rhineland with the central government of Germany by means of claims for money, which only the central government could honor. Relying on the brief of the Judge Advocate, the Commander in Chief asked that the monies paid over by Germany be considered as a war levy and therefore be handled exclusively by the army.

The Assistant Comptroller replied to the Commander in Chief in a letter in which he waived these points as academic, and held that, in his opinion, the above-quoted clauses of the Constitution, together with sections 3617, 3618 and 3619 of the Revised Statutes, settled that the monies paid over by the Germans should be turned into the Treasury of the United States. This decision of the Assistant Comptroller became a constant source of embarrassment to the military authorities in their administration, as will be pointed out in this report hereafter.

Method by which partial payment was obtained.—As early as December 1918, the Commanding General of the Third Army made his first requisition on the Berlin government for money with which to pay the troops. The money was deposited by the German government in the Coblenz branch of the Reichsbank. That part of it not used was immediately stored in the vaults of the headquarters building. Other requisitions were made from time to time until, by April 1919, 308,000,000 marks had been paid over by the German government to the American Army.

In accordance with the decision of the Assistant Comptroller, the money so paid by the Germans was transferred by a voucher to the United States Treasury as miscellaneous receipts and immediately invoiced back to the army under special appropriations. For purposes of accounting, the value of the mark was computed in dollars, and the amounts were thereafter carried in dollars on the books.

As explained later in this chapter, the rate of exchange was in a constant state of fluctuation, the mark generally falling in value each month. As the army kept its books in dollars and its funds in mark notes, this variation in the rate of exchange made bookkeeping very difficult. As a practical measure, the books were balanced each month and the loss or gain caused by the decrease or increase in the value of the marks on hand was debited or credited to the German account for the cost of the occupation.

In the month of March, the United States commenced to sell its war material, and also the abandoned German war material seized during the occupation, to civilian purchasers. The proceeds of these sales were received in marks and turned into the United States Treasury. Inasmuch as the Treasury had practically no use for marks at home, the money was returned to the army for necessary disbursements, such as the pay of troops. So much money was actually obtained from these sources and from the funds of Provost Courts, that it was no longer necessary for the army to requisition marks from the German government, and all such requisitions ceased after April, 1919. The great bulk of sales of property took place in the spring of 1919, and a great store of marks accumulated in the army vaults during the subsequent summer and autumn. However, the value of the mark decreased rapidly during these months, and a heavy loss to the United States resulted as a consequence. The army continued to keep its books in dollars, balancing them at the end of each month and charging off the loss each month against the German government as part of the cost of the occupation.

The expenses of the occupation will undoubtedly be a very difficult item for the United States to collect from Germany. It is apparent today that that country is practically bankrupt, as its entire gold reserves are estimated in January, 1920, not to exceed 300,000,000 marks. It will probably be many years before the United States government will be able to collect the cost of the occupation which Germany agreed to pay under the terms of the Armistice. Whether, in the light of history, this was a wise policy, only time can tell, but one thing is certain, and that is, that for the time being, Germany was enabled to escape one of the just consequences of waging an indefensible war and was permitted to violate one of the most vital terms of the Armistice through the decision of a Treasury official.

Requisition of billets and supplies.—Before the advancing army had set foot on German soil, the Army Commander published as orders the regulations which should govern requisitions. In brief, they required that requisitions should be made only on authority of the officer commanding a locality or by his direct representative, and then only on the local Burgomaster or some other civil official. They also required that receipts be given whenever property was requisitioned. This placed the burden on German officials of collecting material requisitioned by the army, as well as responsibility for an equal distribution among the people of such requisitions.

From the very day it entered Germany, requisition of foodstuffs by the American army was prohibited. This was because of the clause in the Armistice which stated that the Allies would consider the provisioning of Germany during the armistice to the extent recognized as necessary. The Americans realized that, under this clause, they would later be obliged to restore in kind the food requisitioned and that, by requisitioning food, they would give the German government grounds for claiming that the food shortage was due partly to the fact that the Allied armies had greatly depleted the available stores. The Americans realized further that any food which in future should be supplied to Germany would have to come from the United States, as the supplies available in Allied countries for this purpose could be considered as practically negligible. Requisitioning of food, by the American army, was therefore prohibited, except in especially urgent cases. On the other hand, requisitioning of fodder was allowed, provided of course that it was made by the local commanding officer or his representative and on the proper German civil authority. Requisitioning of forage was continued until the lines of communication were able to supply the army's needs, but as such requisitioning so depleted available supplies that farmers had to kill off their cattle, thereby producing for the time being a surplus of meat, but ultimately reducing the vital milk and butter supply, such requisitioning was discontinued at the earliest possible date. This order also established that billets should be requisitioned through the Burgomaster or other responsible civil official, in accordance with the French billeting law. That law did not require householders to prepare food for troops billeted upon them. Little or no difficulty was experienced at any time in billeting the army, inasmuch as the local officials simply utilized the billeting lists which had been prepared for the German troops. Billeting actually proceeded more smoothly in Germany than in either France or Luxemburg.

During the advance to the Rhine, the troops were billeted in houses or barns, or wherever shelter was available. Accommodations were not all that could be desired, but since the army was concentrated along a few roads, it was impossible to provide quarters for all. During the march, however, such hardships as had to be endured made very little difference to the individual soldiers, inasmuch as most of them had become so inured to privation during the summer campaigns that even a hayloft seemed ideal sleeping quarters. As soon as the units were settled in their permanent stations, the Germans were required to furnish them with proper sleeping accommodations. This was because the troops were then spread out all over the country and were occupying practically every town, village and hamlet in the area. (At one time there were nearly 250,000 soldiers billeted on less than 800,000 citizens. Such close billeting was not required in any other area than that occupied by American troops.) The Commanding General also directed that every male German between the ages of fifteen and sixty, who was not in ill-health, should give up his bed if American soldiers billeted in his house lacked them.

Payment for billets and supplies.—The American orders, following the instructions received from Marshal Foch, at first forbade the armies to pay for billets and requisitions, and provided merely that the civil official on whom the requisition was drawn should be given a receipt for all supplies, services or accommodations furnished. Copy of the receipt was to be kept by the officer making the requisition. In the early part of the occupation, the receipts were merely an informal acknowledgment of the transaction, and were turned over by the commanding officer to the Burgomaster as the former was leaving the locality. As conditions became stabilized, however, an efficient administrative system was gradually developed. Town majors were appointed for every city or village; forms for requisitions and receipts were printed and distributed; a central office of rents, requisitions and claims was established and the procurement of billets handled in as simple a manner as if they were provided by one of the supply departments of the army, instead of by the German civil authorities. Under this system, one printed voucher was given to the Burgomaster or Vorsteher, one was kept in the Town Major's office and one forwarded to the Rents, Requisitions and Claims Officer. It was realized that payment for billets would have to be made promptly, if friction with the civil population was to be avoided. The Commanding General consequently directed the Disbursing Quartermaster to pay for billeting, both out of the funds received from the German government and from such other funds as had accumulated from the sale of enemy war material and from that of our own supplies.

Payment for requisitions of other supplies, such as fodder, fuel, rent for land used as drill grounds and target ranges, was made in the same manner as for billets.

Claims.—The Army of Occupation could not avoid damaging or injuring the property of the civil population. Some damage was merely the result of fair wear and tear incident to the occupation, and was not considered as a basis for claims, inasmuch as payment was supposed to be included in the billeting fee. Other damage arose as a necessary incident to the occupation, but was of such a character that it fell with undue weight on particular individuals; for instance, a case in which a stone road was laid across a field in order to bring supplies to a supply dump, or in which fields were cut to pieces by artillery parks. Many claims also arose as the result of accidents or from wilful or negligent acts of individuals belonging to the military forces.

In order to protect the United States from unlawful claims, the American authorities made it an offense to bring a false claim against the United States. An order was soon afterward issued requiring all claims alleging damages by the troops to be submitted to the military authorities within forty-eight hours after the damage had been committed. This order also provided that unless the claims were filed within the time specified, or unless there was a good reason for delay, the claim would be disallowed. This provision was inserted in order to prevent Germans from making a claim months after the damage was committed, when it would have been difficult for the military authorities to secure evidence as to whether or not the damage had actually been committed by American troops.

Claims against the government were eventually divided into three classes: Class A—those incident to the occupation and to the necessary functioning of the army of occupation, and not a result of unlawful acts or negligence on the part of individuals of the American army; Class B—those which were a result of unlawful acts or negligence on the part of individuals of the American army, and in which the circumstances were such that the 105th Article of War could not be applied; Class C—those which were a result of unlawful acts or negligence on the part of individuals of the American army, and in which the circumstances were such that the 105th Article of War could be applied.

A system similar to that used in France for the investigation and filing of claims, for the preparation and submission of vouchers and for the payment of claims, was instituted by the Bureau of Rents, Requisitions and Claims at Headquarters of the Third Army in Coblence. In the settling of Class A claims, after the amount due had been determined and vouchers had been prepared, one voucher was given to the claimant and duplicate vouchers were forwarded to the Officer in Charge of Civil Affairs. The Officer in Charge of Civil Affairs presented these vouchers to the proper officials representing the German government, and directed these officials to pay the claims out of funds at their disposal for that purpose. This roundabout method of payment was necessary to avoid turning in the funds to the United States Treasury under the decision of the Assistant Comptroller, above referred to, for had they been turned into the Treasury, there would have been no way authorized by existing law to secure them again in order to pay claims against the government by enemy nationals. In order to make certain that the German officials actually made the payment, it was directed that the turning over of the money to the German claimant take place in the presence of an American officer, who signed the original voucher as witness and returned it to the American archives for file.

Class B claims were not disposed of so easily. They were certainly not a legitimate cost of the occupation, chargeable to the German government, since they arose from criminal acts of members of the American forces. There was no Congressional provision for payment of this class of claims from money in the Treasury of the United States. No way existed for the time being of settling them. Several authorities on international law argued that the United States was legally financially liable only for such crimes of soldiers as could have been prevented by the exercise of ordinary care on the part of a superior authority. This argument was based on the theory that an employer was not responsible for damage committed by criminal acts of his employes, unless such acts were brought to his attention in time for him to prevent them. Other authorities argued that the government was responsible for all criminal acts of its soldiers, on the theory that as military discipline gave it complete control over their conduct, it was responsible for crimes resulting from relaxation of that control. At any rate, in order to be on the safe side, the Bureau of Rents, Requisitions and Claims instituted a complete investigation of all claims resulting from criminal acts of soldiers and filed a report of the investigating officer, together with the testimony of witnesses in that office, for adjustment after the ratification of peace.

Class C claims were somewhat easier to settle. The 105th Article of War provides that:

> Whenever complaint is made to any commanding officer that damage has been done to the property of any person, or

521572—43——16

> that his property has been wrongfully taken by persons subject to military law, such complaint shall be investigated by a board consisting of any number of officers from one to three, which board shall be convened by the commanding officer and shall have, for the purpose of such investigation, power to summon witnesses and examine them upon oath or affirmation, to receive depositions or other documentary evidence, and to assess the damages sustained against the responsible parties. The assessment of damages made by such board shall be subject to the approval of the commanding officer, and in the amount approved by him shall be stopped against the pay of the offenders. And the order of such commanding officer directing stoppages herein authorized shall be conclusive on any disbursing officer for the payment by him to the injured parties of the stoppages so ordered. Where the offenders cannot be ascertained, but the organization or detachment to which they belong is known, stoppages to the amount of damages inflicted may be made and assessed in such proportion as may be deemed just upon the individual members thereof who are shown to have been present with such organization or detachment at the time the damages complained of were inflicted, as determined by the approved findings of the board.

When claims arose under provisions of the above article, they were promptly settled in accordance therewith, but, in actual practice, the Article could rarely be invoked. When an offense had been committed it was usually very difficult for the Germans to identify the offender, or even his organization. Such claims had therefore to be settled in other ways. Furthermore, if the offense was against the *person* of a German (homicide, rape, etc.), the 105th Article of War could not be invoked, but in such cases disciplinary punishment only could be imposed when the offender could be identified. It is known that claims for damages in practically all such cases (especially those resulting in death) have been submitted to the German authorities, and these will probably be presented as an offset to the cost of occupation, chargeable to Germany. The claim for damages resulting from homicide has usually been made in pounds sterling, the amount being 5,000 pounds.

Rate of exchange.—The Great War radically affected the value of currency throughout the world. The general decrease in money values which resulted therefrom was due largely to the fact that although the world's currency increased somewhat during the five years of hostilities, the commodities which money is used to purchase, had decreased not only by the destruction of war, but also because of the vast amount of labor and energy that was turned from the creation of wealth and supplies to the unproductive work of war and of creating supplies for war. Furthermore, all the great commercial and banking countries of the world had become more or less involved in the war and had inflated their currency in conformity to their needs. Germany, Austria-Hungary, Belgium, France, and Italy had also sent abroad a considerable percentage of their gold reserves, in order to purchase additional war-material. The result was that the paper currency in circulation in these countries had very little actual gold behind it. The enormous reserve wealth of the British Empire enabled it to finance the war without a radical fall in the value of the pound sterling, although it, too, fell to the lowest point in modern times. On the other hand, the currency of the United States, as well as of the commercial neutral nations, rose sharply, resulting in a decrease in the purchasing power of money in the belligerent countries.

Under ordinary circumstances, the mentioned inflation of currency and great exportation of gold by Germany, Austria-Hungary, Belgium, France, and Italy would have caused the value of their respective monetary units to sink in relation to those of the countries still on a sound basis. However, for various reasons, this was not the case. The Allied and Associated Powers pooled their credit and England and America were thereby able to keep up the value of the Belgian and French franc and of the Italian lira. The Central European Powers, on the other hand, maintained the value of the mark by doing most of their purchasing at home and by paying for outside purchases with their normal gold reserves and with the special reserve which Germany had levied from France in 1870 and had stored away against the contingency of another war. This policy of Germany of course diminished her reserves, but as she was completely segregated commercially from most of the rest of the world, it made no particular difference in the value of the mark as long as the blockade continued, the value of the mark being the unit of measure and its value being artificially sustained by law. With no other standard against which to measure it, no one in Germany realized how much it had shrunk. Germany's intention was of course to replenish her gold supply from the indemnities to be paid by her defeated foes.

Hardly had the economic blockade caused by the fighting armies been lifted by the withdrawal of the Germany army, than economic laws began to reassert themselves. During the period of occupation of the Grand Duchy of Luxemburg, Germany had prescribed the rate of exchange between the Luxemburg franc and the mark, at the prewar rate of 1.25 francs=1 mark. When the American army entered Luxemburg, prices were always quoted there in marks, and the average Luxemburger, either because he was very simple or very sharp, demanded 1.25 francs for every mark. This caused a feeling of resentment among the American troops, who were sufficiently good economists to realize that the mark could hardly be worth relatively its pre-war value. However, the banks in Luxemburg soon began to exchange francs for marks at par.

It may be stated parenthetically that there was at this time a tremendous number of marks in circulation in Luxemburg, due of course to the 4¼ years of German occupation. The Luxemburg government ultimately got rid of this unstable currency by buying it back on a certain fixed date at the rate of one mark for 1.25 francs and declaring that after the fixed date, marks would cease to be legal tender. The result of this altruistic action on the part of the government in behalf of its citizens, intended as a compensation for the ruinous rate at which the marks were purchased, was to injure greatly the national credit. In fact, the Luxemburg franc, which would not normally have been affected by the war any more than the Swiss franc, fell as a result of this measure to less than the value of the French franc. The measure was necessary, however, to protect the citizens of the Grand Duchy from great loss, and its people expect the support of the Allied governments in requiring Germany to make good this loss. A similar expedient was resorted to by the French to demonetize the mark in Alsace-Lorraine.

During the advance into Germany, a majority of the American soldiers having only francs, they exchanged or spent them at whatever rate they could obtain. On December 13, pursuant to orders from the Interallied High Command, the Army Commander issued a letter of instructions directing all corps and division Commanders to require the proper local government officials in their areas to publish the official rate of exchange, as fixed by the Allied Commander in Chief, at the rate of 100 francs=142.85 marks, or 1 mark=70 centimes, and to require announcement that to demand or accept a rate more favorable than that prescribed would subject the offenders to trial by military tribunal.

The idea of establishing a rate of exchange was intended to apply to official transactions only. It was not the intention to compel banks or private individuals to exchange marks for francs when they did not wish to do so, but it was the intention to compel them to exchange at the prescribed rate if they elected to exchange at all. There is no doubt that the occupying forces had a legal right to establish an official rate of exchange. The German army itself had always done so, not only in Belgium, France and Luxemburg in the present war, but also in France in the war of 1870. Unfortunately, certain division commanders misinterpreted the policy of the Commanding General and the order of December 13th. This order in part read:

> You will also require the said local officials to include in their proclamation announcing the rate of exchange, a statement to the effect that the rate applies to all civilians as well as to the military and to all classes of currency in circulation within the area of the American occupation.

While this was a correct statement of the policy at the time, it was misconstrued to mean that every person would be *compelled* to exchange money at the official rate, whereas the intention was merely to regulate the rate in commercial transactions. It was not, as has been remarked, intended to make any bank or individual a bureau of exchange against his will.

This course of action gave rise to a great many complaints from banks and other commercial organizations, for, according to a decree enforced since the beginning of 1916, German banks and bankers were prohibited from doing business with neutral banks at their own discretion, in foreign notes, foreign bills, checks, cable transfers or other foreign negotiable instruments, but had to transact all business of this sort through a few central banking institutions in whose hands the whole management of foreign exchange was concentrated. Banks which had exchanged money at the rate set by the Allies had, for that reason, to accept a lower rate in order to transform their francs back into marks. In their complaint, they stated that inasmuch as they had been subjected to considerable financial loss in these transactions, they desired to submit claims for damages against the United States.

These claims caused the American authorities no little embarrassment, because there was no way of determining the amounts that the banks had actually exchanged, except by their unsupported statements, which could not very well be accepted. The American authorities in such cases decided to stand on their undoubted right of imposing a standard rate of exchange on everyone. Fortunately, for this matter, the fall in the value of the mark continued, so that drafts which the German banks had in some instances been compelled to cash at 142 (though usually they cashed them voluntarily), when the current commercial rate was from 135 to 130 were eventually sold at a much higher rate than that at which they had been bought. This somewhat tempered the sting of their loss and the accidental injustice seemed less harsh.

On December 13, 1918, the mark was worth 70 centimes; on December 27th its value had decreased to 60 centimes, and on this date it had also sunk in interior Germany to about 70 centimes. Bankers continued to object and to file protests. Finally, on December 30th, a telegram was received from Advance G. H. Q., stating in substance that it was impossible to fix a rate of exchange for commercial transactions in occupied territory, and that the official rate would apply only to Army disbursing offices. Administrative Bulletin No. 4, January 4, 1919, stated, in Par. 2:

> All officers and men will be instructed that they should not exchange their money with trades-people or banks, but must apply to the office of the appointed Disbursing Quartermaster.

On January 1, 1919, the official rate of the mark rose to 67 centimes; on February 11th it sank to 62 centimes; on March 5th, to 60 centimes; on May 1st, to 50 centimes. From then on, it kept sinking until in January, 1920, seven hundred marks were necessary to purchase one hundred francs.

Inasmuch as the official and the commercial value of the mark were usually at variance, considerable speculation on exchange constantly went on among the members of the army. The Disbursing Quartermaster adopted certain measures to prevent this. Among these was the refusal to change marks into francs unless the person was under orders to enter France, and the refusal to cash officers' personal checks on French banks. Such stringent measures worked a great hardship on many of the officers, who had accumulated their pay checks during the last summer of the war. Shortly after arriving in Germany, an order was issued requiring all officers to cash their pay checks before December 30th. They were given an option of cashing them in francs or marks, and as the mark was so unstable, nearly everyone requested francs. During the early months of the occupation, a great many officers went to France on leave and, fearing to lose their surplus francs, deposited them there in Franco-American banks, like the Guaranty Trust Company, the Farmers Loan & Trust Company, Morgan-Harjes & Company, etc. Upon returning to Germany, they found that neither the Disbursing Quartermaster nor any private banks would cash their checks, so that their money was practically bound up in France. In the early part of the summer, the Allies decided to separate their pooled credit, with the result that the franc promptly dropped from about 5.43 to about 9.53. Many officers thus saw several months' pay melt away while they were earning that for one month, but were able to do nothing about it, on account of the regulations imposed by the army. Thousands of soldiers had likewise attempted to save their surplus cash for use on discharge, and they saw their savings swept away in the same manner. The difficulty was obviated to some extent by the fact that the Post Office Department sold money orders in dollars; but this did not help those who had opened deposits in Paris banks.

The reason for the adoption of this policy was probably to help the French franc by preventing American money from reaching German banks; but the lesson which ought to be learned from this unfortunate procedure is, that the pay of the American Army is expressed in dollars by statute, and that the American soldier has a legal right to demand his pay in dollars wherever he is called upon to serve. If any inconvenience or loss is thereby incurred, such inconvenience or loss should be borne by the Government and not by members of the army.

Allied and neutral securities in enemy hands.—The terms of the Armistice provided that while it lasted, no public securities which could serve as a pledge to the Allies for the recovery of, or reparation for, their losses, should be removed by the enemy. This of course referred solely to securities belonging to the German government and in no way affected those of the German nationals.

Food conditions in Germany became so desparate at the approach of the spring of 1919, that it was found necessary to purchase food from outside sources to feed the population. This brought up the question of payment, which was a factor of great importance, since both the German gold reserve and foreign credits were limited. A conference was held in Brussels on March 13 and 14, 1919, to determine the conditions under which Germany might purchase foodstuffs. It was agreed by all parties at this conference, that Germany's payments should be made in foreign securities, either by their outright sale or by an arrangement through which advances could be made, using the foreign securities as collateral.

In regard to the sale and use of foreign securities as collateral, the Allied and Associated delegates requested the German government to take immediate steps to requisition all foreign bonds, stocks, and securities of every description held by its nationals, with the exception of those of Austria-Hungary, Turkey, Bulgaria and Russia. The German delegates replied that necessary measures would be taken immediately as to bonds, but that the question of the extent to which stocks would be requisitioned would have to be considered. They pointed out that there must be some few exceptions to the rule that *all* securities should be requisitioned, for instance, securities belonging to insurance companies. But, in general, they promised that securities would be requisitioned whether in Germany or not, provided that they were the property of a German resident within Germany.

The Allied and Associated delegates agreed to appoint a technical committee for the purpose of dealing generally with the requisitioned securities and to make arrangements for their sale or their use as collateral. This committee was also to receive information as to the progress made by Germany in the collection of securities.

As a result of this conference, the German government issued an edict on March 26, 1919, ordering all foreign securities, excepting those of Austria-Hungary, Bulgaria, Russia, and Turkey to be immediately placed at the disposal of the state for purchase. This edict was of considerable length and intricacy. Its substance was that all securities as above mentioned must be turned into a bank or banking firm between April 2nd and April 12th and would be then paid for by the government. If not turned in and if later discovered, they would be confiscated. A penalty clause was attached, making failure to declare the securities, punishable by a fine of 10,000 marks or by imprisonment for six months.

Excellent as was the idea underlying this order, it proved practically impossible to enforce, since securities can easily be concealed. The Germans realized very quickly the practical difficulties involved, and, as their interests required it, apparently did their best to enforce the edict. Customs officials on frontiers were directed to prevent the prohibited exportation of securities, co-operation of the banks was obtained and postal censors were required to search registered mail. An "aërial police" was in fact inaugurated, to prevent smuggling the papers out by airplanes. The government was by these means able to collect sufficient securities from its nationals to pay for the cost of the imported food.

The Allied High Command was not satisfied with the German law, because no distinction was made in the treatment of occupied and unoccupied territories. The Allies were selling considerable additional amounts of food to occupied Germany, where the population was being fed in accordance with the ration agreed to at the La Morlaye Conference. This ration was considerably more nourishing than that issued in unoccupied Germany. The Ebert government declined to guarantee these excess purchases of food made by the Rhineland. Marshal Foch, knowing that the Rhineland was well able to pay for the excess out of its own resources, attempted to settle the whole matter by his order issued April 19, 1919. He declined, by this order, to admit the German interpretation of the decisions of the Brussels Conference, and directed that securities in occupied territories should be centralized under control of the Allied military authorities in German banks on the left Rhine bank, and be used under Allied control for the payment of food supplies, or "for any other purpose which the Allies may decide upon." Orders were at once issued by the various occupying armies prohibiting the export of such securities from occupied territories.

The American order on this subject forbade the removal from occupied territory of specie, Allied securities, or Allied bills of exchange, and forbade banks to pay, give or return to anyone, gold coin or bullion, regardless of how held. In order to check the enforcement of this order, banking institutions were required on specified dates to furnish statistics of their financial condition, including lists of foreign securities held by them. The banks at once vigorously protested, but without avail. One private bank, which maintained its protest and refused to furnish the statistics required, was closed by the military authorities until the order had been complied with.

Marshal Foch modified his orders on May 7th, by another, which provided as follows:

> 1. Returns from the sale of securities will be used for the payment of food-stuffs in the proportion and according to the conditions laid down by the Brussels conference. These returns may not be used for any other purpose except by common agreement.
>
> 2. Returns from requisitions on the left bank of the Rhine will not be used for provisioning the local population; but the local population may be provisioned in conformity with the Brussels Agreement.
>
> 3. Securities on the left bank of the Rhine will be centralized and preserved in local banks to be named by the German government in agreement with the Allied governments; for example, at Cologne and Mayence. Necessary instructions will be given, that the decree of the German government, dated March 26th, may be applied immediately in the different zones of occupation, under the understanding that requisitioning will be made on both banks of the Rhine according to the same principles and methods.

Banking institutions.—A complaint was raised by financiers of the various Allied and Associated Powers, immediately after signing of the Armistice, that the financial clauses and restrictions of that document were not sufficiently strict. They said that, under the terms laid down, it would be possible for the Germans to smuggle bullion and securities from their country and thus reduce their wealth to such a point that they could never pay the costs of reparation. The French, since they were most concerned with preventing the bankruptcy of Germany, were particularly nervous about this. They made very elaborate rules for supervision of banking institutions in the territory occupied by them, believing that if they could control the activities of these banks, they would have a good grip on the financial resources of the country. The Americans at first followed the lead of the French in this matter and thoroughly investigated the banks in their own area. The most important of these were:

(1) Deutsche Reichsbank (Branch of the Imperial Bank; headquarters in Berlin. Branches at Coblence and Treves.)
(2) Deutsche Bank (Main office at Berlin; branches at Coblence, Treves, Bernkastel-Cues.)
(3) Disconto Gesellschaft (Main office at Berlin; branches at Coblence and Treves.)
(4) Dresdner Bank (Main office at Berlin; branches at Coblence and Neuwied.)
(5) Bank für Handel und Industrie (Darmstädter Bank. Main offices at Berlin and Darmstadt; branch at Treves.)
(6) Darlehenskasse für Deutschland (Branch at Coblence.)
(7) Landwirtschaftliche Centraldarlehenskasse für Deutschland (Branch at Coblence.)

It was found that the banking laws of Germany were like the corporation laws, far less strict than in America. With exception of the Reichsbank, all the above were joint stock companies doing a large general banking business, each with a special *clientèle* among various industries and trades. The Reichsbank is the chief bank of Germany and is very closely allied with the Treasury. It was organized on January 1, 1876, as the successor of the Prussische Bank. Founded exclusively on private capital, divided into shares, it is according to the German law, a public institution of the government, and not subject to rules governing stock companies. It is, however, placed under direction of the government, as represented by the Chancellor and the directorate of the Reichsbank, the latter consisting of a representative of the Chancellor and four other members. Its functions are to regulate circulation, facilitate exchange and payments and make possible the use of available capital. It has the right to issue notes and script, as have a number of other individual state banks, such as those of Bavaria and Saxony. It stands in close relation to private banks, every important one of which maintains a checking account with it. It has the right to issue loans, using gold, consols and state municipal securities, known as Lombards, as collateral.

Prior to the war, the Reichsbank and the state banks alone had the right to issue notes and script, but when the war began, this privilege was extended to a great number of cities, districts, provinces, chambers of commerce and private firms. The Reichsbank at the same time opened a Darlehenskasse or loan bureau, which lent money on gold, bills of exchange, merchandise and stocks and bonds which were listed in the Berlin Stock Exchange. It was permitted also to issue notes for general circulation, marked "Darlehenskassenscheine," which had deposits of merchandise, including copper, zinc and silver, as security. A vast amount of these latter notes was issued up to December 1918.

The following facts are interesting as showing the great inflation of the Reichsbank's credit during the war. On December 31, 1918, the statement of the Reichsbank showed: Capital 180,000,000 marks; surplus 99,496,423 marks; bank notes outstanding 24,855,257,180, of which about 17,500,000,000 had been issued since August 1, 1914, and against which there were 1,858,577,256 marks in gold bullion and gold coins. Prior to the war, all Reichsbank notes were secured by precious metals to the extent of 30% and the balance by bills of exchange, checks and Treasury notes. On January 1, 1919, the gold reserve behind Reichsbank notes had shrunk to 7½% of the issue. The credit of the Reichsbank is represented principally by the balance of the outstanding notes.

Under the law, the German banks were required to balance their books on December 31st of each year, but the American authorities required them to be balanced at the end of every month in order to furnish the monthly report demanded from each bank in the area.

Most of the banks protested against this, but were reassured when the authorities agreed to keep secret the data disclosed by these reports. One banker, the dean of the Rhineland bankers, who had been a supervisor of French banks during the occupation of 1870, persisted in his refusal to render the required accounts. His bank was closed, and permitted to reopen only when he agreed to fulfill the American orders.

This monthly report required a statement of the total assets of banks, the amount of the total deposits under three headings: (1) the money actually on hand; loans due to be returned within seven days; (2) loans due within three months, and (3) loans not due until three months. A report was also required on the specie, German paper, gold, French paper, French specie and other foreign monies in the banks. Savings banks were required to state the total number of their accounts. All were required to give a list of securities owned and a statement as to where they were held, and, finally, all changes of securities during the month had to be specially noted. These reports were required with a view, as has been said, of keeping check on German finances, especially in the matter of securities. After three months, at the end of April 1919, as they were of no value to the American army and caused endless difficulties, they were discontinued.

Curiously enough, instead of injuring the commercial prosperity of the Rhineland, the American occupation greatly benefited it. The American soldier nominally received his pay in dollars, but as it was given him in the equivalent value in marks, and as the marks had greatly depreciated, the pay even of privates was prodigious. For instance, in the month of October 1919, the rate of enchange being about 28 marks to $1.00, the private, who normally received $30.00 per month, was paid 840 marks, or about 50% more than the best paid skilled German laborer. From the point of view of the Germans, the American army seemed an army of millionaires. Soldiers proverbially spend money like water, and the American soldiers in Germany were no exception. A good 70% of their pay must have found its way into the pockets of eager German merchants and tavern keepers.

As a matter of curiosity, the American authorities caused the German banks to state the amount of their deposits on June 30, 1918, and again on June 30, 1919. The increase of deposits was figured on a per cent basis. The average increase in the deposits of ten banks picked out at random throughout the area, was 41.9%. It is curious to note that, although the increase was general, as no bank showed a decrease, yet the city banks' deposits increased at a ratio far greater than that of country banks. This was probably due to the fact that it was to the cities that the soldier went for his cherished "souvenirs".

Taxation and budgets.—Although under the usages of war, the United States had the undoubted right to control both the budgets and the money collected by taxation within so much of occupied Germany as was under its control, it was decided not to make use of this right. The Officer in Charge of Civil Affairs at Advance G. H. Q., both demanded and received from the German authorities in December, 1918. a list of taxes previously collected, together with a statement of the amounts actually received from each tax.

German wine and luxury taxes.—The members of the occupying forces were of course not required to pay direct German taxes. In December 1918, however, the question arose whether the luxury taxes imposed as a war-measure by the German government could be avoided. These taxes were collected by the government from all merchants who sold luxuries, and the latter of course included them in the prices for their merchandise. It was consequently very difficult to prevent their collecting the tax from members of the occupying forces who desired to purchase the articles in question. The tax amounted in most cases to 15% of the selling price. An order was issued from Advance G. H. Q., on December 28th, which provided that no luxury tax should be charged to any member of the American or Allied forces in the territory occupied by the American army. This order caused considerable difficulty to the German Hauptzollamt, or Chief Tax Office, since there was no way to determine which taxable article had been purchased by Germans and which by members of the occupying forces. Wine was probably the principal article purchased by Americans which was subject to taxation. Hotel and tavern keepers early began to avoid payment of their wine taxes by claiming that their sales had been to Americans. The German authorities inquired, in the latter part of January whether or not Americans were to be required to pay the wine tax. The American authorities replied by merely sending a copy of the order relating to luxury taxes which has been mentioned above as being the only order until then issued on the subject. From the manner of this reply, the Germans assumed that the Americans regarded the wine tax as a luxury tax, and the matter was dropped.

On April 25, 1919, an order was issued from the headquarters of Marshal Foch which provided that the purchase of wine for military or officers' messes should be exempt from taxation, but that wine purchased by individual members of the Allied forces was liable to it. This was in order to assist the German Chief Tax Office. Purchases made for messes, canteens, etc., were to be accompanied by a stamped statement to the effect that they were not liable for the tax, which statement would be accepted by the German authorities in lieu of payment of the tax. Nothing was at first done about this order in the American area, but the Germans learned of its existence and brought it to the attention of the Officer in Charge of Civil Affairs, Third Army, with a request for information as to how the matter was to be treated.

This compelled the American authorities to decide upon a definite policy. It was very desirable that the German government should collect its wine tax unhampered, as the tax is a very considerable source of income, and the German income was to be used largely to pay indemnities. It was well known that the German wine merchants and tavern keepers were using the American order regarding luxury taxes as a ground on which to refuse to pay their own taxes. It seemed, however, impracticable to print tax coupons similar to those used in France, and to distribute them throughout the army, for it was felt that the Army of Occupation was not likely to continue to exist long enough to justify such trouble and expense. The American authorities were well aware that tavern keepers were generally charging the wine tax to soldiers on the same basis as to civilians. Members of the army generally paid what the merchant said the wine had cost, tax included, so that the order of December 28th did not, as a matter of fact, protect them at all, but served merely as an excuse to German wine dealers to avoid their just taxes. No prosecution of dealers for attempting to collect the luxury or wine tax from members of the Army of Occupation ever took place. When the request was received from the German authorities for information as to how the Americans proposed to treat the wine tax in view of Marshal Foch's order of April 25th, the whole matter was carefully considered again, and it was decided that it would be better to let things continue as they were, with the soldiers actually paying the tax, than to change the order and lay down a new policy at so late a date. On July 22nd, the Officer in Charge of Civil Affairs wrote to the Oberzolldirektion at Cologne, saying that since the wine tax was not a luxury tax, the whole matter was thereby settled, because the American authorities had issued no order prohibiting the tax on wine. He went on to point out again that the American authorities had never insisted that a wine tax should not be collected from members of their army. The Germans answered with a request to be allowed to publish a notice in the papers to the effect that the American army was no longer exempt from paying the wine tax. This request was granted on August 20, 1919, after considerable correspondence. The effect of the announcement was, that all tavern keepers, who had been previously including the wine tax in the price of their wine, immediately took occasion to raise their prices 20%, that is, to the amount of the tax. This caused great discontent among the troops and it was therefore decided definitely to exempt the military from payment of the wine tax. All restaurants were at the same time required to post their bills of fare and wine lists, together with prices in a conspicuous place, and it was further required that the prices on the wine list should be indicated both with and without the addition of the wine tax. The difficulty which the German tax officials had had in collecting the just wine and luxury taxes from the merchants, was met by a German order, which provided that upon sale of wine or of an article of luxury to a member of the American army, a coupon was to be presented to the purchaser, on which he should sign his name, state the article purchased, the price paid and the date of purchase. This coupon was to be accepted by the Chief Tax Office in lieu of payment of the tax. This order of September 15, 1919, effectually disposed of the question of taxing members of the American forces.

A considerable number of Allied nationals, businessmen and tourists came into the area in November 1919, and made a practice of refusing to pay the luxury tax. German businessmen were in doubt as to what course to pursue in these cases and, the question being submitted direct to the Office of Civil Affairs, they were informed that in conformity with a decision of the Interallied Rhineland Commission, exemption from the luxury and wine taxes applied only to members of the occupying forces, their families, and to members of the Interallied Rhineland Commission and their families. The matter was then definitely settled.

CHAPTER NO. 13

ARMISTICE INTERPRETATIONS AFFECTING AMERICAN OCCUPIED TERRITORY

Origin and Duties of the Armistice Commission; Release of Prisoners of War; Abandoned Enemy War Material; Material Surrendered in Accordance with the Armistice; the Neutral Zone; Restitution of French and Belgian Property.

Origin and duties of the Armistice Commission.—The third paragraph of the last article of the Armistice is:

"To assure the best execution of the present agreement, the principle of a Permanent International Armistice Commission is admitted. This Commission will perform its duties under the high authority of the military and naval Commander in Chief of the Allied Armies."

This paragraph provided for an Armistice Commission, but in no way defined the duties of this Commission. Clearly, however, it was to be the main liaison agency between the Allies and the Germans. All communications between the hostile powers passed through its hands, all complaints were taken up by it, all requests were submitted through it, and all interpretations of the Armistice were made by it. The Germans very early began to use the Commission as if it were an agency through which new agreements, modifying the terms of the armistice, might be made, and they submitted a number of requests to that end. The Allies did not agree with the German conception of the functions of the Armistice Commission, and in order that a clear understanding should be had, General Nudant, the French representative on the Commission, made the following statement to the German delegates on January 24, 1919:

"The German view-point of the offices of the Permanent Inter-Allied Armistice Commission is, that they are a place for agreements, whereas, my understanding is, that they are simply a place for arranging the detatils for compliance with agreements already entered into.

"Since the Armistice as originally drawn, and as extended on two occasions, is quite general in its terms, there arise many questions as to details.

"The Germans frequently bring up questions which involve a distinct departure from the armistice conditions as laid down. One instance of this was their urgent request for the return of German prisoners.

"It is readily understood that their arguments and discussions on such points are for the purpose of getting what they can, or saving what they can. At the same time, it is readily understood that conditions in Germany have changed greatly since November 11th, when the first armistice agreement was entered into. Most of the German pleas for consideration because of failure to comply strictly with the prescribed conditions are based on the impossibility of so doing, owing to the revolution."

The interpretations of the Allies of the duties of the Armistice Commission were of course accepted without more ado. As a matter of fact, there can be no doubt that General Nudant's explanation was a correct one, for it is inconceivable that the Allies should allow the Armistice Commission to modify the terms of the Armistice. At any rate, the Armistice Commission never was anything more than a liaison agency for the purpose of arranging as expeditiously as possible the details involved in carrying out the armistice. From time to time they took notice of German protests and rectified mistakes or abuses of subordinates, but this part of their duties occupied only a small portion of their time.

Release of prisoners of war.—Article X of the Armistice provided for "the immediate repatriation without reciprocity . . . of all prisoners of war, including persons under trial and persons convicted, belonging to the Allies and the United States."

Hardly had the guns ceased firing on the morning of November 11th, when Allied prisoners began to straggle over from the German line. These returning prisoners were in a pitiful condition. They were all ravenously hungry, and most of them in rags and indescribably filth. This exodus lasted for about a week, and the sight of the deplorable condition of the prisoners caused bitter resentment among the Allied troops. Some of these prisoners brought stories of terrible conditions of hunger in the prison camps from which they had been released.

The Allied members of the Armistice Commission protested vigorously against this treatment of their prisoners by the Germans and threatened reprisals. The Germans explained the condition of the prisoners by stating that as soon as the Armistice was signed, the Allied prisoners had practically mutinied, that they had demanded immediate release and refused to obey orders, but without waiting for proper transportation for their return, had escaped and made their way back to Allied lines. This, they said, accounted for the famished and ragged appearance of the prisoners.

As the weeks passed, other stories began to be heard about the starvation of prisoners who had not escaped but were still in German camps in the interior. New protests were made to the Germans by the Armistice Commission. This time the reply was received that the Germans were doing all in their power to feed and care for Allied prisoners, but that the utmost difficulty was being encountered, partly because the German government had, to a very large extent, lost its control over the country as a result of the revolution, and partly because of the conduct of the prisoners themselves, who refused any longer to obey orders or to submit to the regulation necessary for their proper maintenance and orderly release.

Some stories were also received to the effect that American prisoners had been brutally treated by German guards after the signing of the Armistice. These also were duly protested against in the Armistice Commission. It was not possible for the Allies to investigate the charges made nor to determine the truth about the condition of prisoners, until about the middle of December 1918. It was then discovered that the statements made by the Germans were, in fact, true. The stories of brutalities to American prisoners were found to be without foundation. The American representative on the Armistice Commission stated this openly in the Commission's meeting of December 17, 1918, in order to show the Germans that the Americans intended to deal with them honorably and to give them credit when it was due. General von Winterfeldt, the German representative, expressed himself as indebted to the Americans for this communication.

It was learned later that the Allied prisoners who, at the time of the Armistice, were interned near the front, had, in fact, broken out of the prison camps and returned to the Allies on their own initiative and without resources. They had for a more or less prolonged period been living on the German ration, which at that time was exceedingly scant for all except the fighting forces, and the additional strain of hunger and exhaustion which they underwent in their dash for freedom, was sufficient to account for the deplorable condition in which they arrived.

In regard to the shortage of food in the camps of interior Germany, it was found that the reports from German sources were again substantially true. At the time of the revolution, the greater part of all German organization broke down. This was particularly true of railways, and also of a good many of the German military organizations, where Soldiers Councils attempted to supersede the officers, with a consequent loss of efficiency. Furthermore, in a number of instances, food trains were broken into and robbed by the revolutionists in the

521572—43——17

first days of disorder. As a matter of fact, it has been established that the American prisoners were, on the whole, well treated in the German internment camps. To be sure, their ration was far from sufficient, but most of them agree that it was the same ration on which the German civil population lived, and many have said that, with the addition of the weekly packages received from the Red Cross, they actually fared better than the German troops who were guarding them.

Abandoned enemy war material.—The second paragraph of the sixth article of the Armistice is:

"No destruction of any kind shall be permitted, military establishments of all kinds shall be delivered intact; so, also, military stores, food, munitions and equipment, not removed during the period fixed for evacuation."

This clause clearly established that property abandoned by the German army in its retreat from the battle line to interior Germany was to become the property of the Allied armies as spoils of war. The property received in this manner was not considered as part of the material which the fourth article of the armistice required the German army to surrender to the Allies unless it was officially turned over and accepted. The abandoned property was in addition to such property so surrendered.

As the Third Army advanced into the Rhineland, it found along the roads, German equipment, wagons, trucks, etc., that had broken down during the enemy's retreat. This material, of course, became the property of the Third Army, but was of little value, since it was generally worn out. However, shortly after the American Army of Occupation settled within its area, it came to the attention of the military authorities that there was a substantial amount of war material, which obviously would come within the description of military stores, food, munitions and equipment, still within the American territory and in use in towns, hospitals, etc. An investigation was ordered by the Army Commander on December 12, 1918, for the purpose of determining how these military stores came into the possession of private citizens and public civil officials. These investigations developed the fact that the German army had discovered, during its retreat, that it would not be able, within the thirty-six days allowed for its evacuation, to get all its material out of the area to be occupied. Acting on instructions from Berlin, German unit commanders had sold whatever property they thought it impossible to take with them, to private individuals and public corporations for whatever price they could get, rather than suffer the property to fall into the hands of the Allies.

Probably the underlying motive of the Germans in thus disposing of this property, was simply to evade the terms of the Armistice.

Their official reason was quite different. They contended that they were merely giving the inhabitants of the Rhineland an opportunity to reap a share of benefits from liquidation of military stores—benefits which the rest of Germany would enjoy later when general demobilization was effected. They held that military stores which they sold to *bona fide* purchasers thereby ceased to become military stores and instead became private property and were therefore immune from seizure. They argued that there was nothing in the Armistice forbidding the German Army to sell its property during the period fixed for evacuation, and that therefore, if such a sale was made for value and not for the express purpose of defrauding execution of the Armistice, a valid title was passed to the vendee.

Had the Americans accepted the German point of view on this matter, every individual sale of military stores would have had to be made the subject of an exhaustive, critical examination, to determine whether it was made in good faith, or whether merely to avoid fulfilling the terms of the Armistice. The American authorities wished to live up both to the letter and the spirit of the Armistice, but they were determined not to allow the Germans to impose upon them. Therefore, on December 27, 1918, the Army Commander issued an order directing a careful investigation to be made to discover abandoned enemy war material. This was to be placed under guard and inspected by technical experts. As a result of the investigation made under this order, it was discovered that a large amount of property in the area apparently was of the sort described in Article 6 of the Armistice, and that a great deal of it was in the possession of persons and corporations, many of whom held bills of sale for it from the German military authorities. It was found that all the war material in the area could be divided into three classes:

1. That which had been abandoned by the German army;
2. That which had been transferred by the German army to a third party through the medium of a legal sale effected previous to November 11, 1918;
3. That which had been transferred by the German army to a third party through the medium of a sale effected on or subsequent to November 11, 1918.

Clearly the first class mentioned above became the property of the United States under the Armistice. Equally clearly the property mentioned in the second class was not transferred to the United States by the Armistice. Presumably the German government had a good title to it, and certainly the government had a perfect right to convey that title to a private individual, before it became bound by the armistice agreement. At the time of such a legal conveyance, the goods in question ceased to be military stores and became the property of an individual, which could not legally be touched. The type of property of the third class was more difficult to determine. The Armistice did not prohibit Germans from selling their military stores directly. They always contended that if there was any ambiguity in the terms of the Armistice, it should be decided in their favor, because during the Armistice of 1871 between the French and the Prussians, all points in which the terms of that Agreement were ambiguous, were decided in favor of the French. This was, however, because of a clause in the armistice of 1871 which specifically provided that any doubt as to terms should be decided in favor of the French; from this, the Germans argued that according to precedent, doubtful points should be settled in favor of the defeated side, whether specifically so provided or not. On the other hand, the Armistice of November 11, 1918, contained nothing at all about abandoned military stores, but merely spoke of "military stores, food, munitions and equipment not removed during the period fixed for the evacuation." Clearly, the property in question had not been removed during the period fixed for evacuation. The Germans argued that since the property had been conveyed to purchasers who acted in good faith, it had become theirs, and it would be wrong to take it from them. America did not accept the German point of view, holding that whatever might be claimed by German law officials as to the good faith of a particular purchaser, such purchaser was legally bound by the Armistice, which had been signed by his representative in his behalf. The Armistice made certain definite disposition of property described as "military stores, food and munitions," which might be in the area of occupation upon a fixed date, and while the American army would not question the right of the German army to remove this property during the period of evacuation, it asserted its right and seized the property after December 8, 1918, in spite of the fact that it had perhaps been sold to a German citizen. The American government maintained that the physical characteristics of the property remained the same and it was war material after the sale just as much as before, because as a matter of law, title to this property could not pass from the German military authorities to a German citizen, since the German government had said that that class of property would belong to the United States after December 8th, if left within the American area. The question of individual good faith on the part of the purchaser therefore became immaterial, as all goods passing under such sale were, by virtue of the Armistice, affected by a subsequent condition which defeated the title of the vendee in favor of the United States.

This interpretation was adopted and published in Letter of Instructions No. 8 of the Third Army on January 10, 1919, as follows:

1. The Commanding General directs that you have the following order published for the guidance of the civilian population of the

districts of Germany occupied by the American forces under your command:

2. Paragraph VI of the armistice, signed upon the 11th day of November, 1918, is in part as follows:

a. "Military establishments of all kinds shall be delivered intact; so, also, military stores, food, munitions and equipment not removed during the periods fixed for evacuation."

b. Pursuant to the above provisions of the Armistice, all military stores, food, munitions and equipment which belonged to the German Army upon the 11th day of November, 1918, the day of the signing of the Armistice, and which had not been removed from the zone of the American Army of Occupation during the period of time fixed by the Armistice for evacuation, are hereby declared forfeited under the terms of the Armistice, to the American Army of Occupation.

c. Considerable quantities of material of the kind set forth in paragraph *a*, and especially clothing, leather, iron, blankets, motor trucks, etc., were abandoned by the German Army prior to its evacuation or sold by it to individuals and cities after the Armistice went into effect. All such sales are hereby declared null and void. The title to all such property so abandoned or sold vests in the United States.

d. All individuals, corporations and municipalities having possession or custody of any property described in paragraphs *a* and *b* are hereby directed to turn over the same to the nearest military commander or to the Burgomaster, on or before noon of the 20th of January, 1919.

e. Any property described in paragraphs *a* and *b* of this order in the possession of any individual, corporation or municipality after noon on the 20th of January 1919, will be seized by the American Army and the owner and holder thereof brought to trial forthwith before a military court for violation of this order.

The Germans protested vigorously against this order, at first, to the Commander of the Third Army, from whom they received no response, and subsequently the Armistice Commission and to Marshal Foch himself. Both the Commission and the Marshal endorsed the decision of the Army Commander, and the Germans had nothing to do but submit. A good many individuals attempted to evade the necessity of turning over property and disobeyed the order, although by far the greater majority of them, realizing that the Americans had access to German papers which recorded the sales and consequently would be able to ascertain who had the property, complied with it.

Those who disobeyed the order were tried by provost courts or commissions and were sentenced. The property was confiscated. Many Germans thought that these persons were tried because of their purchase of German war material in the period between the Armistice and Allied occupation of the left Rhine bank, which purchase they believed the Americans had decided was criminally illegal. As a matter of fact, the Americans never prosecuted any one for purchasing war material; what they tried them for, was for failure to obey Letter of Instructions No. 8, directing them to turn over the property they had purchased. Nevertheless, acting on their false assumption that the Americans were trying German citizens because they had purchased war material prior to the occupation, the Germans filed a number of protests, not only with the Army Commander, but with the Armistice Commission and even with Marshal Foch, against the alleged illegal trials. It took several weeks and considerable correspondence before this misunderstanding was cleared up. Finally, to make our position in the matter perfectly clear, General Pershing issued Orders No. 22, Advance G. H. Q., March 30, 1919, as follows:

> Civilian inhabitants of occupied territory will not be prosecuted for retaining or removing German military material between November 11, 1918, and the date of the evacuation of the German forces from territory now occupied by United States troops, when it can be established that such action on their part was in obedience to the orders of the German government, or when they have been merely holders of German war-material found in their possession.

Disposal of this enemy war material, collected as a result of the foregoing orders, proved no small undertaking. A special board called "The Enemy War Material Board," operating under G–4 of the Army, took the matter in hand.

A great mass of material of all descriptions and value was reported throughout the area. It soon became apparent that to collect all this from many different places, would prove a complicated matter and not in the interests of economy. For example, much of the property was of such little value that the cost of moving and guarding it would offset any benefits which might accrue to the army through its collection. Large numbers of horses, in such poor condition as to forbid any further service, had been left behind by the German army in its retreat. They had in some cases been abandoned, in others, sold, to the civilian population, mostly the farmers, who nursed them back to strength and put them to work on their farms. To have taken charge of these horses, for example, would have meant the cost of feeding them, the necessity of caring for them, etc., but further and more important, it would have deprived the farmers of necessary assistance in their efforts to produce those foodstuffs which were needed in Germany.

In view of the above and because it was the intention of ultimately disposing of most of this property through sale, it was decided in most cases, particularly with regard to horses and property of little real value, to permit the holders of them to retain possession, after declaring them to the Army authorities and receiving instructions to care for and guard them. In order to protect individuals who retained their possession from trial by the Army for failure to comply with

Letter of Instructions No. 8, papers were left in their hands as evidence that, for the particular property mentioned, possession by them was authorized, pending its final disposition by the Army. Lists were then compiled in each Kreis or Division area of this class of enemy war-material and forwarded to the Enemy War Material Board in Coblence.

There were, then, three classes of this enemy war material; that actually collected in dumps by the Army, that actually in possession of private parties under the conditions mentioned in the last paragraph, and that not reported nor turned in according to Letter of Instructions No. 8.

One source of information with regard to the latter class proved to be the records of sales of enemy war material kept by various agents of the Berlin government, many of whom were civil government officials, such as Landräte and Burgomasters. By far the greatest amount of information, however, was obtained from the records of the VIII German Army Corps, whose headquarters had been in Coblence. Inasmuch as most of these records, however, were discovered previous to the date of issue of Letter of Instructions No. 8, it is needless to say that the amount of property disclosed by them was not so great as otherwise might have been expected. The mere knowledge on the part of the civilian population that the Army had obtained access to some or all of these records undoubtedly led to voluntary reports which otherwise might not have been made. These records also proved particularly valuable in that they contained, in addition to the date of sale, (a vital matter in itself), the amount of money realized by the German army from each article sold.

With the latter information at hand, it was but one step further to seize the money which these agents had collected in lieu of the property itself. When this was done, the private purchaser was at the same time given a title to the property by the Army. In other words, the Army by this means did not disturb the existing possession of the property, but did receive the money derived from the sale, instead of the German government. Although this policy had advantages, it also had disadvantages. For example, "A" in good faith bought through his Burgomaster a broken-down wagon from the German army. He sold it to "B." "B" sold it to "C." Now in going over the records of the Burgomaster, it was found that "A" had bought this wagon, consequently the money was seized. Also "C" had by now reported his wagon and no records disclosed the fact that "C" had bought the wagon from the German army, consequently "C's" wagon was seized. Here, then, the army had seized *both* the property and the money. There proved to be many such cases and they were difficult to handle. In order to straighten this matter out, the following order was issued:

ENEMY WAR MATERIAL

(a) Where material, previously purchased from the German Army or any regularly constituted government or municipal authority acting for, or in behalf of, the German Army, within the American zone of the occupied area, has been confiscated from any person or corporation, and where the American Army has also taken the money which was paid for the said material from the Kreis concerned, and where the property has not been returned or resold to the purchaser, the person or corporation from whom the property was confiscated may present a claim for the amount paid for the said material. The claims, together with proper receipts and certificates, must be presented to the Officer in Charge of Civil Affairs in the Kreis concerned, within ten (10) days from the date of this notice.

(b) This notice will be brought to the attention of the civil population by the Officer in Charge of Civil Affairs through the Landräte concerned.

(c) All such claims will be investigated by the O. C. C. A., and if considered just, forwarded within 48 hours to G-4, (Munitions), 3rd Army.

The American authorities disposed of the property abandoned or sold by the retreating German army as rapidly as possible by sales. The money received as proceeds therefrom was stored in the vaults of the Headquarters building at Coblence and after being deposited to the credit of the United States, was reinvoiced back to the army and used to pay the troops, as already described in Chapter XII. By the middle of August 1919, all of the enemy war material had been sold or otherwise disposed of by the army, and a receipt was in the hands of every authorized possessor of this class of property. The magnitude of the task accomplished may be realized from the fact that about 10,000,000 marks were obtained from these sales.

Material surrendered in accordance with the Armistice.—Article IV of the Armistice Agreement of November 11, 1918, is as follows:

Abandonment by the German Army of the following equipment in good condition:

5,000 guns (2500 heavy; 2500 field)
25,000 machine guns
3,000 Minenwerfer
1,700 fighting and bombing planes.

First, all D-7's and night bombing planes, the above to be delivered *in situ* to the troops of the Allies and of the United States in accordance with the detailed conditions laid down in Supplementary Note No. 1, fixed at the moment of the signing of the armistice.

Paragraph 3, of Article VII, states:

> There shall be delivered to the Allied and associated powers:
>
> 5,000 locomotives and 150,000 cars in good working order, with all necessary spare parts and fittings, in the periods fixed in detail by Supplementary Note No. 3.

There were in addition many clauses dealing with the surrender of ships and naval material which did not concern the Army of Occupation directly.

Article IV, Supplementary Note No. 2, of the Armistice Agreement regarding material, contained the following:

> (a) *Rolling-stock.*—The rolling-stock delivered to the Allied Armies in the zone included between the present front and line No. 3, not including Alsace-Lorraine, shall be in importance at least equal to:
>
> 5,000 locomotives
> 150,000 cars.
>
> This surrender shall be made in the periods fixed by clause 3, Article VII, of the armistice, and under detailed conditions which shall be decided upon by the Permanent International Armistice Commission.
>
> All this material shall be in good usable condition, and provided with all the usual spare parts and fittings. It may be used with its own or other personnel at any part of the railway systems of the Allied Armies.
>
> The material necessary for the maintenance of the railway system of Alsace-Lorraine shall be placed at the disposal of the French Army.
>
> The material left in place in the countries on the left bank of the Rhine up to the bridgeheads on the other side, must be sufficient to permit the normal operation of the railways in these territories.

Paragraph IV, Article VII, of the armistice follows:

> There shall likewise be delivered 5,000 motor-lorries in good condition within a period of 36 days. The railways of Alsace-Lorraine shall be handed over in a period of 31 days, together with all personnel and material organically connected with this system.

The Germans were remarkably slow in carrying out the above quoted clauses of the armistice, and continually attempted to foist material on the Allies that was not in good condition. On the other hand, the Allies subjected every piece of material to a rigid inspection and accepted nothing that was not in first class condition. As a result of this, on December 16, 1918, which was the last day of the 31-day period agreed upon on November 11th, the Allies had accepted only 2,000 locomotives out of the 5,000 due, and 20,709 cars, instead of 150,000. The Germans claimed that, as a result of the war, practically none of their locomotives were in such condition that the Allies would accept them. They stated that the difficulties of the revolution prevented their delivering the material on time; that they could not find men to do repair work; that the receiving facilities of the Allies were not adequate; that the personnel who delivered the surrendered material to the Allies were so badly treated that it was almost impossible to induce men to undertake its transport.

Not only were the Germans dilatory in delivering railway transportation, but they were equally so in delivering other kinds of war material. At times, they even endeavored to surrender obsolete cannon without recoil cylinders. They failed to turn over the required number of aeroplanes, and a compromise was offered, in which the Allies agreed to accept twenty heavy draught horses for every aeroplane not surrendered. The delivery of motor lorries was particularly delayed. This was probably because most of the German army lorries belonged to private German business firms which the government had subsidized in time of peace and requisitioned during the war. As soon as the war was over, a good many of the firms got their lorries back, and it became very difficult for the government to reacquire them in order to surrender them.

By February 6, 1919, the Allies had received and accepted 3,861 locomotives and 118,698 cars. The Germans asserted that in order to get this number accepted, they had offered 9,000 locomotives and 200,000 cars, and that the locomotives which had been accepted, represented two-fifths of their entire stock of serviceable engines. They also said that the delay was in part due to the fact that the different Allied committees for the reception of surrendered war material demanded such different standards of excellence, that they were never certain whether an object would be accepted or not.

As the full quota of material to be surrendered by the Germans was approached, the Allies made an arrangement among themselves for division of the property. Under this arrangement, Belgium was to receive one-tenth of all the war material surrendered; America was to receive two-tenths; England, three-tenths, and France, four-tenths.

By February 13, 1919, the Americans had received their full share of guns and machine guns, and by February 19th, of motor lorries. By March 1st, all the railway material, including that necessary for maintenance of the Alsace-Lorraine line, had been delivered, and by March 13th, all the motor lorries due.

The neutral zone.—Paragraph 3, Article V, of the Armistice prescribes that:

> A neutral zone shall be set up on the right bank of the Rhine between the river and a line drawn parallel to the bridgehead and the river and ten kilometers distant, starting from the Dutch frontier and extending to the Swiss frontier.

The Armistice also provided in Article V that the countries on the west of the Rhine should be assured by garrisons holding the principal crossings of the river (Mayence, Coblence and Cologne), together with bridgeheads at these points of a thirty-kilometer radius on the right bank. The bridgeheads were thus originally prescribed as segments of circles, whose centers were the centers of the towns mentioned and whose radii were thirty kilometers in length. It was discovered shortly after the Armistice that if these bridgeheads were occupied as thus provided for, considerable administrative difficulty would arise. For instance, the Mayence bridgehead would cut into the Stadtkreis of Frankfort on Main in such a way as to paralyze the efficient administration of that city. Consequently, the perimeters of the bridgeheads were modified somewhat to conform to the expediency of local administration. An indentation was made in the Mayence bridgehead excluding the Stadkreis of Frankfort, and the perimeter of the Cologne bridgehead was modified to conform as nearly as possible with the administrative boundaries of the Kreise and Bürgermeistereien in the area. The Coblence bridgehead, however, was maintained as laid down in the Armistice, its border remaining the arc of the circle.

In th early part of 1920, when it seemed likely that peace would soon be ratified by the Senate, it was decided to modify the boundary of the American portion of the Coblence bridgehead. This was because of the fact that a good many Bürgermeistereien and several Kreise were arbitrarily cut by the boundary of the occupied zone as laid down in the armistice, and because the administration of these political areas was consequently rendered exceedingly difficult. The Army therefore arranged to modify the boundary of the bridgehead by ceding portions of the area to the Germans and accepting portions in return which had formerly been in the neutral zone, so that the new boundary, which would be in existence during the ten years of occupation provided by the peace treaty, would conform to the boundaries of German administrative districts. This alteration, however, had not been accomplished when the Inter-Allied Rhineland High Commission assumed control.

The question of control of the neutral zone lying between occupied territory and unoccupied Germany was not taken up in the Armistice. As soon as the occupation had been established, Marshal Pétain, Commander of the French Armies, submitted the following memorandum on the subject to Marshal Foch, Commander in Chief of the Allied Armies. The suggestions contained therein were adopted in their entirety, and the neutral zone was thenceforth governed under their provisions.

SUGGESTION FOR A NOTE RELATING TO CONTROL IN NEUTRAL ZONE

Execution of Note No. 7281, from the Marshal Commander in Chief of the Allied Armies, November 29, 1918.

I. GENERAL ARRANGEMENTS

1. In each army and in accordance with the general conditions fixed by the above mentioned note No. 7281, control will be enforced by officers provided with an order, either permanent or special, issued by a General in command of an army, or on his authority.

These officers may be accompanied by the personnel necessary for the accomplishment of their task. They may present themselves without being announced in advance, and will do so under a flag of truce. Orders will be given by the German military authorities to allow them to move about freely upon presentation of their passes.

II. The number and the duration of missions are unlimited.

III. In the exercise of control, the officers entrusted with these missions will appeal to the civil authorities and to the commanders of police sectors, to be mentioned later in Article VIII. These agents will offer every facility necessary for the accomplishment of the missions of these officers.

IV. The officers entrusted with missions may ask any one of the authorities mentioned in the preceding article for any papers that they judge necessary. These papers will be signed, and responsibility will rest with the signer. Officers on missions may also ask for certain other papers, to be mentioned later. Lastly, they may require that all necessary verifications, identifications and instructions be carried out according to their orders and in their presence, even if individual dwellings are concerned, and that no objection be opposed in these matters.

V. In case a part of the community lies in the neutral zone, and the capital of the community is outside of that zone, officers on missions may proceed freely to the capital under the conditions fixed in Article I, above.

II. LIMITS AND ORGANIZATION OF TERRITORY IN NEUTRAL ZONE

VI. The eastern limit of the neutral zone is fixed by the attached map.

VII. The neutral zone is divided into police sectors outlined in the attached map.

VIII. Command in each sector is exercised by a field officer, who is assisted by an interpreter. In everything concerning control on behalf of the French Armies, commanders of the police sectors will correspond direct with the French Command.

III. MILITARY GOVERNMENT IN NEUTRAL ZONE

IX. No modification may be made without previous authorization of the French Command. As soon as the rules for government of the neutral zone come into force, commanders of police sectors will make known the distribution of troops of their sectors in the different towns and villages. Later modification will be made the subject of a special report. A report will likewise be drawn up in case of a relief. In case use is made of the police forces for the maintenance or reestablishment of order, a report should be sent without delay to the commander of the sector.

X. Inasmuch as the German troops that are allowed in the neutral zone are retained there for policing it, steps will be taken without delay to dissolve civil guards. However, municipalities are authorized to maintain police agents to the limit of the personnel as it existed on August 1, 1914, on condition that a list of the agents be sent without delay to the local French command, and that each agent be provided with a regular uniform bearing distinctive and indubitable insignia. In case of necessity, local civil authorities may ask the French command to consent to abrogation of the present article.

XI. German troops in the neutral zone having merely police duties may not perform any military manoeuvers. They may not practice shooting, even at barracks, or throwing grenades or any other weapons.

XII. No work shall be done in the neutral zone that may be considered as contributing in any way, even indirectly, to the offensive or defensive equipment of the zone, not excepting work done by civil workers. Especially there shall be no placing nor removal of railways, no change in the existing systems of railways, even of narrow-gauge systems. In case work appears necessary or urgent, an authorization for it may be asked of the local French command. No beginning of such work may take place before the authorization is granted.

XIII. Supplies of all kinds existing in the neutral zone, without distinction as to whether they belong to the State or to individuals, will be immediately listed with the local French command, who may demand that they be given up if they are capable of use for military purposes. The French command reserves to itself the right to determine these matters.

XIV. The use of optical or other signals is prohibited in the neutral zone. This prohibition applies both to soldiers and civilians. No wireless station will be maintained or established in the neutral zone without authorization of the local French command.

XV. Pigeon-cotes, either civilian or military, are prohibited.

XVI. Aviators are prohibited from flying over the neutral zone.

XVII. No camouflage may be carried out in the neutral zone, not even for practice.

IV. CONTROL OF POLICE GARRISONS

XVIII. The commander of each police sector will make known in writing within twenty-four hours after these regulations take effect, a list of personnel and the distribution of local police forces in the sector. These forces may not exceed the normal strength of the territorial police in time of peace. No increase in the personnel may be made without previous authorization from the local French command.

XIX. The commander of the sector must have a list in his possession, kept accurately up to date, of all soldiers composing each garrison. This list must be presented to the officers of missions upon demand. Officers of missions may also ask to see all food tickets, lists of ration strength or any other administrative tables whatever.

XX. Within the limits fixed by Article XVIII, the commander of each sector will make known in writing the civilian population in each of the communities of his sector. The inhabitants of each community will be listed according to profession and name. This list will be kept up to date and presented to officers of missions whenever they see fit to require it.

XXI. Any increase in population of a community will be brought to the knowledge of the local French command as soon as that increase passes the number of 50. A list of names of transients will be kept in each community. This list will be presented whenever required by officers of missions.

XXII. Any increase in the population considered as unusual by the French command, may be construed by him as constituting a violation of Note No. 323–2 of the Marshal Commander in Chief of the Allied Armies.

XXIII. Officers on missions may make investigations in industrial establishments to find out whether the persons listed as workmen may not be soldiers brought into the neutral zone in excess of the numbers fixed by Note No. 323–2, previously cited.

XXIV. Demobilized German soldiers may not be authorized to enter the neutral zone nor to take up their residence there, unless they had lived there before August 1, 1914. Return of every demobilized soldier should be immediately made known by the commander of the sector to the local French command. The same applies to the departure of any demobilized soldier whose residence has been in the community. A list of demobilized men will be kept in each community, as well as of soldiers who have been demobilized and whose legal residence is in the community.

XXV. German soldiers coming into the neutral zone on leave will present themselves upon arrival to the commander of the police sector. The names of these soldiers will be written on the list of transients mentioned in Article XXI, above. Rank, organization and length of stay will be stated opposite each name.

V. ARMS AND AMMUNITION

XXVI. Importation of arms and ammunitions into the neutral zone is prohibited. This prohibition applies to arms and ammunition of all kinds; side arms, fire arms, artillery ammunition of all calibres, cartridges, grenades, fireworks, signalling flares, etc.

XXVII. Arms and ammunition for use of the army will be evacuated to the rear of each community of the neutral zone without delay. Arms and ammunition belonging to civilians must be given up to the authorities and kept in store by the authorities. A table of the arms and ammunition in storage in the community will be drawn up and given to the local French command, who may order their evacuation to the rear.

XXVIII. Arms and ammunition belonging to police troops will be kept by them. However, these troops will keep only their regular arms and the ammunition forming the normal supply, i. e., ammunition carried by a man and that carried by a combat train. A list of the arms and ammunition kept for this purpose will be drawn up by the commanders of police sectors.

VI. GENERAL ARRANGEMENTS

XXIX. The local French commander must be kept informed of any indication or incident that might presage excitement or conflict, and of all preventive or repressive measures taken by German authorities.

XXX. Hunting in any form is prohibited in the neutral zone. Should it become necessary to take steps to destroy predatory animals, these steps can be done only upon authorization of the local French command and under conditions fixed by him.

XXXI. Fireworks are prohibited.

XXXII. Passage of boats from one bank of the Rhine to the other is prohibited for civilians, other than regular boatmen.

XXXIII. Telegraphic or telephonic communication between the neutral zone and territories occupied by the Allied Armies may take place only upon authorization of the local French command. The French command reserves the right to control the communications above mentioned. Authorizations already granted may likewise be revoked.

As might have been expected, the Germans protested against these regulations. They objected to control of the circulation of German civilians, on the ground that that was "none of the Allies' business." Marshal Foch insisted on it, however, pointing out that it was necessary for him to have a control on immigration into the neutral zone in order that he might prevent the bringing in of German soldiers in civilian clothes in sufficient numbers to endanger the safety of the bridgeheads in the event of the reopening of hostilities. The Germans also protested against the unannounced arrival of Allied officers on tours of inspection. Marshal Foch replied that this was necessary in order to detect possible construction of fortifications, etc., because if the visits were not made unexpectedly, it would be possible for the Germans to camouflage any military works on which they might be engaged.

The Allies, and particularly the French, were caused a great deal of inconvenience by the small strip of neutral zone, not more than a kilometer wide in some places, which lay between the Coblence and the Mayence bridgeheads. The Regierungspräsident of Wiesbaden and some of the inhabitants of this strip of neutral zone petitioned the Allies to occupy it, in order to eliminate the difficulty that was had in the distribution of food, fuel, etc. The Allies did bring up the question several times before the Armistice Commission and requested that this narrow strip be surrendered to them. The German representatives, however, consistently refused to do so, on the ground that the inhabitants of the district did not wish it occupied. As the terms of the Armistice on this point were explicitly in favor of the German point of view, the latter carried the day.

Early in March 1919, the industrial situation in unoccupied Germany had become very critical and Bolshevism was spreading rapidly, particularly in Düsseldorf and the adjacent Ruhr district. In order to combat this menace, the Allies permitted the Germans to strengthen the forces maintained by them in the neutral zone. A month later, when order had been restored, the Germans petitioned that they be allowed to increase their garrisons in unoccupied territory to equal the number of troops that had been maintained there prior to the war. This petition was denied as unnecessary and possibly dangerous. It was found later that the German forces in the neutral zone, though less than the number allowed by the Allies, were composed almost entirely of non-commissioned officers acting as privates. When the Allies protested against this, the Germans replied by saying that they were unable to recruit sufficient soldiers from the ranks, that since service as non-commissioned officers was a profession in Germany, it would be unfair to allow men who had devoted their life to the work to lose their occupation as a result of the country's defeat.

Restitution of French and Belgian property.—Germany, during the war, despoiled many factories and manufacturing plants in northern France and Belgium of their machinery, tools, material, etc. This was done with a double purpose; first, to obtain material necessary for war work in Germany as cheaply as possible; second, so to injure the industrial efficiency of France and Belgium, that when peace should come, neither of these countries would be able to compete successfully with Germany in commerce. The extent to which this exploitation was carried on, may be gathered from the following letter of the American representative on the Commission of Industrial Restitution:

> 1. *Evolution.* There were two main periods of the exploitation of Northern France and Belgium,—that in the early part of the war, when requisitions were made by various military units to meet immediate needs, and a later period beginning 1915, but only completely organized in 1917, when requisitions were made to meet more general needs. In the latter part of 1915 there were a number of Industrial Committees (Wirtschaftsausschüsse) attached to the higher staffs to exploit the (a) agriculture, (b) forests, (c) and the industries of the occupied regions. These had the power of requisition, and exercised it to meet the demands of German industry.
>
> 2. Alongside of these committees worked delegates from the Ministry of War who had about the same powers. The demand for materials from the occupied regions became so great, that in 1917 it was decided to create a higher organism which would tend to centralize the whole. By the spring of 1918, it may be said that the system was completed which, based on the carefully compiled lists of materials available, was destined to permit full exploitation.
>
> 3. *Organization.* The listing and requisitioning was made by (1) the Ministry of War, (BDKM) within Germany, by the Quartermaster General at the front, and by the military governor of Belgium in that country.
>
> 4. The Ministry of War had a number of organs charged with this work, the principal of which was the *Wumba* (Waffen- und Munitionsbeschaffungsamt). This society was divided into sections, one for handling plants and large machines, another for tools and electrical machinery, a third for raw materials, and another for booty of war.
>
> 5. The service of the Q. M. General was divided by regions—five for the entire front, which also included the rear (Etappen). At the head of this system was Colonel von Mueller of the Wako. In each region was a representative of the Q. M., another from the Ministry of War and finally a number of military organizations (aviation and artillery), which had limited powers of requisition.
>
> 6. *Its Functioning.* Wumba was the point of contract between the German manufacturer and the Government. According to a document signed by Moeller (one of its members), "Wumba is simply an intermediary with powers to meet the needs of German industry and at the same time to fix prices." It was organized as a commercial firm, touching a 5% commission on all transactions, with smaller profits from sales of iron in bulk and from the use of prisoners of war.
>
> Wumba had complete lists of machinery available and put these at the disposition of the German buyer. He picked the machines he wanted, settled on the price with Wumba and then entered into relations with the military organization which had powers to requisition them.
>
> 7. The demounting or demolition was done by one of five firms recognized by Wumba. These firms had expert personnel

comprising German engineers, adjusters, and as many prisoners of war as they needed. To one of these firms, the same engineer Moeller wrote on November 17th:

"We have recently become more and more inclined to carry out the complete demolition of Belgium. I feel no hesitation in energetically exploiting Belgian territory at present. We are now in position to accept large orders for demolishing industries in Belgium. I leave you therefore quite free to receive orders."

8. The value of requisition of machinery was fixed at what it had been in July, 1914, provided it had undergone no depreciation. The price was always arbitrarily fixed and was always far below the real value of the machine.

9. The profits realized by the demolition firms, as their books show, averaged 20%.

10. These totals are but partially complete. In fact, the Germans themselves do not know the full extent of the exploitation in Northern France and Belgium. It was not until March, 1918, that the Wumba really centralized the demands from the German buyers and worked in complete coordination with the organs, with powers of requisition. The result is, that data in regard to the machines is scattered in a dozen different offices in Berlin. There are machines which were turned over to manufacturers without record having been taken of the transaction, and there are manufacturers who, as the big iron manufacturer, used their influence to secure machines outside regular channels.

The people of the Allied countries, especially France and Belgium, always desired that Germany should be made to pay damages for the destruction wrought in France and Belgium. Nothing concerning this was mentioned in the Armistice Agreement of November 11, 1918. All that appears in this document concerning the payment of damages, is contained in Article XIX, which follows:

XIX. *Financial Clauses.* Without prejudice to future claims and demands on the part of the Allies and of the United States:

REPARATION OF DAMAGES

While the armistice lasts, no public securities which can serve as a pledge to the Allies for the recovery of or reparation for war losses, shall be removed by the enemy.

Immediate restitution of the cash deposit in the National Bank of Belgium, and a general immediate return of all documents, money, securities, personal and fiduciary, together with facilities for the issue thereof, belonging to public or private interests in the invaded countries. Restitution of the Russian and Rumanian gold taken by the Germans or delivered to them by Russia or Roumania. This gold will be taken in charge by the Allies until the signing of peace.

The plan of restoring the material taken from France and Belgium was first officially broached to the special Economic Commission at Luxemburg, December 22, 1918, by Marshal Foch. The American delegate on this Commission asked the Marshal under what clause of the Armistice such a restitution could be compelled. The Marshal replied that the phrase "reparation of damages" in Article XIX, above, was sufficiently broad to cover the matter.

On January 16, 1919, the protocol extending the Armistice for a second time was signed by the Germans and the Allies. The Allies at that time insisted on inserting clauses requiring specifically that Germany should restore what she had taken. The Germans objected. They offered to pay an indemnity for the property removed, but desired to keep the machinery as necessary to their commerce and industries. The Allies were obdurate on the matter, however, and the following paragraphs were finally included in the protocol of January 16, 1919.

RESTITUTION OF MATERIAL REMOVED FROM BELGIAN AND FRENCH TERRITORY

1. The restitution of material removed from Belgian and French territory being indispensable to the starting up of factories, the following measures will be taken:

2. Machines, pieces of machinery, tools, industrial and agricultural, parts of every kind, and, in general, every industrial or agricultural tool removed from territory occupied by the German Army on the western front, under whatever pretext it may be, by German authority, military or civilian, or by German individuals, will be held at the disposal of the Allies, to be sent back to the place of origin, if the French or Belgian Governments so desire. These objects will not be subjected to any changes or damage.

3. To prepare for this restitution, the German Government will immediately give to the Armistice Commission, all official accounts and memoranda relative to the objects, all contracts of sale, contracts of lease or otherwise, all correspondence which has reference to them, statements and other papers of value in regard to their existence, origin, transformation, actual condition and present location.

4. Delegates of the French and Belgian Governments will proceed to Germany for inspection and examination on the spot of the specified objects, if this appears to them to be necessary.

5. The return of these goods will be carried out according to specified instructions, which will be given by the French or Belgian authorities as they shall decide.

6. In particular, there shall be declared, with a view to immediate restitution, depots of all sorts in parks, at anchor, in boats or in factories, also belts, electric motors, parts of motors, or parts of sails taken away from France or Belgium.

7. The accounts mentioned under #3 and #6 will be begun after eight days, dating from January 20, 1919, and will be furnished by the 1st of April, 1919.

As soon as the protocol of January 16th had been agreed upon, a subcommission was appointed by the Inter-Allied Armistice Commission, known as the "Service of Industrial Restitution", or "S. I. R." The members of the S. I. R. held their first meeting at Spa on January 22, 1919, and proceeded at once to organize for their tremendous task. On January 24th, they required the Germans to issue a decree ordering all German citizens in possession of any machinery, parts of machinery, or agricultural implements, requisitioned in France or Belgium, to report them to the German Government authorities. These instructions were at once complied with, and a decree was issued February 1, 1919, which called for all reports to be made before February 20th.

Carrying out the provisions of the protocol of January 16th, the S. I. R., early in February, passed a resolution requiring the German government to furnish the Commission with forms made from official records, showing the exact place in which requisitioned French and Belgian material was situated. They also provided at the same time French and Belgian representatives of the Commission should make tours of inspection throughout Germany, seeking for machinery which had been removed from their countries. The expense of these tours of inspections was to be borne by the German Government.

The S. I. R. finished its work of organization on February 5th, and moved to Wiesbaden to commence arbitrations. Simultaneously, a corresponding German committee was organized at Frankfort. The duties of this latter committee were to facilitate the collection by the German Government of French and Belgian machinery from individual Germans.

There were many delays in restoring this property, and the date on which it had to be declared was extended from time to time until it was April 20th, instead of February 20th. The German excuses claimed that the delays were due to the lack of governmental organization resulting from the revolution, lack of transportation, etc. The French and Belgians were inclined to give the Germans a taste of their own medicine in the matter, and at first insisted on immediate return of the property, regardless of possible effects on industrial conditions in Germany. By the beginning of March, it was seen that the restoration of the property was not a matter of concern for Americans, but rather for the French and Belgians alone; and the American representative was accordingly withdrawn from the S. I. R. Thereafter, he made occasional trips to Wiesbaden, but more as a liaison agent than as an actual member of the commission.

Shortly after the commission was organized, a number of Germans who, in accordance with instructions received from their government, had been actively engaged in removal of material from France and Belgium during the war, were arrested. The Germans protested vigorously against their arrest, claiming that it was a violation of that clause in the armistice which provided that no one should be punished for an act of war perpetrated before November 11, 1918. They claimed that the removal of French and Belgian property was necessary as a military measure and consequently was an act of war. On April 8th, the Allies definitely pledged themselves not to prosecute any one for having removed property in pursuance of orders, and to release any German individuals who might be in captivity awaiting trial on such a charge. This agreement, of course, did not protect Germans who, on their own initiative, and for their own benefit, had taken property from France or Belgium while members of the occupying forces. The conduct of these latter was not an act of war, but common pillage, which is not permitted under the laws of civilized warfare.

During the period of Allied occupation of the Rhineland, two Germans living in the American area were accused by the French of specific acts of pillage. In both these cases, the American authorities decided, after proper evidence sustaining the accusations had been submitted, to permit the houses of the accused to be searched. In one case, none of the property claimed to have been stolen, was found, and no further action was taken. In the other, some of the stolen property was discovered, and the thief, a Captain, was delivered to the French military authorities for trial on the charge of violating the laws of war. He committed suicide in jail before trial.

The Germans entered frequent protests that the French and Belgian officers, while on tours of inspection in Germany to discover stolen machinery, were using their official position to obtain trade secrets. The American authorities thereupon issued strict orders requiring these inspectors to notify the Americans before they made a tour of inspection in the American area, and requiring them to refrain from prying into business secrets.

When the Armistice was again extended by the protocol of February 16, 1919, it was provided that the Germans should return the railway fixtures which they had taken from France and Belgium during the war. At the same time, an active campaign was started to recover live stock, horses, cattle and chickens, which had been taken away. The Germans protested against this latter action, claiming that the live stock was properly requisitioned as a necessity of war, and maintaining that as it had not been specifically agreed that it should be returned, they had a right to keep it. No consideration, however, was paid to this point of view.

About this time, the Germans entered still another protest against their designation by the Allied press as "thieves," "robbers," etc. They claimed that inasmuch as the blockade of the Allies had prevented their purchasing material from abroad, they had been compelled by military necessity to requisition property from the Allied lands in which they had had control, and that they were justified under military law in so doing. As might have been expected, the Allies sharply disagreed with this point of view.

On April 11, 1919, four sub-committees were attached to the S. I. R. at Wiesbaden and placed under its control. These committees were:

(a) Committee for Return of Industrial Machinery.
(b) Committee for Return of Railway Material.
(c) Committee for Return of Agricultural Machinery, Implements, Horses, Cattle, etc.
(d) Committee for Return of Furniture and Furnishings of all kinds.

The Allies caused a census to be taken of the horses and live stock in the occupied area. The Germans objected very strongly to this, maintaining once more that live stock could properly be requisitioned in war time, and arguing that the removal of such live stock from private persons in occupied territory was a violation of the right of private ownership guaranteed in the Armistice. The work of recovering live stock progressed very unsatisfactorily. For instance, 80,000 head of cattle had been requisioned in Belgium, and of these, only 8,000 could be recovered. This was because many had been slaughtered, and also because they had been so mixed in removal from one district to another, that they could not be identified.

The Germans were accused by the French of intentionally delivering harvest machinery first and planting machinery later, too late, in fact, for the planting of the spring crops. The Germans denied this, claiming that they sent forward what they could get.

It was soon found that the Germans would be able to restore only a part of the machinery removed, and the question arose of how they should pay for the rest. It was impossible to estimate the cost of each machine, partly because of the enormity of such a task, and partly because the price of all kinds of machinery had very greatly increased between the time of their purchase and the spring of 1919. It was finally decided that the Germans should pay on the basis of the cost of an average machine on August 1, 1914, averaged with the cost of the same machine on April 1, 1919.

The French and Belgians demanded immediate return of their machinery. German workmen, on several occasions, threatened to demolish the machines rather than see them removed. Such threats as these induced the Allies to modify their policy. Even from the very beginning they had been much more yielding in their demand for the return of machinery which had been set up in operation in Alsace-Lorraine and the Saar districts, where the French were anxious to conciliate the population. Machinery in those districts was permitted to remain *in situ* until some business arrangement could be arrived at which would hurt both parties as little as possible. The French realized that although the machinery had been unlawfully removed, it would be unwise to remove it in an arbitrary manner, since it was now in the hands of virtually innocent purchasers.

A very curious situation arose with regard to property in Alsace-Lorraine which had been requisitioned by the German government during the war. Since these provinces were at that time German territory, the whole matter appeared to be a problem of internal German administration. Nevertheless, when Alsace-Lorraine had passed to France, the Allies required that this property be restored. Fortunately the larger part of it was church bells which had been taken into Germany and stored as a metal reserve. Inasmuch as these bells had not yet been melted up, it was easily agreed that the Germans should return them upon receipt of the price paid for them when they were requisitioned.

Gradually, the lenient and businesslike system of restitution which the French had inaugurated with respect to property in Alsace-Lorraine and the Saar valley, was extended to that in Germany itself. The French permitted the German firm in possession of the machinery and the French firm who owned it, to make agreements as to its restoration to their mutual advantage. An example of such an agreement is given below:

A French firm agreed that the Germans might keep their machinery, provided that the Germans deposited money to the value of the machinery in a French bank; that the Germans should supply equivalent machinery within eight months; that the French firm should not be obliged to show the Germans their plans for reestablishing their factory; that until the machinery arrived, the Germans should supply the French with a certain portion of their produce at cost.

Such agreements as these were not disadvantageous to Allied businessmen, were even an advantage to the Germans and at the same time tended to stabilize the economic conditions in Europe, which after the war, were very precarious.

CHAPTER NO. 14

AMERICAN SUPERVISION OF GERMAN CIVIL GOVERNMENT

I. Supervision of Civil Legislative Organs: National and Prussian Elections; Interdiction of Local Elections; Prohibition of Attendance of Deputies from Occupied Territory at the Provinziallandtag; Summary of American Policy towards German Legislative Bodies.

II. Supervision of Executive Officials: Supervision of Oberpräsidium and Regierung; Supervision of Inferior German Officials; Supervision of German Police; Creation of Police Reserves; Policy Regarding Changes of Officials; Disciplinary Actions against Officials; Character of German Bureaucracy.

III. Supervision of German Judiciary.

Paragraph V of the Armistice Agreement of November 11, 1918, states:

> "The countries of the left bank of the Rhine shall be administered by the local authorities under the control of the troops of occupation of the Allies and the United States."

This paragraph exercised an immense and far reaching influence over the character of the succeeding military government of the Rhineland. History records very few instances of an invading army stabilizing its control under conditions so favorable to the task in hand. The surrender to the Allies by the German government of the whole administrative, legislative and judicial machinery of the territories to be occupied, rendered the problems of civil control which confronted the Allied armies comparatively easy of solution. Supervision requires tact and judgment rather than the display of constructive genius. One is spared the necessity of erecting courts, appointing officials or providing intricate systems of education or finances. The most critical problem that can confront an occupying army was solved when Germany's representatives placed their signatures on the Armistice in the forest of Compiègne. In Cuba and the Philippines we had striven to revise native methods, and with results which fully justified the amount of labor expended. During the whole occupation of Germany, legislatures, executives, judges, customs officials, the most complicated and reputedly most effective civil organization in the world, were placed under the direction of American military government.

The contrast with earlier occupations becomes most striking if we compare the various causes which led to our presence on foreign soil. The military occupations of Cuba and the Philippines had been begun with the definite purpose of rehabilitating the corrupt and decadent governments of those lands. Whether the magnitude of the task undertaken was realized at the time is perhaps doubtful but, once begun, we did realize that we could not be content with reorganizing the government, but would have to revivify the whole civilization. Compared to these difficulties, our task in Germany was simple and direct. We had no intention of attempting to influence Germany to accept our form of government. Our enemy was a nation as highly organized in many ways as were we in others. Her people, finally awake to the danger of delegating power to a small group of autocratic noblemen, were far more capable of erecting a new government, suitable to their temperament and development, than were officers of a foreign army. The troops of occupation had but to maintain themselves on foreign soil, and their task, that of constituting a weapon to prevent renewal of hostilities and to force compliance with the peace terms, would be accomplished. Our supervision of the civil administration need only suffice to insure the maximum security of troops. America coveted no Rhenish territory. It has never been an American policy to stoop to maneuvering to obtain commercial or political advantages, nor to carry on propaganda. Supervision of the civil government might be either strict or loose, in accordance with the docility or hostility of the population, but there was no intention to interfere with German customs or institutions. As a practical proposition, the American army found it impossible to avoid interference in matters not strictly within this definition of "security". Army Headquarters found much difficulty in restraining subordinate officers from interfering too much in civil affairs, and from accepting responsibility which properly should have fallen on German shoulders.

There was another factor in the wording of Paragraph V. which rendered the task of military government comparatively simple. In December 1918, unoccupied Germany was still suffering from the effects of the revolution of November 9th. The overthrow of a régime, so long established and so well recognized as that of Germany, cannot be accomplished without the dislocation of the machinery of government. In revolutions in the past, there has usually resulted a complete change of administrative personnel. In unoccupied Germany, many officials either resigned or were displaced, a step which inevitably affected the efficiency of the governmental services. Paragraph V. of the Armistice, however, was so interpreted by both Germany and the Allies that "local authorities" meant the officials of the old régime. As a result, the armies of occupation had a thoroughly capable group of officials to carry out their orders and wishes, and were spared recognizing and dealing with the chance appointees of Workmen's Councils whose selections were made during the hysteria of revolution.

This chapter, dealing with American supervision of the German organs of civil government, may logically be divided under three headings:

1. Supervision of legislative organs of government;
2. Of the executive administration;
3. Of the judiciary.

The American policy towards each of these governmental functions will be treated chronologically and in some detail. The political events and life of Western Germany during the occupation will be treated only incidentally, and then, with a view of emphasizing those events which influenced the decisions of the Army of Occupation.

I. SUPERVISION OF CIVIL LEGISLATIVE ORGANS

National and Prussian elections.—The revolution of November 9th naturally had its effect on the political and administrative systems of Germany. Prussia in particular had been governed for centuries by a King and a Council of Ministers, whose decrees had the force of law. A well organized administrative system subordinate to the council secured the enforcement of the royal laws and regulations. Local government was still in an embryonic stage of development. The Diets, from the Prussian Landtag to the Village Council, were chosen by an elective system which granted two thirds of all delegates to the propertied and governing classes. A somewhat more liberal system of elections prevailed in the states of South Germany, but even there, Kings and Councils had managed to retain decisive authority in their own hands.

The revolution of November 9th was essentially a protest against an outworn system. Its leaders, having overthrown the old régime, appreciated the necessity of obtaining immediate reorganization of the form of both Prussian and local governments. Ebert and Scheidemann were no hairbrained fanatics, aiming to secure the Socialist party in power at the expense of Germany. A counter-revolution by the monarchic elements seemed a distinct probability and its success possible, if a combination of Socialist and Bourgeois parties was not promptly effected to defend the republic. The adhesion of the latter parties to the government could be secured at that time only by a promise from the Socialist leader to proceed to election of the German and Prussian assemblies. It was promised that these assemblies, when chosen, should have the power to formulate constitutions for the respective states, and in the meantime, to organize governments of a representative and democratic character. It was accordingly decided to hold elections for the National Assembly on January 19th, 1919, and for the Prussian Assembly on January 26th, 1919. No provision was made to exclude those portions of the country occupied by foreign armies from participation in these elections. Alsace-Lorraine was, however, excluded, as it was a foregone conclusion that those provinces would be ceded to France.

The decision whether to authorize or to refuse permission in occupied territory for these elections, was a matter properly pertaining to Allied G. H. Q. In view of the necessity of permitting Germany and Prussia to reorganize their governments sufficiently to sign and abide by the peace terms, Marshal Foch decided to accord the German authorities the necessary authority to hold the elections. This decision was thoroughly in accord with French sentiment at the time, there being no expressed desire to annex the Rhineland to France.

The American army particularly desired a free and fair expression of the will of the German people. The war which had just ended, had, on our side, been a struggle against an autocratic government. These elections therefore constituted the symbol of our victory and, it was hoped, the dawn of a better day for Germany.

Instructions to corps commanders defining the attitude of the American army during the elections were sent out in telegraphic form by Army Headquarters on January 15th. This telegram laid the responsibility for proper conduct of the elections squarely on the shoulders of the duly constituted civil authorities. Members of the American Army were prohibited from expressing any preference for, or opposition against, any German political principle. The last sentence of this telegram was:

> "They (the civil authorities) will take such steps as may be necessary to secure a full, free and fair expression of the will of the voters of Germany in order that the constitution now about to be written and the government now about to be established may be founded upon the immovable foundation of the will of the people."

Advance G. H. Q. informed the army on January 6th that similar facilities should be given the German civil government for the elections of January 26th for the Prussian Assembly. It is worthy of record that no communication was ever received at the Office of Civil Affairs, Third Army, complaining of interference in these elections by any individual of the American forces.

Despite intense popular interest, the elections passed off without disorder. The results for the American area are noted below, as they indicate the comparative strength of the various political parties in American occupied territory.

NATIONAL ELECTIONS

January 19th, 1919

Centrum (Catholic Party)	277,811
National Peoples Party (Reactionary)	2,831
Democratic (Bourgeois)	46,243
Social Democrats (Workingmen)	58,361
Miscellaneous	3,223
Total	388,469

PRUSSIAN ELECTIONS

January 26th, 1919

Centrum (Catholic Party)	267,840
National Peoples Party (Reactionary)	4,152
Democratic (Bourgeois)	36,852
Social Democrats (Workingmen)	51,561
Miscellaneous	2,895
Total	363,300

Interdiction of local elections.—The National and Prussian Assemblies, having met at Weimar and Berlin, respectively, proceeded to the organization of representative governments for Germany and Prussia. Both governments, when organized, developed radical tendencies, the majority Socialist party having polled nearly half the voting strength of the nation.

The local diets, assemblies and councils throughout the country had been chosen by the then prevailing three class franchise. Most of them had not been renewed since 1914. During the five year interim, great changes in public sentiment had developed and these bodies no longer represented the wishes of their constituents. Among the first acts of the new Prussian government was the passage of laws dissolving the old Kreistage and councils and directing new elections by universal suffrage. These laws, published on January 24th and February 18th, required the elections for municipal councils to be completed by March 2nd, and those for the Kreistage by May 4th.

When the American authorities learned that the Oberpräsident of the Rhine Province had received instructions from Berlin to hold these elections, they believed it good policy to withhold their consent until knowledge of the wishes of the High Command could be obtained, as uniformity of policy on such a vital matter was imperative. It was known that the advisability of permitting these elections was then under consideration at Allied G. H. Q.

A telegram from Marshal Foch on February 17th announced that it had been decided to forbid execution of the decrees of the Prussian government pertaining to elections for municipal councils. The telegram stated in explanation that it seemed clearly unwise for the time being to overturn existing local authorities. Billeting and local requisition of the armies had been effected by order on the German civil authorities. A wholesale replacement of these experienced officials by new and untried men, a step which would logically have followed the choosing of new councils, could only prove harmful to the interests of the armies. It also seemed inadvisable to begin another series of elections and subject the temper of the people to the partisan issues of politics. Third Army Headquarters transmitted the instructions of Allied G. H. Q. to all corps and division commanders. Each of these generals was directed to take such steps as he deemed necessary to secure the enforcement of the Marshal's orders.

It will be noted that this decision of Allied G. H. Q. contained no reference to the Prussian law of February 18, 1919, directing new elections for the Kreistage. On March 21st the Officer in Charge of Civil Affairs, 3rd Division, informed the Army Command that the Landrat of Mayen had received orders from the German government to hold elections for a new Kreistag. This official begged that we prohibit the election, stating that inasmuch as the members of the Kreistag were chosen by the communal councils, and as the old councils were still in existence, the new diet would be no more representative of the will of the people than was the old. The army authorities did not feel themselves competent to decide questions of policy of this nature, and referred the matter to Advanced G. H. Q. A reply was received on March 21st from the Officer in Charge of Civil Affairs in occupied territory, which stated that the interdiction of municipal elections had been considered as including Kreistag elections. The Landrat of Kreis Mayen was therefore so informed.

On April 19, the Marshal, Commander in Chief, informed the commanders of the various Allied armies that local elections might take place in the neutral zone. The German authorities, however, were warned that if disorder resulted, the Allied Command would entertain no request from German sources for reinforcement of their garrisons in the neutral zone. This was a matter of very little importance to the American army as they at no time attempted to exercise any jurisdiction in unoccupied territory.

The Landrat of Berncastel informed the Officer in Charge of Civil Affairs on May 26th that elections for the local Kreistag had been held in his Kreis on May 3rd. He further stated that he had since learned that the American army had prohibited these elections, and now requested an American decision as to which of the Kreistage, the old or the new, should be recognized by him as legal. The mere fact that such elections had taken place appeared on its face a gross violation of orders. It was not surprising, however, that election had taken place without knowledge of the American forces. Members of the Kreistag are chosen by the various Bürgermeistereiversammlungen of the Kreis. The meetings of these assemblies, in view of their local character and because of the small importance of the matters therein discussed, were not attended by army representatives. The army, like Advanced G. H. Q., had taken the point of view, that the directions conveyed to the local civil authorities and unit commanders, in Letters of Instruction No. 13, February 19th, 1919, (prohibition of municipal elections) was a sufficient statement of American policy to prevent holding of elections without previous authorization. In this case, however, the Officer in Charge of Civil Affairs, Third Army, decided that a careful investigation should be made before instituting disciplinary proceedings. This investigation disclosed the fact that, through an oversight, neither the Oberpräsident of the Rhine Province nor the Regierungspräsidenten of Coblence or Treves had been informed of a specific American policy in respect to Kreistag elections. All idea of disciplining the officials was at once dismissed, but at the same time it was felt that the higher officials had taken an improper stand in not consulting the American authorities beforehand in regard to such elections. The elections which had already been held in the Kreise were declared null and void and the Oberprasident notified on June 4th that elections of any character must thenceforth be authorized officially by the American army.

The signing of peace at Versailles on June 28th and its subsequent ratification by Germany rendered it desirable for the army to relax some of the restrictions which had been imposed on the civil population.

Marshal Foch had granted the armies permission on August 10th to hold such elections as they deemed proper. The American army believed the moment opportune to grant the German government permission to renew the local councils and assemblies. The previous interdiction of these elections had been wholly due to the military situation, and, while appreciating the difficulties placed in the way of the growth of local democracy, it was not thought advisable to depart in this instance from an Inter-allied policy. Now that the factors which had worked for the prohibition no longer existed, the Omerpräsident was informed on August 28th that Kreis, municipal and communal elections might be held. In order to avoid disorder and to provide necessary police safeguards, the German authorities were directed to inform the Office of Civil Affairs at least five days in advance of the dates and places of the various elections.

The elections took place during October and November, and their orderliness and quietness reflected credit on the local population. A marked diminution of popular interest was noted in comparison with the National and Prussian elections earlier in the year. The results, however, indicated decided losses for the Centrum party, with corresponding gains to the majority Socialists. In some cities, as in Andernach, the Independent Socialist party (an organization with Sparticist tendencies), placed candidates in the field and polled a considerable vote, although the results did not endanger the position of the more conservative parties.

Prohibition of attendance of deputies from occupied territory at the provinziallandtag.—The self-governing organ of the Prussian province is the Landtag. This body ordinarily meets every year, the session lasting about a month. The seat of the Landtag of the Rhine Province is at Düsseldorf, lying in unoccupied Germany, while the offices of the government appointed functionaries are at Coblence. The annual session of the Landtag was summoned by its President to meet in March 1919, and notices to that effect were sent to the deputies on both banks of the Rhine. In order to obtain the consent of the occupying armies, the German government, at the same time, approached the Allied High Command through the Armistice Commission at Spa, with a request that permission be granted deputies within the area under Allied control to attend the session of the Landtag.

Marshal Foch resolved not to grant this request. An examination of the business to come before the Landtag had disclosed little of importance, and it was found that much of it could be dealt with elsewhere. Furthermore, the meeting of the representative body of the Landtag in a city outside Allied jurisdiction might easily have led to acts of subversion of the military governments by members of that body. The decision of Marshal Foch appeared wise and was approved by the American command. In consequence, the G–2 section of the General Staff, Third Army, refused to issue circulation passes for Düsseldorf to deputies living in American territory.

Summary of American policy towards German legislative bodies.—The supervision of German legislative bodies in occupied territory was a matter which good policy dictated should be regulated alike in all Allied zones. If permission had been granted Germany to hold elections in one zone and withheld in another, an appearance of weakness would have been created and a weapon placed in the hands of our enemies to use in future dealings with the Allies; and Berlin was already aware of minor differences of policy among the armies on the Rhine. Reconstruction of the German government was a matter of such prime importance that it was necessary for the Allies, in this case at least, to act as a unit.

The interdiction of local elections by the American army was, as has been shown, due directly to instructions of the Allied High Command. The refusal to issue passports to Landtag deputies residing in the American area, also originated in the same source. Action in both cases was dictated by military necessity. It was nevertheless felt to be regrettable that the United States Army should be placed in a position in which it was forced to prevent the orderly growth of

German democracy. The legislatures and assemblies which had been found in occupied territory were wholly out of touch with the new republican sentiments of the people. Only the requirements of order and the welfare of the army could have forced us to take the decided stand which we temporarily did take against the renewal of these bodies.

The American army never supervised the sessions of a local legislative body, nor took steps to influence the passage or defeat of laws. Many German and Prussian laws and regulations were forbidden in American territory, and local communities often permitted to work out their salvation as best they could. Matters within the jurisdiction of local assemblies were so limited and of such slight importance, that the very existence of these bodies could ordinarily be disregarded. In one instance, however, attention was drawn to the privileges of the Kreistage. Kreis Mayen had been without a Landrat from January 1919, until October of the same year. In September, the attention of the Office of Civil Affairs was directed to the matter, as the temporary appointee showed a lack of energy in solving food and fuel problems. The American military authorities believed that the official's timidity was primarily due to his not holding a regular appointment. It was therefore to military interests to insist that the German government immediately fill the post. However, on further investigation, it was ascertained that the temporary incumbent had already been appointed by Berlin, but was awaiting confirmation by the Kreistag. This body had the right to express approval or disapproval of the government appointee, although it might not select a man of its own choice. No steps were therefore taken by the Office of Civil Affairs to fill the post, prior to an expression of local sentiment.

II. SUPERVISION OF EXECUTIVE OFFICIALS

Military government inevitably tends to concentrate its energy on the supervision of civil executives. The wishes of the civil population, as exemplified by the laws and decrees of their legislative bodies, are usually held in abeyance whenever they run contrary to the wishes of the occupying forces, and the orders of the military commander become temporarily the law of the land. The enforcement of these orders therefore is the foremost function of military government.

The Armistice had required Germany to surrender her entire administrative machinery of government in Rhenish territory to the Allies. This machinery offered an efficient and easy procedure of transmitting military orders to the civil population. Paragraph V. of the Armistice even stipulated that the exercise of control by the occupying armies should be effected in this manner. As a practical proposition, the German civil administration would have had either to be utilized as it was, or provision made for the erection of an entirely new one. The latter course was never for a moment considered.

Once having accepted the German official as a representative of American military government, there remained but to organize an effective system of supervision. We could not be content with issuing orders and leaving their contents to become known automatically to every civilian within the territory. The official had to be controlled at every step. Measures had also to be taken to learn of all laws and regulations issued by the Imperial and Prussian governments, for the Rhineland was still a province of Prussia. Publication of such of these laws and regulations as would adversely affect the occupying armies, had to be forbidden in the territory. Finally, we had to observe each official, his efficiency and his willingness to cooperate with the American army. If recalcitrant or lazy, he was disciplined or removed from office. Inefficiency of officials in executing their own laws was not a matter for our concern, but slackness and carelessness in the transmission of American orders could not be tolerated, and the responsible official usually suffered.

Reference was made in the introductory chapter of this report to the facility with which German civil administration could be adopted to a military government. This is a factor which should be emphasized at this point. The average German official has served his allotted time in the army. During this service, he has learned to recognize that implicit obedience to military orders is expected as a matter of course. Many of the officials had served in Belgium and had experienced at first hand the strict discipline demanded by the German régime in that country. They now realized that failure to render implicit obedience to the conqueror would only bring punishment on their heads. Being essentially soldiers themselves, they fully recognized the authority and power of the occupying forces.

The German conception of civil government, moreover, is itself a military one. All officials are primarily concerned with transmitting the orders of the central government, and exercising such supervision as is necessary to assure their execution. Important officials are appointed by Berlin and their tenure of office does not in any way depend on local popularity. In consequence, the transfer of supreme authority from Berlin to the Allied Commands did not alter their fundamental position.

Supervision of Oberpräsidium and Regierung.—Supervision of civil officials began with the Office of Civil Affairs of the army and extended down to regiments and lesser units. We were very fortunate in having Army Headquarters in the political capital of the Rhine Province. The jurisdiction of the Oberpräsident extended over the entire American area, with the exception of Kreis Montabaur.

The German government, however, early in the occupation, with a view to obviate administrative difficulties, gave the Oberpräsident authority to transmit communications of the American army to the Landrat of that Kreis. This put that Kreis under our control as a compact political unit, and made it unnecessary for the Army Command to transmit communications to Regierungspräsidenten or Landräte. An order to the Oberpräsident accomplished identical results and placed the burden of responsibility for its transmittal on the shoulders of the civil authorities. In comparison with our Allies, whose territories included a number of political units without a central control, the advantages of the compact American area are obvious. It greatly simplified correspondence, as well as the necessity for close supervision of the Regierung. In certain matters, however, it was still necessary to inquire into the business of these latter bodies. The Regierung received many laws and regulations directly from Berlin, not by way of the Oberpräsidium. The Office of Civil Affairs believed it necessary to examine this correspondence to ascertain that no orders went into effect which were inimical to the Armies of Occupation.

The quantity of correspondence between the Office of Civil Affairs and the Oberpräsidium was so large that it was found necessary to appoint an American officer as a representative with the latter. This officer's duties included, in addition, such supervision of the Regierung of Coblence as appeared necessary. The higher German officials were still granted the privilege of appealing to or personally interviewing the Officer in Charge of Civil Affairs, but routine matters were invariably left to his subordinate. The Oberpräsidium and the Regierung in their turn appointed officials possessing a knowledge of the English language to attend to correspondence with the American Army. The arrangement proving satisfactory, the O. C. C. A. was thus relieved of the necessity of constantly interviewing civil officials, and could devote himself to questions of policy.

The duties of the American representative with the Oberpräsident were fairly numerous.

I. He transmitted and received all communications passing between the military and the civil governments:

II. He was expected to familiarize himself thoroughly with the workings of provincial and local government:

III. He was to study the personalities of the various German officials and report any failures to comply with American orders:

IV. He was to inspect the correspondence of the German government offices daily and ascertain that no propaganda was being made against the occupying forces:

V. He was to examine carefully all new Imperial and Prussian laws, copies of which were furnished him weekly by the Regierung, in order to discover possible interference with the rights of the American Army. Circulation thereof was accordingly either authorized or forbidden.

A large amount of correspondence between the Tenth French Army and the Regierung of Coblence also passed through the hands of this officer. This was due to the fact that a number of Kreise of Regierungsbezirk Coblence were under French control. This method of transmission was a necessity, as the American military authorities alone had power to discipline civil officials in its area. French requests, when submitted by the Office of Civil Affairs to the Regierung thereby received the force of orders, and infractions could be punished legitimately by American provost courts. An American representative with the Regierung of Treves was also appointed by the Officer in Charge of Civil Affairs in American occupied territory. Matters handled by this officer, however, were largely local in nature, as army orders were transmitted to the Oberpräsident and by him to the Regierungspräsident.

Normal communications between the German and the military authorities were written and formal. The terseness so striking in American military letters was for a long time commented on adversely by the Germans, who deemed it either a want of courtesy or an attempt to wound their pride. This was due to ignorance of American character and customs, for German officials were uniformly treated by the Office of Civil Affairs with the utmost consideration. On occasion, when attempting to formulate policies affecting the life of the community, written communications failed to effect a satisfactory interchange of views. In food and fuel matters, in particular, the personal expressions of the officials were desired, rather than official statements in writing. Conferences were therefore held, and facilities given the Germans to state their practical difficulties. These were considered and, when possible, smoothed over. Such conferences were not held at regular intervals, but only when the purpose of the Army Command could be served in no other way. During the session, both sides maintained a formal and dignified attitude, but as long as the civil authorities did not adopt an antagonistic tone, they were permitted and frequently invited to criticise military regulations, both those already in effect and proposed ones.

In the first conference with higher German officials in Coblence on January 11, 1919, the Officer in Charge of Civil Affairs warned them, that, in view of Germany's methods in initiating and carrying on the war, they must not expect a sympathetic attitude on our part. They were further informed that they should never forget that we came among them as a conquering army and therefore as unwelcome guests; we proposed to deal with them strictly but impartially and their future peace of mind depended wholly on whether they were willing to accept the fact of military occupation and the obligations it laid upon all

Germans. They were also advised that infractions of, or any failure to carry out exactly and promptly military orders would be punished. In later conferences in the autumn of 1919, when the coal question had become critical, both sides met in a less formal manner to seek a solution of the difficult situation in which the local population found itself. In the light of fifteen months experience and in all fairness, it must be recorded that German officials with few exceptions fully performed their duties toward the American forces of occupation.

Supervision of inferior German officials.—This varied greatly during the occupation. While tactical units still controlled civil matters, such supervision as took place was loosely organized, and often indifferently carried out. This was perhaps unavoidable, in view of the fact that the boundaries of divisional areas and German political subdivisions were not correlated. In Regierungsbezirk Treves, the 90th Division occupied Kreise Daun, Wittlich and Berncastel, while the 89th Division was billeted in Kreise Prüm, Bitburg, Trier-Land and Saarburg. It was manifestly impossible for a single divisional civil affairs officer to supervise, as in case of the 89th Division, four separate civil administrative bodies, separated by many kilometers of territory. The ordinary solution adopted was, for civil affairs officers to appoint subordinates to supervise the execution of American orders in each Kreis. This was never entirely successful, and the very essential supervision of German officials was sometimes entirely lacking. Another failure of control by tactical units to meet the exigencies of the situation arose from the locations chosen for Divisional Headquarters. The city usually selected was the largest town in the area or, in some instances, the town with the most ample hotel accommodations, rather than the political capital. In consequence, the activities of the O. C. C. A. were handicapped by the necessity of travelling to another town when he desired to interview the Landrat, as the latter could seldom travel except by rail on account of lack of German transportation facilities. Actual supervision of the duties of the Landrat's office was thus out of the question.

With the introduction of territorial control in May 1919, the advantages of the new system soon made themselves apparent. The Kreis O. C. C. A. ordinarily had his offices in the Kreishaus adjoining the office of the Landrat. He could inspect daily the latter's correspondence, verify his receipt of army orders from the Oberpräsident, and generally study problems confronting the local administration. It was natural in these circumstances that a spirit of cooperation should develop between the O. C. C. A's and the Landräte. Both were concerned with the solution of identical problems and both felt a sense of responsibility to the population they were governing. It was very shortly apparent under this system that American orders were receiving more attention at the hands of the German officials. and at the same time the O. C. C. A's found many opportunities to assist the Landräte in food and fuel difficulties. In order that the Army representatives should thoroughly inform themselves as to the operations of Prussian local government, each O. C. C. A. was required to submit an essay on the organization of his Kreis to the Office of Civil Affairs. These reports were of only incidental value in themselves, but stimulated the O. C. C. A's to a thorough investigation of the laws and customs of the country which they were administering.

Supervision of German police.—The methods employed by the army to control the activities of the German civil police varied considerably during the course of the occupation. Police were authorized to carry arms by the provisions of the "Anordnungen" of Dec. 9, 1918, and, from then on, came to be recognized as an adjunct to the military forces for the preservation of public order. Their authority, however, was limited strictly to the civilian population, and under no circumstance were they permitted to arrest or even touch a member of the Allied forces. Supervision over local police and gendarmes was exercised during the first five months of 1919 by the troop commander having jurisdiction over the town or district to which the police belonged. In most cases, this officer delegated the actual responsibility to one of his staff officers, commonly the assistant Provost Marshal. No very effective results were obtained by this arrangement, and the civil police, whose morale had already seriously deteriorated during the war and revolution, sank to an even lower level of efficiency. It must be admitted that their position was not one to carry prestige with the civil population, as the actual maintenance of public order was the duty, pure and simple, of the military police of the army. In a great number of cases, the army found it necessary to overrule their actions, and in some cases to bring policemen before provost courts. Disciplinary punishment was entirely necessary under the circumstances, but it nevertheless tended to weaken the prestige and *esprit de corps* of the police.

The transformation in organization of the military government which followed the return of most combat divisions to the United States, brought with it a change in attitude on the part of the army towards the civil police. In the larger part of the American area, troop garrisons were confined to Kreis capitals, and the maintenance of order in the outlying sections of the Kreise thereby devolved on the civil police. The Kreis Officers in Charge of Civil Affairs, while mainly depending on the military police companies assigned to them for the preservation of public order, nevertheless began to take an interest in the German police and to place more responsibility in them. The withdrawal of the military police companies towards the end of the summer of 1919 from the outlying Kreise, forced the Officers in Charge of Civil Affairs to go even further, and after the concentration of the American garrison in Coblence, the civil police remained in the Kreise as the sole surety for the maintenance of public order.

A much closer supervision of the city police had always been exercised by the army than in the outlying districts. The Assistant Provost Marshal of the city, who was directly responsible to the Garrison Commander, maintained a close *liaison* with the civil officials who directed the police forces, and in many instances directed the activities of the civil police himself. With the abolition of the office of Garrison Commander, in August 1919, the Provost Marshal's Department was placed under the G-1 section of the Army, which, appreciating the service that the civil police, if under proper direction, could render to the army, at once took steps to improve their morale and efficiency. A superviser of German police was appointed for the American area, whose duties were to control their activities and at the same time bring about a close liaison with the American military police. The results of this arrangement were astonishingly good, and by December 1919, a great change, both in efficiency and appearance, had taken place in the Coblence force. German police took over almost exclusive direction of traffic, using the methods and signals which they had learned from the Americans. At the same time, there was an undeniable improvement of morale, which fully justified the action of the army in what was, properly, a strictly German internal matter.

However satisfactory the results which the appointment of a German police supervisor may have achieved in Coblence, it nevertheless somewhat complicated the situation in the outlying districts. Until then, responsibility for the maintenance of public order in a Kreis had rested on the shoulder of the local Officer in Charge of Civil Affairs. Extension of the authority of the new office into the Kreise, now created a dual responsibility, which at times gave rise to friction. Trouble which arose in Ahrweiler over arms permits for police reserves brought the matter to a head. On December 6, 1919, the Officer in Charge of Civil Affairs and G-1 section came to an agreement, that the authority of the German Police Supervisor should be confined to the Coblence area, and that responsibility for the preservation of order in the Kreise should be placed on the Civil Affairs officer. The Provost Marshal's Department was, however, to be permitted to make such inspections of the police of these districts as were deemed necessary, in order that the Commanding General might be informed as to their reliability. Such inspections were to be made in the company of the local Officer in Charge of Civil Affairs.

Police reserves.—As early as the middle of July 1919, the American authorities commenced work on a scheme by which they hoped to make the maintenance of public order a duty of each town and village. In Germany, before the war, police activity was centralized at Berlin, much to the detriment of democracy. This prevented the growth of a feeling that the preservation of public order was the duty of each individual city and town. The army desired, as far as was possible, to avoid interfering in the strikes and industrial disputes which became so common. The German regular police forces, however, were very small and it was impossible to increase them, because of the financial burden which such an act would place on the community. The military authorities therefore turned to the idea of a police reserve, as exemplified in New York City, for a solution of the difficulty. Every care was to be taken to prevent such police reserves from assuming a military character, and so conflicting with the treaty of Versailles. Nevertheless, the army did not desire to be forced into a position in which it would be constantly forced to side with either capital or labor, and believed that, whenever possible, public order in such disputes ought to be maintained by the Germans themselves.

On November 24th, therefore, the Army Commander authorized the creation of police reserves. Units were to be established in every community, at the rate of ten reserves to every active policeman. Supervision of these units was to be exercised by the Kreis Officers in Charge of Civil Affairs, who were to inquire carefully into the character of the personnel selected by the Burgomaster. The police reserves were to be armed with pistols, but were under no conditions to perform any military drill.

The creation of these police reserves resulted at once in proof of the wisdom of the plan. A food riot in November at Ahrweiler was quelled by the civil authorities without resort to the American authorities, largely by the prompt response of the police reserves to the call of the Burgomaster. Reserves in other Kreise also did good service in assisting the regular police to enforce the food laws in 1920. It is, however, too early to determine clearly whether the theory of local responsibility will take a firm hold on the German people, or whether they will revert to their conception that police activity is a matter strictly pertaining to the central government.

Policy regarding changes of officials by legal and revolutionary means.—The importance of the American policy regarding changes of officials cannot be emphasized too much. Its effect on German politics has been dealt with at length in another chapter of this report, that on the Rhenish movement.

When American troops entered Germany in December 1918, they found a considerable number of Workmen's and Soldier's Councils in existence. These organizations had displaced the duly constituted civil authorities, and were performing all ordinary governmental duties. The Allied forces were therefore called upon to decide who were the "local authorities" referred to in Paragraph V. of the Armistice. It was unanimously agreed that these self-appointed committees could not be recognized without serious complications. The councils were, as their title indicates, composed of workmen and soldiers. They contained very few individuals who had had any experience in civil government, and the majority were poorly educated and drafted from the lowest strata of society. It would have been a difficult matter to attempt to billet our army or to requisition supplies, using such committees as agents. It was therefore decided to accept the officials of the old regime as "local authorities", whether those antedating the revolution or those duly appointed thereafter by Berlin. This policy as then adopted was adhered to by the American army during the whole occupation. Even when the position of the old officials was threatened by Dr. Dorten's revolutionary propaganda, the American army adhered to its earlier policy and refused to recognize the Wiesbaden "government". Meetings whose announced purpose was to discuss the overthrow of Prussian rule, were absolutely forbidden. The fate of Germany was a question which, we felt, the peace conference was alone called upon to solve. Ill advised solution of the Rhenish problem in the meantime by military men could never have assisted President Wilson to solve the many tangles of European politics.

When the Rhenish agitation was at its height, the Prussian government made an attempt to relieve the existing officials and appoint new men, who were conspicuous for their popularity in the Rhineland, but at the same time loyal to Prussia. As is noted in Chapter XV., the American army could not accept these appointments of the Prussian government as binding in occupied territory any more than the self constituted Soldiers and Workmen's Councils. The Oberprasident was consequently so advised by the Office of Civil Affairs in a letter announcing that we could recognize neither these spurious appointments of Berlin nor any further important appointments without previous consultation.

In part, this letter read:

> "I wish to call your attention to the fact that the American authorities will not permit any important official to be removed from office or any one to be installed in office, whether by yourself, any of your subordinates or by the order of the Central Prussian government, without first bringing the matter to the attention of the American authorities and obtaining consent to the change."

This policy necessitated a considerable amount of work for the Office of Civil Affairs. It was nearly impossible to exactly interpret what was meant by "Important Officials". The provincial government decided to be on the safe side, and thereafter submitted the appointment of many minor officials to the army for approval. The Office of Civil Affairs agreed, though this practice had never been contemplated, and took no steps to alter it. Changes of minor officials were approved by the American representative with the Oberpräsident. Appointments of Landräte were transmitted to the Office in Charge of Civil Affairs, A. F. in G. Submission by the German government of the appointments of new officials in American territory was in most cases merely formal. The American authorities desired to interfere as little as possible in German government and had no intention under ordinary conditions of refusing recognition to duly appointed officials. The cases of Herr von Groote, Oberpräsident of the Rhine Province, and Dr. Momm, Regierungspräsident of Treves, were exceptional ones, and our action was due to the desirability of quelling a political agitation which might well have assumed the greatest consequence. In case of transfer of officials from French or Belgian territory to ours, the German request was sent by the army to the Ally concerned, for an expression of approval or disapproval. Only two such transfers were made.

Disciplinary actions against officials.—A number of important policies were involved in awarding suitable punishments to negligent or recalcitrant officials. This was an important matter, and one which, it was realized, would have to be carefully considered from all viewpoints. Offenses by officials are bound to occur under any military government and must be promptly punished if the latter is to maintain a firm hand. The choice of suitable penalties in Germany, however, was felt to be a difficult matter. Officials ought not to be placed in positions which would cause them to loose the respect of their subordinates or the people at large. If punishments of such nature had been awarded, the dignity of the civil service would have suffered and the individual concerned rendered useless thereafter as a servant of the military government. Imprisonment was the particular punishment which the Office of Civil Affairs felt should be avoided. Fines, suspension from office or deportation from occupied territory appeared the only logical penalties which we could inflict without injuring ourselves. Deportation appeared such a severe punishment, depriving the official, as it did, of his home as well as his office, that it was decided to reserve it for very grave offenses only, particularly for acts which threatened the security of the army. Notwithstanding the efficacy of this form of punishment on the individual, it was realized that its constant use would cripple the civil machinery and eventually react on the army. It is gratifying to note that on no single occasion was an offense of an official judged serious enough to warrant such punishment.

In striking contrast to our actions was the French policy in regard to deportations. The French Command evidently believed that officials who adopted a hostile attitude should be rigorously punished and that deportation alone could accomplish that purpose. In one

Kreis of the Saar district alone, orders of banishment were issued up to October 15, 1919, against twenty-three officials. In the Rhine Province, instances of its use are not so numerous, but are still sufficient to warrant the belief that it was the normal French method of punishing officials. The British policy was to banish only in extreme cases. Fines and suspension from office were the normal disciplinary measures resorted to by their army as well as ours.

Disciplinary punishments inflicted on officials in American occupied territory were for the following offenses:

Failure to report sales of German war material	2
Disobedience or inefficiency in execution of American orders	34
Failure to prevent German patriotic demonstrations	3
Smuggling	6
Negligence in carrying out American sanitary regulations	7
Offensive remarks about American Army	2
Homicide	1
Total	55

Imprisonment was awarded only in those cases in which the offense was strictly criminal, such as the killing of an American soldier by a German policeman. In many cases, the sentences imposed by the court were afterward suspended during good behavior. Of the higher officials, records only show four instances in which punishment or reprimand was necessary. Two of these were for offenses of a minor nature, the officials concerned being the Landräte of Mayen and Wittlich. The former, Herr Walter Theodor Losen, had attempted to alter an American order, but as there were mitigating circumstances, he was merely given a reprimand. In the second case, in which Dr. Simmons, Landrat of Wittlich, was the accused, he was sentenced by a provost court to pay a fine of 500 marks for failing to publish an American order in due time. This fine, however, was later remitted. A more serious case centered around the Landrat of Mayen, Herr von Lettow-Vorbeck, who was tried by a military commission and fined 5,000 marks for failing to report the sale of foodstuffs, which had belonged to the German army, to the people of his Kreis. The fourth instance occurred on October 6, 1919, when the Officer in Charge of Civil Affairs, A. F. in G., suspended Regierungsrat Herr Braun, an official of the Regierung of Coblence, for one month from office for gross carelessness in transmitting American orders.

The Acting Regierungspräsident of Coblence, Herr Jürgenson, narrowly escaped punishment by the American authorities in March 1919. He had been duly served with a requisition for 200 laborers, who were to report to an officer of the American army on the morning of the 17th. At the appointed hour, only a very few of the required 200 appeared. An investigation revealed the fact that there had been a striking lack of attention to the order, if not a wilful defiance, in the

Regierungspräsident's office. Herr Jürgenson was placed in arrest and charges against him were drawn up, which ordinarily would have resulted in trial by a military commission. Shortly afterward, however, he became seriously ill and, as he was not expected to recover, the charges were withheld. On April 16th, he was released from arrest and the case dropped, but he was soon thereafter retired by the German government.

Character of German bureaucracy.—The efficiency of the American military government, for reasons explained in preceding pages, depended to a large extent on the capabilities and cooperation of the civil officials. Like most bureaucracies, the German civil service is cumbersome and slow moving. There are innumerable departments, in some instances entirely independent of the local authorities and serving directly under Berlin. The railways and customs services are instances of such independent services. This somewhat compromised the desire of the Office of Civil Affairs to utilize the Oberpräsidium as a medium for the transmission of communications. Prussian officialdom has in many respects, however, decided advantages over governmental services in other countries. The service has always been a profession, membership to which was attained by long study, followed by a series of severe examinations. When once admittance had been gained, the official was certain of life service unless found guilty of misdemeanor or misuse of office. While pay was never high, the service was recruited among the sons of the wealthy, and membership in it assured the individual a standing in society. In consequence, corruption and bribery were so rare before the war as to be almost unknown. With the coming of the revolution, the morale of officials was much shaken, and their sense of responsibility to the republican *régime* became more nominal than real. Still, the American army found no evidence of misdemeanors of this sort among the higher officials. A few cases of bribery, smuggling, etc., came to light among customs guards, railway employees, etc.; but it must be borne in mind that there had been a complete change of these minor employes during the period of demobilization.

The local government, however, was in a difficult position as a result of the occupation. The officials were largely of Prussian extraction and were looked upon as foreigners by the Rhinelanders. Their authority before and during the war had been maintained by the police power of Germany and Prussia. With the crumbling of the old *régime* in November 1918, the German army could no longer be looked to as a prop of their authority. The Rhineland passed into enemy-hands, and foreign bayonets and rifles effectually secured the old officials in office. As far as Rhenish officialdom was concerned, a hundred revolutions might have come and gone at Berlin, so few were the changes of personnel that occurred. The people, however, were alive to the new feelings of democracy, and looked askance on the old officials, Germans though they were, who continued to govern them, by the grace of the enemy. The officials themselves could no longer look to Berlin for advice or comfort, for Social Democracy there sat enthroned, just as anxious to please the Rhenish population, as the people themselves desired to select governors of their own choosing. These representatives of old Germany saw in the future, when Allied troops would have departed, that they would have to render an account to the people. Either they would have then to swim with the tide, or abandon their lifelong profession. So, inevitably, they turned to political intrigue as a solution for their difficulties. Government law and regulations, which displeased the local population, were carried out halfheartedly. This was particularly evident in the case of food laws, the American area being essentially an agricultural country and hostile to the efforts of the German government to regulate prices. Not only were their efforts to enforce the food laws unavailing, but so also was their attempt to check profiteering by the so-called "Schieber". In the last case, the officials were undoubtedly willing enough to act, but felt that they did not have sufficient means at hand to enforce the laws. Under the emperor they had had a well organized police force at their beck and call. This force had been somewhat impaired during the war, and to a much greater extent in the period following the revolution, when gendarmes and policemen had been recruited among returning soldiers. The force, thus newly organized, was very much inferior to the police of the empire. Discipline had been shaken and the government took no steps to check slackness. Profiteers flourished everywhere and flaunted the efforts of the police to arrest them. Eventually, the American authorities, in the interest of public order, had themselves to take the matter in hand. They assured the population in a proclamation that they would try by provost courts any one guilty of offiering violence to a policeman or food official in the execution of his official duties.

Notwithstanding this lack of energy, it must be admitted that the local officials were in a trying position. Their attitude to the occupying authorities was generally correct. Orders of the American Command were punctiliously obeyed, and if disciplinary action was occasionally necessary, the fault generally could be ascribed to ordinary carelessness. No instance has come to light of a clear cut defiance of American regulations by a German official.

The services of the Oberpräsidium to the American military government are especially worthy of comment in this regard. His Excellency Herr von Groote, Oberpräsident of the Rhine Province, and his subordinate officials in the Oberpräsidium, accepted the occupation in a quiet and dignified manner and spared no effort to meet the wishes of the military government. The army did not have to deliver a single reprimand to an official of the Oberpräsidium, and our orders, with a few minor exceptions, were punctiliously and exactly complied with. Oberpräsident von Groote, the Oberpräsidialrat Dr. Brandt, Regierungspräsidenten von Gröning of Coblence and Dr. Momm of Treves, were all men of character and ability. Their path was difficult, the demands of the Army of Occupation were often a heavy burden to their communities, but they put their shoulders to the task, in an effort to push difficulties aside, and to solve the many problems brought about by occupation by a foreign army. The office of Oberbürgermeister of Coblence was a particularly trying one. Dr. Klostermann, the first incumbent, died during the first few days of the American occupation. After a long interval, during which the city, according to the German system, was seeking a new official, Dr. Jansen carried on the exacting work of the office; finally Dr. Russell, an exceptionally able and tactful official, formerly Burgomaster of Buer (Westphalia), was chosen to manage the affairs of the city, which he succeeded in accomplishing in an entirely satisfactory manner.

Supervision of German judiciary.—While the American authorities did not intend to limit the jurisdiction of German courts over German civilians, a certain amount of supervision was nevertheless found necessary. During the first few months of occupation, such supervision as took place was spasmodic and unsatisfactory. The personnel of the divisional staffs for civil affairs were not numerous enough to provide experienced officers for all duties. Examination of the activities of German courts was hence entirely lacking. A change for the better was noted when territorial control was instituted. The Officer in Charge of Civil Affairs, Third Army, required weekly statements from each Kreis Officer of the number and nature of all criminal trials held before civil courts. These reports enabled the army authorities to gain a comprehensive view of civil judicial activities and emphasized any failure of the German officials to enforce their own food laws.

It was found necessary to deny jurisdiction to civil courts in a number of instances. The proclamation on May 28, 1919, of the German government, declared that the leaders of the Rhenish movement were traitors to their country, and should be tried as such. This proclamation was forbidden in the American area, but knowledge of its contents became generally known through news dispatches from unoccupied Germany. On June 15th, the German police of Coblence arrested one Philip Hedwig, of Wiesbaden, discovered in the act of distributing propaganda in favor of a Rhenish republic. The Officer in Charge of

Civil Affairs, Coblence, did not, at the time, receive information of the arrest, but a few days later learned that the civil courts were about to try Hedwig for treason. Such procedure could not be tolerated by the American authorities under any conditions. Coblence was under American jurisdiction and therefore we could not recognize treason against an enemy nation as a crime in our own country. The right of the German courts to try cases under the German and Prussian acts forbidding treason was therefore denied. Hedwig was removed from civil custody and tried before American provost courts for violation of American regulations.

Another somewhat similar case arose in January 1919, in Kreis Ahrweiler. A civilian by the name of Metzler was arrested by the local police on the warrant of a court in Düsseldorf. This warrant charged Metzler with high treason to Germany, the specific acts alleged being that he conveyed military information during the war to the Allied governments. The army could not in this case recognize the right of the German government to extradite Metzler to unoccupied territory, and therefore ordered him released.

An attempt was made in May by the German government to enforce the penalties, imposed by its military courts, on civilians living in occupied territory. In neither France nor Germany is it the custom to require the presence of the defendant in court. Fines and imprisonments may be imposed in his absence, and the civil authorities ordered by the court to carry out the punishment. As many German deserters had fled to occupied territory, their military commanders were at a loss to find a way to enforce the penalties decreed by their courts. They therefore approached the American military authorities with the request that they be permitted to carry out punishments in our zone. Such procedure was repugnant to the American sense of justice. The German military commission was therefore informed that as the United States was still technically at war, it could not permit the punishment of an enemy military court to be carried out in its zone.

A more knotty question arose from the requests of civil courts that American officers and soldiers should appear before them as witnesses. American testimony was in many cases declared by the court to be essential to justice. There did not seem to be good reason why the American army should deny the German courts the right to hear all possible testimony. The question was raised, however, that American soldiers, in appearing before a court as witnesses, placed themselves within its jurisdiction and could be compelled to answer all questions put to them. After considering the question, it was finally decided that the right of the court to summon an American soldier before it could be agreed to only subject to reservations. The summons of the civil court was to be forwarded to the Office of Civil Affairs, accompanied by a statement of the testimony expected. The officer or soldier in every case was to be accompanied by counsel, and was under no conditions to be compelled by the court to incriminate himself. This procedure was adopted, and on several instances American soldiers appeared as witnesses before German courts. Only a very few of the German requests for witnesses, however, were granted and then only because the expected testimony appeared indispensable to the courts of justice.

CHAPTER No. 15

THE RHENISH SEPARATIST MOVEMENT

Historical Basis of the Movement; Outline of the Separatist Agitation; Phase from November 1918, to May 1919; Political Situation at end of May 1919; Agitation in the Palatinate; Dr. Dorten's attempted *Coup d' Etat;* Policy of the American Military Authorities regarding Changes of Officials; Proclamation of the "Republic"; Dr. Dorten fails to gain Popular Support; Attitude of Governments of Germany and Prussia; The Fall of the "Republic" on June 4th; Renewal of the Movement During the Summer; Further Agitation in the Palatinate; Dr. Dorten's Statement of his Policy; Separatist Movement in Birkenfeld; Declaration of Centrum Party; Change in French Rhenish Policy.

Historical basis of the movement.—Probably the most important political manifestation of the people of the Rhineland during the winter and spring of 1918–1919, centered around the agitation for a Rhenish republic. So important would have been the results if this agitation had resulted in the formation of an independent Rhine state, and so momentous, as far as its success was concerned, proved to be the attitude of the occupying military forces, that it is considered wise to treat the movement in this report at some length.

No just appreciation of the causes of such a movement can be had without an examination of the political, geographical and historical background of Western Germany, and its relations with Prussia and with the Bismarckian Empire. While no attempt is made to give a history of Western Germany in a governmental report of this sort, certain points in the relations of Western Germany and Prussia must be mentioned here at least briefly.

During the Middle Ages Western Germany had been part of the Holy Roman Empire. The authority of the Emperor, however, had been more formal than real. The Rhenish territory, which after the World War was occupied by the Allied armies, had been divided into a great number of clerical and lay principalities, practically independent of the Emperor's authority. These states maintained their own military forces and levied their own taxes and customs duties. Such a condition weighed heavily against any great industrial development, and the Rhineland with the adjacent territories of Hesse and the Palatinate, remained up to the nineteenth century primarily an agricultural country. At various times the Palatinate and the district about Saarbrücken were under French rule, the former territory hav-

521572—43——20

ing been terribly devastated during the wars of Louis XIV. That portion of the Rhineland occupied by the American army contained, in those days, three fair-sized states, besides numerous less important ones. These were the Archbishopric of Treves, which included most of the Moselle Valley and the city of Coblence, and the principalities of Sayn and Wied, in what in 1919 formed the Coblence bridgehead. Following the French Revolution, the armies of the new republic carried the frontier of France to the Rhine. Napoleon became the all powerful arbiter of Europe and, at the height of his success, decreed the abolition of the Holy Roman Empire. From the states on the right bank of the Rhine, he formed the Confederation of the Rhine. The area on the left bank, now occupied by the Americans, was formally annexed to France and formed part of the Department of the Eifel. Violent though the wholesale destruction of these petty states was, it proved in the long run beneficial, since tariff frontiers were abolished and the groundwork laid for industrial development.

During the succeeding decade Germany lay dormant; but the national uprising of 1813 and the War of the Allies by ousting the French from the Rhine, paved the way for the unity of the entire German nation, which followed in the middle of the century. By the Treaty of Vienna in 1815, the Rhineland was awarded to Prussia in recognition of that nation's great services in the overthrow of Napoleon. The western portion of Germany received its present political form by this treaty. It is true that Prussia acquired a certain amount of territory in lower Hesse by the war of 1866, but, in a broad sense, it may be stated that the changes in Western Germany since 1815 have been merely changes incident to the unification of the states into the Empire. Alterations of state boundaries have taken place only on rare occasions.

An outline of the political organization of Western Germany of today, is therefore a succinct description of the results of the work of the master diplomatists at the Congress of Vienna.

Two provinces of Prussia are comprised within the territory occupied by the Allied armies at the present time. These are the Rhine Province and the Province of Hesse-Cassel. One other province, that of Westphalia, has, with these, also felt the agitation for a separate Western German State, which forms the subject of this chapter; but this province lies entirely outside of Allied jurisdiction. To the south of the Rhineland, on both banks of the river lies the former Grand Duchy of Hesse-Darmstadt, having within its borders the important cities of Worms and Mayence. Still further south is the Bavarian Palatinate, with the cities of Ludwigshafen, Speyer and Landau. This district, called in German the "Pfalz", is separated from the rest of Bavaria by the intervening states of Baden and Hesse-Darmstadt. One more autonomous state, the small principality of Birkenfeld, lies within the triangle formed by the Moselle, the Rhine and by Lorraine, and is surrounded by Prussia and the newly formed Saar state.

The Rhineland has always been a Catholic country, and in fact much of it was governed during the Middle Ages by the Archbishops of Cologne, Treves and Mayence. The grafting of the richest provinces of Germany to Lutheran Prussia was therefore a task requiring the wisest statecraft. Tact and diplomacy were as necessary for the Prussians to win the Rhinelanders to their side as was the maintenance of armed garrisons. These arts the Prussian lacked, quite as much in the past century as he does in the present. The machinery of Prussian civil service was installed with all its complexity and perfection of method. Whatever may be said against the Prussian system of government, one cannot but recognize the efficiency and faithfulness of its Landräte, councillors and minor officials. Even the most ardent supporter today of an independent Rhenish state recognizes that the Prussian genius for organization alone made possible the industrial and business prosperity which swept over the western portion of the country in the nineteenth century. The growth of a national German spirit, which sprang up in the wars of the Bismarckian Era, must also be attributed to the leadership of Prussia. Yet there was another side to the picture. While the Rhinelander admired the organization and the able administration which Prussia had accorded his country, he never fully reconciled himself to a government alien in religion and temperament. The Rhinelander saw officials come and go, the majority of whom had been born and educated in Brandenburg and East Prussia. Office holders of high rank with local traditions were so rare as to excite comment. As is natural when communities are governed by men from other parts, a belief develops that the interests of the locality are slighted. What few instruments of local self-government did exist in the Rhineland, were merely pawns in the hands of officials of the central Government. Despite the great industrial progress of the Rhineland, it would be vain to deny that, even before 1914, large elements of the population bitterly disliked Prussian rule. Still, under the old, conservative, reactionary government, there was no interference with the religion of the Rhineland and, until the revolution of 1918, the demand for an independent Western German State was so slight that it may be disregarded.

The reign of William II was marked by the great industrial development which converted Germany from an agricultural into a manufacturing country. The Rhineland took such an important share in this economic revolution, that today it is, with Westphalia, the wealthiest region of Germany. Important political events of these years were the formation of the Centrum party and the growth of the Social

Democrats. The former party was welded together by Bismarck's bitter attack on the Catholic Church. The ensuing struggle, usually called the *Kulturkampf*, united Catholics into one political party to maintain the rights and privileges of the Church. The indecisive results of the conflict in no way impaired the strength of the Centrum, which secured and thereafter maintained the balance of power in the Reichstag. Down to the outbreak of the World War, a period which witnessed the phenomenal growth of the Socialist party, the Centrum became absolutely indispensable to the Government. The voting strength of the party lay principally in Westphalia, the Rhineland and Bavaria.

Outline of separatist agitation.—As has been stated, the desires and longings of Western Germany for self government had remained dormant until the downfall of the Empire and the disaster to Prussian military power. The appearance of a separatist sentiment may even be traced definitely to the first days of the revolution. Since that event, the agitation has assumed a number of different forms, embracing now an entire territory, now an individual district; at one moment violent, at another quiescent. Whatever form is assumed, it was quick to incur the vituperation of all elements of the population which desired a strong German state. The first phase of the agitation lasted from November 1918, until the middle of May 1919. In this period, separatist adherents were moderate in their demands, and advocated separation of the Rhineland from Prussia and the retention of the new state within Germany.

This early phase was followed by another, lasting from the middle of May until the middle of June. This was marked by the appearance at Wiesbaden and in the Palatinate of extreme revolutionary movements. Dr. Dorten proclaimed a Rhenish republic at Wiesbaden, and advocated immediate separation from Prussia. Although much of the Allied press represented his *coup d'etat* as aiming at complete separation from Germany, Dr. Dorten's public statements always were moderate in tone, and emphasized the desire of the Rhineland to remain a state within Germany. On the other hand, the supporters of the movement in the Palatinate openly proclaimed themselves as favoring a neutral state, with, however, economic union with the Saar basin.

The complete failure of the leaders of these movements to gain popular backing was followed by still another phase, marked by the return of the Separatists to constitutional methods. Many influential Rhinelanders and Hessians believe firmly in the necessity of more self government, and in 1920, it appears certain that the western parts of Germany will eventually receive some measure of autonomy at the hands of the central government. The American military authorities at all times carefully observed the political situation and desiring to remain neutral in a purely German question, took no decided stand, until the authority of the regularly appointed German officials was threatened by Dr. Dorten's *coup d'etat.* This phase of the separatist movement will therefore be treated in this report in somewhat greater detail than the events themselves actually deserve, in order to explain the standpoint of the American military authorities.

First phase; November 1918—May 1919.—When the Emperor abdicated in November 1918, power passed to the Social Democrats, who organized a cabinet with Ebert as provisional President, and issued a statement proclaiming their intentions of shortly holding a national election. The men comprising the Ebert government had always been members of the opposition party, who had favored the extreme left, and, consequently it is not surprising that they developed radical tendencies at once. In the first few months of the new régime all established institutions were violently criticized and attacked, and both the Lutheran and Catholic Churches received their share of abuse. Not only did the Catholic Church face disestablishment, but loss of its revenue and private property as well. The Centrum Party naturally became alarmed, and its leaders in the Rhineland began to consider, as an alternative to the disestablishment of their church, the creation of a separate Rhenish state, in which the predominantly Catholic interests of the section might be safeguarded. The Ebert government, as it began to taste the responsibilities of Government, however, soon cooled down, and, as the Spartacist danger grew, began to realize the necessity of a unified front of all moderate elements. The results in the Rhine Provinces of the January election for the National Assembly were quite satisfactory for the Centrum party. The votes by parties were:

Centrum	1,665,027
Social-Democratic party	864,554
Democratic party	381,822
German National Peoples party	238,508
Independent Socialist party	188,733
German Peoples party	100,055

It will, however, be noted that although the Centrum had a considerable plurality in the province, it did not have an absolute majority of the voting strength. When the National assembly was convened at Weimar in March 1919, Ebert was confirmed as provisional President, and a cabinet formed with Scheidemann as Premier. The Centrum Party, which, shortly before, had made the statement that it had preserved citizenship, the Church and faith against the assaults of Social Democrats, now decided to coperate with the latter in the work of reconstruction. It even agreed to accept the separation of Church and State, provided the government safeguarded the financial interests of the Church; and it was on the basis of such a compromise, that leading members of the party consented to accept positions of responsibility in the cabinet. The formation of such a coalition government at this critical period, materially strengthened the country, assuring, as it did, the support of the moderate elements to the Social Democrats. While the Centrum party during this period never officially stated its position in regard to Rhenish separation from Prussia, it was generally believed that it was favorably disposed towards the idea, and desired the Rhineland to take its place in the nation alongside of Bavaria, Saxony and Würtemberg. As a party, however, they repeatedly denied their intention of forming a neutral state, independent of the rest of Germany, or an autonomous one, dominated by a foreign interest.

After the national elections, separatist agitation for a time died out, and what little discussion of the matter existed was generally confined to the columns of the "Kölnische Volkszeitung." The editor of this paper, Herr Frohberger, is an Alsatian by birth, and a Catholic priest. During this period he was probably the most active participant in the separatist movement.

Political situation at the end of May 1919.—May and June proved critical months for Germany. National unity was unquestionably threatened by Dr. Dorten's attempted revolution, and the Separatist movement in the Palatinate. Why, the end of May was chosen by the Separatists as the opportune moment to present their program to the people, is still a somewhat obscure point. It is significant to note, however, that the middle of May marked the awakening of the German people to the extent of their defeat. The announcement of the peace terms at Versailles was followed by a wave of discouragement which penetrated all classes, rich and poor. It was evidently felt by those who favored separation, that a Rhine republic would appear at this moment to the people as the panacea of all their ills. A few industrial magnates also may have viewed it as a possibility for preservation of their business and for a lighter indemnity, if the Rhineland became a neutral state. The support given the movement from such sources, accounts in large measure for the apparently unlimited funds at the disposal of the Separatists, and must not be underestimated. Yet, on the other hand, the presence of these elements among the Separatists seemed as counter propaganda for the government among the working-classes, and eventually proved one of the most effective weapons in the hands of the Unionists, in lining up the Social Democrats against the Rhenish movement. It is also noteworthy that a large group in France took this moment to express themselves as disappointed with the peace terms. Their opinions, as voiced by the "Echo de Paris" and other conservative papers, had constantly presented the necessity of the Rhine as the national French frontier. Probably such views were held by a minority of the French people, but at this time Germany was still feared in France, which again saw a future enemy in her. A benevolent attitude towards movements tending to weaken Germany, and the creation of buffer states, appeared natural to French generals, statesmen, politicians and to a very large group of the people as well.

The dispair of the Germans of their future, and the anxiety of a large number of the French for their own safety, made the moment appear an auspicious one to the Separatists to make a bid for popular support, using the desire of France for annexation as a means to their ends. They realized, of course, that the backing of the Allied armies was absolutely necessary to the success of the movement. Whether they felt at this time or not that the Centrists, would follow their lead, is a matter of conjecture. Leaders of that party had never expressed themselves publicly as favoring the movement. There is, however, reason to believe that at one time they secretly favored the scheme, in so far as it related to separation from Prussia. No German source has, however, suggested that the Centrum party stood behind the Separatists of the Palatinate. In regard to their participation in Dr. Dorten's agitation, conservative papers have claimed repeatedly, since its failure, that its scheme was originally one of the Centrum party.

May agitation in the Palatinate.—France has for centuries had an active interest in the border country called the Palatinate. Landau and Speyer were under French sovereignty, in the days of the Holy Roman Empire, and it was natural that France should believe that less antipathy existed to her there than in other parts of Germany. The friendship of certain individual Germans for France was noticeable there even in the early days of the occupation, and resulted in friction between the loyal Germans and the Francophiles. On May 22nd, General Gérard, commanding the Eighth Army, found it necessary to issue a proclamation forbidding the molestation by Germans of their fellow citizens who showed sympathy for "*la France, victorieuse et bienveillante*". In May, moreover, certain political happenings in Bavaria had produced a strong undercurrent of feeling not unfavorable to the idea of separation from Germany. Munich had been seized by local Spartacists and was in the throes of a reign of terror which threw all Bavaria into a turmoil. The conservative elements of the population naturally viewed these events with extreme distaste, and preferred separation from Germany to a continued union with an anarchistic state.

During the night of May 21st, unsigned proclamations appeared throughout the Palatinate, calling upon citizens to form a neutral republic, with economic union with the Saar basin. The following quotations from the proclamation are of interest as being the frankest and boldest expressions heard at any time during the agitation, for separation from Germany:

"MEN AND WOMEN OF THE PALATINATE

"As a result of the hard peace terms imposed by the Allied powers, Germany is not only politically and militarily rendered helpless, but also comes into an undesirable economic situation. It is not to be hoped that any essential amelioration can be got in the peace terms.

"The economic situation of the Palatinate, in comparison with the country on the other side of the Rhine, will become still worse, cut off as it is from the coal and industrial districts of the Saar, deprived of its own mines, diminished in extent by loss of almost three Bezirke, with their flourishing trade. Enclosed upon almost all sides and economically isolated, the Palatinate is to keep a military occupation for fifteen years duration, with attendant continual difficulties of intercourse across the Rhine. There will be three frontiers—a political frontier, a military frontier, a tariff frontier. There is but one road to safety from this situation of despair:

"FORMATION OF AN INDEPENDENT NEUTRAL STATE WITHOUT LOSS OF TERRITORY, WITH ECONOMIC UNION WITH THE SAAR BASIN

"We desire to remain German throughout—German institutions, German customs, German administration,—but we wish to insure to ourselves the advantages of economic co-operation with the Saar Basin, by which alone the Palatinate may be saved from otherwise inevitable ruin.

. . . .

"Citizens of the Palatinate! Let us have no false shame! Save our dear Fatherland and DEMAND THE FORMATION OF THE PALATINATE AS AN INDEPENDENT NEUTRAL STATE."

Several days later there were demonstrations in Speyer, Landau and Zweibrücken, but generally popular approval of the movement seems to have been withheld. The leaders presented a petition to Regierungspräsident von Winterstein, urging him to recognize an independent republic. The statement was also made that General Gérard would recognize, protect and support an independent neutral Palatinate. The German authorities promptly arrested the petitioners. Shortly after these events, Regierungspräsident von Winterstein was expelled from occupied territory and other government officials in Speyer and Landau, who had shown themselves particularly hostile to the Separatists, were arrested by French troops.

The action of the French military authorities in removing von Winterstein was the occasion for a protest on the part of General von Hammerstein, Chief of the German delegation on the Armistice Commission at Spa, which is of interest as a concrete expression of the views of the German government. In part it reads as follows:

> The French Commander in Chief of the Palatinate, General Gérard, and the French authorities under him, have undertaken by misuse of their rights under the Armistice Agreement, to promote and support the revolution which seeks the separation of the Palatinate from Bavaria and from Germany. The attempt to proclaim an independent republic of the Palatinate and the forceful separation of this political unit from Bavaria and from Germany, has been initiated by about twenty-one persons. The great majority of the inhabitants of the Palatinate, stirred and excited by the treasonable acts of these persons, are convinced that their leaders have been influenced by low and materialistic motives.
>
> At the present time, when the plenipotentiaries of the German government and the Allied and Associated governments are assembled in Versailles with the purpose of bringing about the conclusion of peace, the German government cannot conceive of so flagrant a violation of the right of self-determination, as that of General Gérard, in which he has the approval of the Commander in Chief of the Allied Armies, Marshal Foch, and that of the French government.
>
> The German government therefore believes that it may justly appeal to the principles of the future League of Nations, as approved by all peoples and states, and to righteousness and international morals, in expecting the immediate recall of the French Commander in Chief of the Palatinate and the immediate cessation of his acts.

The movement, however, despite the apparent anxiety of the German representatives on the Armistice Commission, never assumed proportions which warranted attempt by the agitators to remove the existing government, and Bavarian officials, except for a few individuals, dismissed by the French military authorities, continued in office. The agitation, however, did not abate, and in June an illustrated paper appeared, called the "*Pfälzische Woche*," which because of its advocacy of a neutral Palatinate was considered in many circles to be French propaganda.

Dr. Dorten's attempted "Coup d'etat."—Dr. Dorten's attempt at a revolution, complete fiasco though it was, had considerable comment in the Allied press, and particularly in Paris journals. More important for the success or failure of the movement, was the immense amount of discussion on the future of the Rhineland which it aroused in Germany. If France had seriously believed that Western Germany was ripe for separation, her hopes must have been shattered. She was perfectly aware that numerous groups in Hesse-Nassau and the Rhineland desired separation from Prussia, which, in reality, would give her the entering wedge she coveted in German affairs. The benevolent attitude which she adopted toward Dr. Dorten, far from aiding him, seriously compromised his position, and threw Western Germany, temporarily at least, back into the arms of Prussia. Not less important among the causes of the fiasco, were the personalities of the persons who appeared as leaders of the movement. Their obscurity forbade the press and public from treating them with the seriousness which the importance of their cause warranted. These facts alone probably doomed the movement to failure, quite apart from the decision of the British and American military authorities not to adopt a benevolent attitude towards Dr. Dorten and his followers.

The American zone lay like a strategic wedge between the areas of the Rhineland occupied by the other Allied armies. Its geographic position was such therefore, that a real impression of Rhenish unity for a separate state could not be achieved without at least some popular backing in that region. Furthermore, it contained the capital of the Rhine Province, Coblence, with the government buildings and high officials. Any revolution would be an unreality, in which the officials of the central government continued to act at Coblence.

Policy of American military authorities regarding changes of civil officials.—On May 22, 1919, the Commanding General, Third Army, received information from French sources that a Rhenish republic would shortly be proclaimed in Wiesbaden. General Mangin was interested in its success and begged for the assistance of the American Army in furthering it. The new republic was to be of an essentially conservative character, with Catholic and agricultural interests predominating. Although it was contemplated that for the present it would remain a sovereign state within the German confederation, later it would be entirely independent. About fifty backers of the movement were to come to Coblence the next day and begin an active propaganda throughout the American territory.

The request of the Commanding General of the Tenth French Army, and the arrival in Coblence the next day of supporters of the "Republic", obliged the American military authorities to come to a prompt decision as to the policy to be adopted towards the movement.

Paragraph V. of the Armistice provided in part as follows:

> The countries on the left bank of the Rhine shall be administered by the local authorities under the control of the troops of occupation of the Allies and of the United States.

When the Allies entered Germany in December 1918, they found that the duly constituted local authorities had, in many places, been Soldiers and Workingmen's Councils, which had sprung up during the revolution. It was then and there decided by the military authorities that the "local authorities" referred to in the Armistice were the officials chosen under the then existing German law. The Soldiers and Workingmen's Councils were therefore not recognized.

Under authority of Marshal Foch, elections were permitted to be held throughout occupied territory on January 19, 1919, for the German National Assembly. Similar elections were permitted on January 26 for the Prussian National Assembly. This permission was granted in order to enable Prussia and Germany to organize their governments on a constitutional basis, the Reichstag having been swept out of existence during the revolution. By the end of May, both assemblies had been constituted. That of Prussia was sitting at Berlin, framing a new constitution, while the National Assembly was at Weimar performing the same task for the nation. The Prussian government, on January 24th, issued a proclamation directing the holding of municipal elections in all Prussian territory, including the Rhine Province. The attention of the American military authorities having been drawn to this proclamation, the Oberpräsident was notified that the elections would not be permitted without approval of the competent Allied authorities. The matter was referred to Marshal Foch, who, in due course of time, notified the American Commander in Chief as follows of his decision:

> Upon proposal of the Allied authorities of occupied territory, I have decided not to grant the order of the Prussian Government calling an election for the entire renewal of the municipal councils of Rhenish Prussia.

The local authorities were accordingly notified that these elections would not be permitted.

In searching for precedents which should govern American policy in regard to the Rhenish republic, it was apparent that although granting Germany sufficient authority to hold elections to reorganize her government, we had consistently refused to permit her to hold local elections to change existing officials. As a revolution which would proclaim a state independent of Prussia and yet permit Prussian officials to remain in office, would be an absurdity, any official recognition of the movement would require a drastic change of American policy. Since, under the Armistice conditions, we were charged with the responsibility of administering the government through local officials, it was decided that the trained Prussian officials, who had satisfactorily performed their duties, would be preferable to a group of untrained political adventurers. Having refused the German people the right to change their officials by orderly elections, there was still more reason to prevent such change by revolutionary means.

Pursuant to this policy and having requested the officials to remain in office, General Mangin's representative was informed that we must refuse to recognize revolutionary movements of any character, and that if agitators for a Rhenish republic entered the American area, no matter whether of French or German nationality, they would be treated on the same basis as other agitators. The Army Commander added that he had no authority to deviate from these instructions, which had been laid down for him by higher authority.

This action of the Commanding General, Third Army, was sustained by the Commander in Chief of the A. E. F., who, on May 22nd, telegraphed as follows:

HAEF May 22, 1919.

C. G. Third Army,

With reference to telephone conversation re possible political agitation by outsiders in territory occupied by the Army of Occupation, the Commander in Chief approves the action already taken by you and he further directs that you refuse to allow French or any other political agitators to operate in the territory controlled by you no matter by whose orders they may claim to be operating or desire to be permitted to operate.

McANDREW
Chief of Staff.

Proclamation of the "Republic."—During the night of May 31, 1919, the following proclamation was posted in Wiesbaden, Mayence, and other towns of French occupied territory:

PROCLAMATION OF THE RHENISH REPUBLIC

At this critical time, when decisions of importance to the Rhenish people are about to be made, the people itself desires an opportunity for a hearing. It wishes to appeal to the right of self-determination—a right which has been generally recognized throughout the world and to which all opposing forces must give way. The Rhenish people desires a peace which will be the basis for a reconciliation of all nations. It wishes for that reason to dissociate itself completely from the evils which have caused many wars: namely, feudalism and militarism. After these evils have been abolished, the difficulties in the way of a lasting peace will be forever removed. The draft of the peace treaty contains provisions intended to compensate France and Belgium for the tremendous losses sustained by them during the war. The German government has recognized the justice of these provisions. Adequate guarantees are also provided against the resumption of hostilities. These terms involve a terrific burden upon the German people. The most important duty of the Rhenish people is to do everything in its power to bring about a universal and permanent international reconciliation.

We therefore make the following proclamation: An independent Rhenish Republic within the boundaries of the German nation is hereby erected. It is to be a republic of peace and will be composed of the Rhine Province, Old Nassau, Rhenish Hesse and the Bavarian Palatinate. Its establishment will rest upon the following basis: I. The boundaries will remain the same as heretofore; *Birkenfeld will be included.* II. Changes in boundaries may be made only with consent of the inhabitants of the territory affected, which must be obtained by a referendum. The government of the Rhenish Republic will be provisionally conducted by delegates selected by the undersigned committees. Permission will be obtained at once for the holding of elections to select members of the Rhenish State Assembly. These elections will conform in all respects with those at which members of the German National Assembly were elected. The seat of government and the place at which the State Assembly will assemble will be Wiesbaden. Provincial and local officials will continue to perform their duties as usual until further notice. The provisional government of the Rhenish Republic will take the place of the Central Governments of Prussia, Hesse and Bavaria, respectively.

Long live the Rhenish Republic!

Aix-la-Chapelle, Mayence, Speyer, Wiesbaden, June 1, 1919.

The Labor Committee of Aix-La-Chapelle,
The Nassau-Rhenish-Hessian Labor Company,
The Wiesbaden Labor Committee.

The posting of this proclamation met with prompt resistance from disorderly elements in both Mayence and Wiesbaden. Many of the sheets were torn from the buildings on which they had been posted, but reappeared the following night. The people generally claimed never to have even heard of the Labor Committees whose names were appended to the proclamation. The press, however, announced on the next day the composition of the cabinet which was to govern the newborn state. Its membership was as follows:

Dr. Dorten; Premier and Minister of Foreign Affairs;
Lawyer Eckermann; Justice;
Schoolmaster Klingelschmitt; Art and Science;
Schoolmaster Kremer; Education;
Workmen's Secretary Kraemer; General Welfare;
Dr. Liebling; Finance;
Monikes; Interior;
Salm; Agriculture and Food.

None of these persons was a man of prominence, and their names in no way tended to gather public opinion around the standard of the new republic.

During the afternoon of June 2nd, the Acting Regierungspräsident of Wiesbaden, Herr Meister, was summoned to French Headquarters and informed by the commandant of the city that he should submit to the orders of the provisionally established Rhenish state. The commandant used the words "Je vous prie," in transmitting the order, and as this phrase aroused some uncertainty in the mind of the Regierungspräsident, he requested an explanation. On being told that it was an order, Herr Meister answered that, as a Prussian official, he could not observe an order which would force him to commit treachery against his own people, and begged to be relieved from office. This request was granted, and Oberreigierungsrat Springorum, one of his subordinates, relieved him as Acting Regierungspräsident. The French, however, failed to renew their demands for recognition, so that at the end of the day, the old government still maintained its position.

During the evening, a telegram was sent out from Wiesbaden over the French military wires, announcing the formation of the "Rhenish Republic." This telegram aroused considerable discussion and was widely commented on by the Allied as well as by the German press.

> Mayence No. 11523 B This morning Rhine Republic was proclaimed in all cities without difficulty. Provisional government presided over by Doctor Dorten is at present installed at Wiesbaden and is obeyed. This event which ends annoying uncertainty appears desirable to majority of population who remain very calm. Doctor Dorten has addressed message to Marshal Foch, to the Conference, and to all Generals commanding Armies of Occupation for their respective governments.

Dr. Dorten fails to gain popular support.—The Rhineland seethed during the next week with discussions of Dr. Dorten's *coup d'etat*. The events following close on one another, rendered success of the movement impossible, at least in so far as it ever depended on popular support. The attitude of the Social Democrats was already a foregone conclusion. Its strength in the January elections lay chiefly in the populous industrial towns of the lower Rhine. The party organizations of Cologne, Crefeld, Solingen and Essen came out immediately in opposition to Dr. Dorten. In Mayence, on June 4th, shopkeepers and workers struck, in order to signify their disapproval of the movement, and simultaneously in Coblence the American military authorities only narrowly averted a strike of railway employes, who appeared to want to join the demonstration. When, however, the strikers were informed that the Army would not permit a strike, inasmuch as it would hinder the supply of the army, they contented themselves with sending a telegram of protest to Premier Scheidemann of the German government. As the telegram probably faithfully presented the ideas of the majority of the working-classes in the American territory, the following sentiments are significant:

> The Coblence directors of the Railway Employees Association, representing the majority of the employees in the Coblence district, vehemently protest against the traitorous act of a small minority who propose to establish an independent Rhenish republic.
>
> We urge the national government to bring the leaders of the movement to trial on the charge of treason.
>
> We, Rhenish railway employees, cleave to a united Germany and solemnly give our oath of fealty to our German brethren beyond the Rhine.

These demonstrations left only the attitude of the Centrum as an uncertain factor in the situation. All eyes watched its leaders. On

June 3rd a proclamation appeared in the *Coblenzer Zeitung* which relieved the prevailing uncertainty. It follows in full:

RHENISH MEN AND WOMEN.

> We hereby raise sharpest protest against the discord sown in the ranks of Rhinelanders in this, the hardest hour of the German nation, by the summons for formation of a Rhenish republic. Never was unity more necessary than at this moment. He does not deserve the name of a German who in the hour of greatest common need prefers personal, political and individual ends to the welfare of the whole Fatherland.
>
> We believe the great majority of the Rhenish population condemn the doings of the so-called Working Committee. We appeal most earnestly to all of our Rhenish countrymen in the critical hours before us, to remain ever true to our beloved German Fatherland and not allow themselves to be led astray by promises or threats of any kind.
>
> COBLENCE, June 3, 1919.
> (Signed) Centrum Party,
> German Democratic Party,
> German Peoples Party,
> German National Peoples Party,
> Rhenish Peoples Party,
> Social Democratic Party.

The signature of the Centrum, heading the list of parties, even taking into consideration that the proclamation was purely a local act, was an all-important indication of public opinion. Either opposition to Dr. Dorten was overwhelming, or a disastrous rift in the Centrum party had already arisen. The appearance of similar protests in other cities, however, showed the party to be unanimous in its disapproval.

It is not difficult to sum up the reasons for the Wiesbaden fiasco. Primarily, it had the support of none of the important parties of the Rhineland. Failing in this, only popular heroes or chaos could have afforded the movement a chance of winning popular favor. Certainly Dr. Dorten and his associates were not the type of men to launch a great popular movement, and the disorganization of the country was still not so complete as to utterly destroy the morale of the nation.

There was, however, another factor in the situation which must not be neglected. The Rhinelanders were a Germanic people, who, according to their lights, had fought for their Fatherland through four years of strife. They had finally seen the debacle of their nation and the invasion of the homeland by enemy armies. From school days on, they had been taught to believe that France was a determined enemy of their nation. Now, rightly or wrongly, they believed that France was interfering in a strictly internal matter of their own, and was fostering national disunion for her own purposes. All classes, whether they had formerly advocated more extensive self government for the

Rhineland or had believed in continued Prussian sovereignty, now united in protest. At this moment it was the unanimous German belief, that the French military power stood behind Dr. Dorten.

Attitude of German and Prussian governments.—Discussion of the movement was as violent and bitter in Berlin and Weimar as in the Rhineland. Practically every official or politician of prominence came out in public utterances, condemning the movement. Premier Hirsch of Prussia, in fact, used such strong language in condemning the previous attitude of the Centrum as to bring on himself the criticism of every Rhenish newspaper. Herr Kastert and Herr Kückhoff, prominent Deputies of the Centrum at Weimar, found their continued presence at Weimar untenable, on the mere suspicion that they had a conference with General Mangin at Mayence. In Cologne, Herr Frohberger was obliged to retire from the editorial staff of the *Kölnische Volkszeitung.* Ministers and Deputies generally took an exaggerated view of the incidents at Wiesbaden. The Coblence newspapers of June 3rd contained a dispatch from Berlin, stating that the Prussian government was meditating a general transfer of higher officials of the Rhine Province. It stated that Herr Trimborn was being considered as a successor to Oberpräsident von Groote, and a Herr Fuchs to Regierungspräsident Dr. Momm of Treves. Other dispatches from Berlin emphasized the undue alarm which had seized the government on learning of the events at Wiesbaden. In desperation at what was considered a critical situation, the government now attempted to strengthen their position in the Rhineland by appointing officials, loyal to Prussia but recognized as having the sympathy of the local population. However, the policy of the Third Army in regard to changes of civil officials remained as firm against changes by edict of Berlin as by revolution. The Oberpräsident and the Regierungspräsident of Coblence and Treves were informed that they were to disregard orders of transfer, unless they had first obtained the approval of the army authorities. The feverish haste of Berlin to strengthen its position in the Rhineland could not but appear amusing to both Germans and Americans, who were in a position to estimate rightly the importance of Dr. Dorten's manifestoes. The proposed transfers were clearly inadvisable both for military and civil interests. Herr von Groote was a Rhinelander by birth, who had won the Oberpräsidency for himself by over twenty years service as a Landrat. He had served the American army with faithfulness, tact and due regard to the interests of his people. His transfer at such a moment would have been equally harmful to Germans and to Americans.

The fall of the "Republic".—During the morning of June 4th, Dr. Dorten, accompanied by his "cabinet" and Col. Pinot, the French administrator, appeared at the government building in Wiesbaden. Acting Regierungspräsident Springorum was summoned to meet the party, and at once scenting danger, asked Col. Pinot to define the attitude of the French government. On being told that France would preserve a neutral attitude in a purely German question, and that his presence was merely to assure public order, Herr Springorum personally "escorted" Dr. Dorten to the street. The cabinet members were ejected at the same time, in accordance with Herr Springorum's instructions. The "Ministers" appear to have been roughly handled by the populace, which had gathered outside the building. Herr Eckermann, Minister of Justice, and Herr Kremer, Minister of Instruction, were so severely bruised that they had to be taken to hospital. The story of this incident appeared in the press and a storm of ridicule descended on the head of Dr. Dorten and his government, nearly all newspapers taking this occasion to affirm the loyalty of the Rhine Province to Germany. Dr. Dorten, confronted with failure in his attempt to take possession of the Government Building, issued a statement that the old officials would remain in office for the present. The new republic, once an object of fear on the part of Prussia, now appeared merely ridiculous.

On June 11th, General Mangin took occasion to assure the Bürgermeister of Mayence of the neutrality of the French army in domestic German affairs. Count Brockdorf-Rantzau, the German delegate at the Peace Conference at Versailles, paid a flying visit to Cologne for a conference with Cardinal v. Hartmann, the Archbishop, Oberbürgermeister Adenauer and other prominent Centrists. On his return to Versailles, he expressed himself as assured of the impotency of the separatist movement. These events had fully shown the wisdom of the American policy of retaining the old officials in office. Any other course than that adopted, would have involved us in the support of a group of adventurers, who would have controlled the government without the shadow of a popular following. Such a situation would have spelled untold difficulties for the American army.

Renewal of movement during the summer.—Dr. Dorten, following his summary expulsion from the Wiesbaden Government Building, disappeared also from the front pages of the press. It was realized that in the Rhineland, however, he was merely biding his time, and working to create a public sentiment favorable to his cause. Information reached American Headquarters, from time to time, that his propaganda was in full swing in French occupied territory, but it was difficult to ascertain what its success was. Dr. Dorten's agents gave the American area a wide berth, for a few weeks, but gradually evidence cropped out that his propaganda was being carried on in the vicinity of Coblence.

The German police on June 15th arrested Philip Hellwig, his wife and Benedikt Wirth, who had come from Wiesbaden, and had been discovered in the act of distributing pamphlets which advocated an independent Rhineland. They were lodged in the city jail and proceedings were begun by the civil officials to try them for high treason. The American army had never permitted the civil tribunals to bring action against German citizens for political offenses. The act of Dr. Dorten's agents, however, in scattering propaganda which had been unauthorized by the army, constituted a distinct breach of American regulations. The three offenders were therefore withdrawn from civil jurisdiction and brought to trial before an American provost court. A fine of 5000 marks was imposed on Hellwig by this court, which fine was paid by a French officer. The military commander of Coblence took this occasion to warn the civil authorities that they must not presume in future to try individuals for any political offense whatever, without previous sanction of the American authorities. This warning was considered very necessary in view of the high tension existing at the time between various groups of political adherents. Our actions in trying Hellwig before an American court undoubtedly impressed all German parties with our complete neutrality in the Rhenish question.

On July 15th, the *Coblenzer Zeitung* reported a movement as being on foot to form a Greater Hessian State within the German nation. The boundaries of the new republic were to include both Hesses (Hesse-Darmstadt and Hesse-Nassau) and the Palatinate. Ministerpräsident Ulrich of Hesse-Darmstadt was declared to be the prime mover in this new movement. The subject was discussed at Weimar, and reference made to a reported visit of Ministerpräsident Ulrich to General Mangin at Mayence. Ministerpräsident Ulrich replied personally to the accusation and stated that his Hessian policy was thoroughly in accord with that of the national Ministers. He added that his visit to General Mangin at Mayence had been merely for the purpose of adjusting certain administrative difficulties.

On August 13th, Dr. Dorten was arrested at Cologne by the German police. An explanation was at once demanded of the Oberbürgermeister by the British military authorities. It developed that Dr. Dorten had been arrested and placed in what is known as "protective custody". The police stated that there were no charges of high treason involved, but that it was necessary to restrict Dr. Dorton's movement to protect him from the mob. The British authorities, however, considered the action of the police as unwarranted, and gave orders that Dr. Dorten be given the freedom of the city, on condition that he pledge his word not to leave without previously obtaining permission from the British. Later in the day there was a hostile demonstration before the house of Herr Frohberger, editor of the *Kölnische Volks-*

zeitung, with whom Dr. Dorten was staying at the time. The mob, however, was easily dispersed by the German police. The next day Dr. Dorten was directed by the British to leave Cologne, and to return only on permission of the military governor. This incident, though of relatively slight importance, probably induced the British to issue a proclamation which warned the civil population that no revolutionary changes of government officials would be permitted. Their policy was thoroughly in accord with that of the American army, both insisting that a Rhenish republic, aside from the question of its desirability, could be brought about only by the Germans themselves and then in a constitutional manner.

The National Assembly at Weimer finally ratified the new German constitution, on August 3rd. In the debates which preceded the adoption, the danger of separating the Rhineland from Germany had been repeatedly emphasized by members of all parties. Although the Centrum had very strongly urged the creation of the Rhineland as a sovereign state, it, too, shared the prevailing German belief that Dr. Dorten's movement savored entirely too much of foreign intrigue. The constitution made provision in Article 18 for the creation of new states, either by the separation of a province or district, or by the union of petty states, providing the people of the territory concerned had expressed themselves by a plebiscite as so disposed. In order to ward off the immediate danger of disunion and to calm the excited state of public feeling in the Rhine Province, the party leaders finally agreed to postpone the effectiveness of Article 18 for two years. This decision, spoken of in German as the *Sperrfrist*, was directly incorporated into the constitution as Article 167. By so doing, Dr. Dorten's movement was outlawed for the time being.

Further agitation in the Palatinate.—During the early summer of 1919, the Separatists of the Palatinate had been carefully paving the ground for a new attempt to seize the reins of government. After their unsuccessful attempt in May to enlist popular backing, they realized that they must secure the support of at least one political party to attain even a measure of success. The parties of the right and the majority Socialists were bitterly hostile; the Centrum was lukewarm. Only the Independent Socialists, ever on the look-out to stir up trouble, fell in line with their schemes. Haase, the leader of the May upheaval, now found an able, if unsavory, assistant in a man named Eggelsdorf. The latter individual had a long criminal record and was reputed to have been one of the Spartacist leaders in Munich.

A large crowd attempted, on the night of August 27th, to storm the post office building at Ludwigshafen. For a while, the employes succeeded in holding off the mob, but the attack was renewed at about 4:00 a. m., with hand grenades, and all resistance was overpowered.

The manager of the post office and one of his assistants were killed in the melee. French troops were called out and the city was policed by them. The attempt of the Separatists to proclaim an independent Palatinate came to a prompt and final conclusion, but not before a profound impression had been made on all classes in occupied and unoccupied Germany. The French military government investigated the occurrences and placed the blame on the murdered officials, who were declared to have been overzealous and untactful in dealing with the mob. Regret was expressed, however, for the occurrences. In view of the repeated charges from German sources that General Gérard had instigated the outbreak, it is only proper to quote the reply of the Commander in Chief of the Allied Armies to the protest of the German delegate on the Armistice Commission:

> The French military authorities of the Palatinate endeavor to observe scrupulous and strict neutrality between the different German political parties in occupied Rhenish territory. The French military authorities are guided by the sole desire to maintain order. Its relations with the German population are guided by this consideration alone.

Dr. Dorten's statement.—The Officer in Charge of Civil Affairs in Kreis Bitburg received information, on the evening of August 24th, that Dr. Dorten was in his Kreis and had that afternoon addressed a meeting of adherents of the Rhenish movement in the town of Kylburg. Efforts to find Dr. Dorten were unavailing, but on the next day, his motor car was seized by the military authorities, together with a man who appeared to have been with him. On orders of the army authorities, the car, together with Dr. Dorten's colleague, Eugen Hauptmann, a journalist of Bonn, were brought to Coblence. In order to acquaint Dr. Dorten definitely with American regulations on political propaganda and to prevent meetings in future similar to the one in Kylburg, it was thought best to send Herr Hauptmann to Wiesbaden to invite Dr. Dorten to come to Coblence for an interview. A pass was granted Dr. Dorten for this visit.

Dr. Dorten accordingly appeared at American Headquarters on August 30th and on the same day, representatives of G–2 and the Office of Civil Affairs interviewed him in connection with the meeting at Kylburg. He stated that his flight was due to information that the Prussian police were pursuing him, and that he had not learned that he had been wanted by the American military authorities for violation of their regulations. He announced to the authorities that he had no intention of violating the orders of the occupying forces, and begged that he be furnished with a copy of the regulations which he was asked to obey in American territory.

The American representatives then took occasion to question Dr. Dorten on the political platform of his party and the future intentions of the Separatists. Dr. Dorten stated that his movement was purely German in character and aimed solely at separation from Prussia. The proclamation of June 1st had been published in order to stimulate the demand for a plebiscite. He felt it necessary to erect an organ which could speak in the name of the Rhenish people. Feeling that Prussia would never accept the idea of free determination of peoples, he had taken it upon himself to proclaim a republic and to appeal to the Entente to force Prussia to recognize the Rhineland as an independent state. He declared that his movement was perfectly constitutional and that Article 167 could not apply to the Rhenish agitation, as the republic antedated the constitution. On being questioned as to his further plans, he stated that he was collecting signatures to a petition for Rhenish independence. Upon obtaining the names of half the electorate (he claimed then to have 1,400,000), he intended to present the petition to the Entente and beg them to require Prussia to grant a plebiscite to the Rhineland. He commented on the occurrences in the Palatinate, and condemned the leaders of the Separatists in that province. The Rhenish republic desired to include the Palatinate within its boundaries, and he had assisted the opponents of the Separatists in that region. He believed that separation of this border-country from Germany ought to be avoided at all cost. He added that General Mangin, while not personally unfavorable to the Separatists, had officially stated that the French military authorities at Mayence would observe a strict neutrality.

The American representatives then told Dr. Dorten that the United States was governing its portion of the Rhine Province under the provisions of the Armistice, which document required them to exercise control through the existing officials. Change of these officials by revolutionary means could not be contemplated, and propaganda aiming to force such action would not be tolerated in the American zone. In closing the interview, Dr. Dorten was warned that it would be necessary for him to obtain a pass from G–2 before re-entering American territory, a restriction which, however, was shortly afterwards removed.

Separatist movement in Birkenfeld.—Birkenfeld, a small principality of the upper Rhineland, surrounded entirely by territory of the Rhine Province, did not escape its share of the current of Separatism. Birkenfeld was a medieval remnant of the domain of the Grand Duchy of Oldenburg, and before the revolution was governed by the ducal officials with the assistance of a local diet. After the revolution, the Grand Duchy became a republic, but the relation of Oldenburg and Birkenfeld remained practically the same as before. Such a system of control was extremely unsatisfactory to Birkenfeld, and its citizens realized that the only solution of their difficulties would be for Birkenfeld to seek union with a neighboring state. The Rhine Province was naturally favored by a large number of the population, but a certain minority favored union with the Palatinate, while still another group, although a very small one, desired annexation to the newly formed Saar state.

On July 14th, a group of nine men, most of them citizens of Birkenfeld, constituted themselves into a provisional government and declared a republic independent of Oldenburg. Although unrecognized by Germany, they issued a proclamation, guaranteeing that Birkenfeld should remain German, and adding that the old government would remain in force until the Provincial Committee (Landesausschuss) had elected a Constitutional government. This new government would leave the question of incorporation of Birkenfeld into another state or union of states, to a plebiscite.

On July 17, the Provincial Committee, which had in the meantime been elected under French supervision, met in Birkenfeld. In the discussion which ensued, the usurpers proved so weak that Regierungspräsident Hartüng and the officials of the old regime managed to stay in office. It was decided by all parties, however, that secession from Oldenburg was desirable, and a delegation was appointed to proceed to the capital and arrange for the legal secession of the former principality from Oldenburg.

On August 30, the Provincial Committee and the government of nine met once more with Major Bastiani, the French administrator of Birkenfeld, present. The report of the delegation sent to Oldenburg was thoroughly gone over, with particular attention to the resolution of the Oldenburg Landtag, that it would at once begin negotiations with Prussia for the incorporation of Birkenfeld into the Rhine Province. This was bitterly opposed by the revolutionary committee, which claimed that the delegation sent to Oldenburg had exceeded its powers. After more discussion, the Provincial Committee, with the exception of one member endorsed the resolution of the Oldenburg Landtag. The revolutionary committee of nine then announced that they would take over the government, and that the Provincial Committee was dissolved. Under the influence of Major Bastiani, the old government yielded and announced its resignation. Herr Zoller of Zweibrücken, a Palatinate activist, was announced by the revolutionary committee as President of the Birkenfeld Republic, and men by the names of Haupt and Eifel were appointed assistant administrators.

The new President at once telegraphed to the German government, asking for recognition of the independent republic as a federal state of the nation. This request was promptly refused by the German government, which, in its reply, pointed to Articles 18 and 167 of the Constitution, which forbade the creation of new states before the year 1921.

Herr Zoller and his confederates paid no attention to this decision, but proceeded to organize their government. Popular feeling in Birkenfeld, which almost unanimously opposed the new régime, became greatly agitated, and a series of political strikes followed, which came to an end only with the intervention of French troops. Heavy fines were imposed on the agitators of these strikes by the military provost courts, which also ordered a number of the more prominent supporters of the old régime to leave occupied territory.

General Mangin received two delegations from Birkenfeld, on September 17, at Mayence. One represented the old régime, and the other the Zoller government. An exact report of the results of the conference has never been published, but the Zoller government continued in power. General Mangin, however, publicly authorized municipal elections throughout Birkenfeld on September 28th and for the local Diet on October 5. This authorization was subsequently revoked and the municipal elections were postponed until October 20 and those for the Landtag until October 26.

The result of these elections proved an overwhelming disaster to the Zoller government. Although the opposition parties charged Zoller with every sort of chicanery and misuse of power to influence the elections, the vote was so decisively anti-government as to leave no possibility of doubt that Birkenfeld rejected Separatism. The result of the municipal elections on the 20th indicated the state of popular feeling towards Zoller. Birkenfeld, the stronghold of the movement, out of a city council of fifteen, returned only three Separatists. The Landtag election on the 26th was bitterly contested by both parties. The Zoller government received in all but 700 votes, while the opposition rolled up a total of 13,051. The new Landtag as a result, contained only two Zollerites in a membership of twenty-five.

Declaration of Centrum party.—An important convention of the Centrum party in which the Rhineland question was thoroughly gone over, was held in Cologne on September 16th. The principal speaker was a Dr. Lanscher, formerly one of the most outspoken members of the party in favor of a separate Rhineland. He confessed that owing to the failure of Westphalia to join the movement, it could not succeed on economic grounds, especially as all the political parties on the left bank of the Rhine, except the Centrum, and even a part of that party, opposed separation. Finally, there was the danger that the Rhineland, like the Saar district, might fall under control of the Allies. All the succeeding speakers seemed to be animated by the desire of preserving a united Germany. The convention finally adopted the following resolution, with only two dissenting votes. It reads, in part, as follows:

> The Rhenish question has been placed on a totally new basis through new developments, especially through constitutional and legislative enactments of the National Assembly.

Cession of certain sovereign states rights to the national government was necessary in order to stimulate the development of the idea of a united nation. This also establishes a new basis for the efforts for creation of a Rhenish state. Taking the changed conditions into account, the convention of the Centrum party, here assembled, has decided that the following principles shall guide the party in its actions on the Rhenish question:

1. The wish for Rhenish autonomy shall be granted only in connection with a territorial reorganization of the nation which will take into consideration the interests of the entire nation, and must be accomplished in the manner prescribed by the constitution.

2. The final object of territorial reorganization shall be the creation of a united nation, composed of autonomous federal states. In accomplishing this, various states will be partitioned or united with each other, and Prussia in particular will be forced to relinquish her position of preponderance. It is intended to create states having equal rights, upon a basis of tribal affinity, and to secure a maximum of ethical and economic good thereby. All states are to be granted the greatest possible autonomy.

3. In case it be impracticable to realize the plan of a united nation, as outlined in the first and second paragraphs of this resolution, efforts looking towards the creation of new federal states, in accordance with the provisions of Article 18 of the Constitution, may again be instituted.

4. We unreservedly recognize the provisions of the National Assembly, including those providing for the territorial reorganization of Germany. We do, however, expect, in case national necessity and vital interests of the Rhineland should at any time in future demand an abrogation of Article 167, that the national government will declare this article to be without effect. (Article 167 provides that no efforts for separation of the Rhineland from Prussia shall be instituted for two years.)

5. In the interim, we demand the fullest autonomy for all Prussian provinces. The autonomy promised by Prussia is merely a pretense, and will not suffice the Rhenish people.

6. Whosoever takes part in any schemes seeking either directly or indirectly to effect a separation from the nation, harms the national, ethical and economic life of the population of the Rhineland, and is considered as having severed his connection with the German Centrum party.

7. The convention requests the national government to institute a territorial reorganization of the nation, as outlined above, as soon as possible.

Change in French Rhenish policy.—The French government officially announced, on October 8th, the unification of the Fayolle group of armies on the Rhine, the Tenth and Eighth, together with the relief of their commanding Generals, Fayolle, Mangin and Gérard, and the appointment of General Degoutte to command the permanent French Army of the Rhine. The action of the French government in relieving these Generals, who had achieved a decided popularity in the French army and nation for their achievements during the war, was bitterly attacked by a large part of the press, particularly by the conservative royalist papers. The *Echo de Paris* stated that the action of the government in General Mangin's case was due to the latter's activity in furthering the Rhenish movement for independence. It furthermore hinted that the action of the Premier was due to repeated protests by the American delegation to the peace conference on French policy in the Rhineland. This unwarranted charge was at once denied by the American Peace Delegation.

The German press, keenly alive to any possible evidence of dissension in Allied ranks, magnified the relief of the French generals into a matter of national importance. All seemed to take it for granted that the Rhenish movement, without French support, was doomed to speedy death and the unity of German peoples was no longer threatened.

The American army had at all times preserved a disinterested and impersonal attitude in the movement. While the Army Command believed that the creation of an autonomous Rhenish state within the German union would prove in the end desirable both from the German and Allied points of view, it did not feel at liberty to interfere in a purely internal affair of Germany.

The reasons actuating the adoption of the American policy may be summarized as follows:

1. The Rhineland was being governed under the terms of the Armistice. That instrument required government through the medium of existing officials. American recognition of a Separatist government would have involved the wholesale replacement of officials.

2. The American task was primarily that of maintaining order. Revolutionary agitation by parties aiming to create a new nation could lead only to disorder.

3. An autonomous Rhineland cannot be brought about by foreign armies and misuse of military power. The Rhenish question was a peculiarly *German* question. Mere suspicion that a foreign power stood at the elbow of Dr. Dorten was sufficient to prevent the Rhenish people, at heart ardently desiring an autonomous state, coming to the support of a Rhenish republic.

CHAPTER No. 16

PROPOSED MILITARY GOVERNMENT IN CASE OF A FURTHER ADVANCE

Political Situation, Middle of June 1919; Situation of Military Government at Time of Contemplated Further Advance; Strategical Outline of Proposed Advance; Political Geography of Proposed American Advanced Zone on Right Rhine Bank; Plans for Extension of Military Government; Proclamations and Regulations Prepared in Case of Further Advance; Recapitulation.

Political situation middle of June 1919.—The armistice agreement of November 11, 1918, had required Germany to evacuate the left bank of the Rhine, with the adjoining bridgeheads, as a guarantee that she would refrain from further aggressions. The enemy was to be so placed in irons that he must either sign the Allied peace terms, whatever they might be, or consent to renew the war under conditions extremely disadvantageous to himself. In the immediate months following the armistice, there seemed no possibility of a recurrence of hostilities. The Spartacist situation was regarded by the old Junker element, formerly the exponents of Pan-Germanism, as more dangerous to Germany, at least for the time being, than even complete occupation of the country by the Allies. Influential Berlin newspapers even begged that foreign troops garrison the larger cities of Germany in order to end once and for all the anarchy which was threatening to render economic recovery of the country impossible. The government beseeched the Allies to present the peace terms to them at once, in order that the industrial life of the nation could be resumed. Enemy demobilization was apparently proceeding smoothly and rapidly, and the possibility of a resumption of the war did not seem to be seriously entertained either in Germany or in Paris.

The situation, however, had altered so considerably by June that renewal of the war appeared as a distinct possibility. For several weeks following the presentation of the Allied peace terms at Versailles, the question of peace or war seemed to depend on a throw of the die. The voice of the old war party was heard once more in Berlin; the mailed fist reappeared, after not having been seen since November 1918 and the streets once more re-echoed to the tread of disciplined troops.

Complete and striking as was this change in public feeling, it came as no surprise to the armies on the Rhine. The intelligence staffs of the several Allies had carefully followed the ups and downs of German politics. They had minutely scrutinized the reorganization of the army and had noted the re-entry of the Junker element into national affairs. On the Rhine, as early as April one began to feel that Germany's signature to the peace treaty might not be secured so easily as had been supposed in January. The reasons for the change in the temper of the civil population can be ascertained by an examination of the events which occurred in the intervening months.

First and foremost must be reckoned the unexpected delay before presentation of the terms. In January, Germany was still enmeshed in the throes of Workmen's Councils. Her people were suffering from lack of food and all classes hoped for and looked forward to a speedy peace. The government was feeble enough and was to all appearances growing weaker day by day. Much had occurred in the interval between January and June. Order had been re-established, the national elections had been held and the position of the Majority Socialists confirmed by their results. The solidarity of all the conservative elements behind the Ebert government brought about inevitably the re-establishment of an organized military force. In November and December 1918 old army units had either completely disintegrated or become so thoroughly imbued with radical ideals as to be a source of danger rather than strength to the government. The birth of the Reichswehr—the new army—may be traced to the formation of bands of volunteers, enlisted independently by energetic officers of the old army. It was recognized by the government when its value was proven in the January riots in Berlin. They were thoroughly organized under Noske and by June the volunteer bands had become an organized force, numbering fully 400,000 men.

Fully as important a factor in raising the morale of the people was the lifting by the Allies of the food-blockade. With food available in quantities larger than had been their lot in the past four years, the spirit of the people inevitably revived. Rumors began to reach the German people in April and May, of the severe peace-terms being prepared in Paris. Coincidently a return of the old feeling of national dignity was noted. Speakers of all parties expressed themselves as ready to ruin Germany rather than submit to her degradation. It must be remembered that Berlin was just as aware as Paris or London that the Allied public were no longer so keen for hostilities as they had been in November 1918. These German politicians who now talked of plunging their country into inevitable disaster, apparently counted on the appearance of Bolshevism in France and England to prevent the Allied nations from exercising their full strength.

But if certain elements in the German state did advocate a last resistance, other and more numerous groups were quite as loud in demanding complete surrender. These were inevitably the wiser, more far-seeing elements. To Germany as a whole, it should have been easily apparent that her loss of munitions and railway material under the armistice had been so serious as to almost preclude the possibilities of a defensive struggle, let alone driving the Allies from German soil. This group of men believed that signs of discontent within enemy states would prove as illusive in the future as in the past. However much they may have disliked to sign the terms, they realized that to force an issue in June 1919 would merely incur the destruction of Germany.

Secretary Dutasta of the Peace Conference presented the reply of the Council of Five to the German counter-proposals to Count Brockdorff-Rantzau, on June 16th. This reply gave Germany five days to signify her willingness to sign the peace-terms. It also warned Germany that in case of a refusal, hostilities would be resumed at once.

Evidence of an approaching governmental crisis appeared in Weimar almost at once, and simultaneously a decline in the chauvinistic spirit of the Assembly could be noted, in striking contrast to the earlier spirit of resistance. By June 20th, competent critics believed that Germany must and would signify her assent to the peace terms. Nevertheless the preparations of the army on the Rhine to advance and occupy further German territory, were in no wise abated, and by June 20th all was in readiness.

Situation of military government at time of contemplated further advance.—The situation of the military government in American occupied territory in the period of June 18th to 23rd, 1919, was briefly as follows:

Territorial governments were in full operation in all the Kreise of Bezirk Treves which were under American control. They had further been instituted in Kreise Adenau, Cochem and Ahrweiler, of Regierungsbezirk Coblence. Briefly, each Kreis was controlled by an Officer in Charge of Civil Affairs, who functioned directly under the control of the Officer in Charge of Civil Affairs, Third Army. In order to maintain order in the districts left without garrisons on the departure of divisions to the United States, companies of military police had been transferred to each Kreis capital. These bodies of troops, small as they were, had effectively maintained the authority of the United States by motorized patrols over a very extended area.

The forward Kreise of the American area on both sides of the Rhine still billeted the regular divisions which had been retained in the Army of Occupation. On the left bank, the Third Division was billeted around Mayen, while the Fourth was further down the Rhine in Kreis Ahrweiler. On the right bank, the First and Second Divisions garrisoned the American section of the bridgehead, dividing between themselves Kreise Unterwesterwald and Neuwied, with adjoining strips of Altenkirchen, Marienburg, Westerburg and Coblence-Land. Coblence-Stadt and that part of Coblence-Land lying on the left bank of the Rhine, held the Headquarters of the Army, together with units of army troops. Control of this district was exercised by the garrison commander of Coblence. In none of these forward Kreise, had territorial governments been constituted. They were controlled, as formerly, by the Commanding General or Officer of the division or army troops which occupied them. Each General had on his staff an officer, designated as an Officer in Charge of Civil Affairs, who acted as the Division Commander's representative in civil matters. It was unfortunate that no uniform method of treating civil matters had ever been instituted in the American area. The existing system was theoretically far from perfect, but since it had developed bit by bit as a result of contact with local conditions and institutions, its imperfections must be considered as inevitable.

Strategical outline of proposed advance.—Upon expiration of the time allowed the German government to signify its willingness to sign the peace-terms, the armistice would automatically come to an end. It was proposed to institute at once a vigorous advance into unoccupied Germany in order to bring the government to terms. This advance would have limited objectives, the supposition being that Germany was ready to sign the peace, provided a sufficient show of force was made by the Allies. The forward movement was to be carried out simultaneously by each of the Allies armies.

The contemplated advance of the American Army was to be in a northeasterly direction in conjunction with the French Tenth Army on the right and the British on the left. The right of the army was to follow a line marked by Oberlahnstein, Limburg, inclusive, Wetzlar, inclusive, Lauterbad, inclusive. The left boundary of the army zone was to be marked by a line, Linz, inclusive, Siegen, exclusive, Brillon, exclusive. The limit of the advance was to be made by the III Corps, composed at this time of the 1st and 2nd Divisions, these two divisions being disposed from right to left. The 3rd Division, which up to this time had belonged to the III Corps, was now detailed and assigned to a special duty. As we had no cavalry at this time in Europe, the French had lent the American Army their Second Cavalry Division. This unit was assigned to the duty of closing the gap on the left flank, which would arise because the British were separated from us by the neutral zone between the two bridgeheads, at least during the initial phases of the forward movement. As soon as the army had reached the designated limits of advance, the III Corps was to garrison as much of the advanced zone as extended from their outposts back to the general line Siegen-Wetzler. In the meantime, the 3rd Division was also to move forward and garrison the remainder of the advance zone, back to the Rhine. A part of the 4th Division was simultaneously to cross from the left to the right bank of the Rhine, and place units in some of the industrial towns along the Rhine, but its main body was to be retained on the left bank. The orders transmitted to this division directed further that it should guard the railways and bridgeheads across the Rhine. It was intended that Army Headquarters should remain in Coblence.

The following chart will illustrate the complex political organization of the territory which the American forces were planning to occupy. With parts of five Bezirke and seventeen Kreise falling under American military government, the task confronting the officers charged with civil administration in the advanced zone was expected to be extremely difficult and one requiring careful preparation to assure success. The zones allotted to each of the Allied Armies were based on strategical rather than political considerations.

POLITICAL GEOGRAPHY OF PROPOSED AMERICAN ADVANCED ZONE ON RIGHT RHINE BANK

	Tactical unit to occupy district
Prussia	
I. *Rhine Province:*	
1. Bezirk *Coblence:*	
a. Kreis Altenkirchen (all)—*Altenkirchen	3rd Div.
b. Kreis Wetzlar (part)—*Wetzlar	3rd Div.
II. *Province Hesse—Nassau:*	
1. Bezirk *Wiesbaden:*	
a. Kreis Westerburg (all)—*Westerburg	3rd Div.
b. Kreis Ober-Westerwald (all)—*Marienberg.	3rd Div.
c. Kreis Unterlahn (part)—*Diez	3rd Div.
d. Kreis Oberlahn (part)—*Weilburg	3rd Div.
e. Kreis Dill (all)—*Dillenberg	3rd Div.
f. Hinterland (all)—*Biedenkopf	III Corps
2. Bezirk *Cassel:*	
a. Kreis Frankenberg (part)—*Frankenburg	III Corps
b. Kreis Marburg (all)—*Marburg	III Corps
c. Kreis Kirchhain (part)—*Kirchhain	III Corps
III. *Province Westphalia:*	
1. Bezirk *Arnsberg:*	
a. Kreis Siegen (part)—Siegen	3rd Div.
b. Kreis Wiltgenstein (all)—*Berleburg	III Corps
c. Kreis Meschede (part)—Meschede	III Corps
d. Kreis Brillon (part)—Brillon	III Corps
Grand Duchy of Hesse	
IV. *Province Oberhesse:*	
1. Bezirk *Oberhesse:*	
a. Kreis Giessen (part)—*Giessen	III Corps
b. Kreis Alsfeld (part)—Alsfeld	III Corps

(*Indicates that Kreis capital is in proposed American zone.)

Organization of military government in the original bridgehead had been prepared for the 4th Division by previous American troops, and no trouble was expected in arranging local affairs. (The political administration of the bridgehead has been discussed elsewhere in this report.)

It is interesting to note in connection with the proposed advance, that the American army was thus planning a return visit to the home of the Hessian mercenaries of American Revolutionary fame. These troops had been raised in the vicinity of Cassel, contingents coming from both Giessen and Marburg. After 150 years, American troops had found it necessary to carry the war into the heart of the very country which had in 1776 lent itself, according to the practice of that period, to further the cause of an alien king.

Plans for extension of military government.—The diversity of German political organizations in the proposed forward zone, together with the lack of unity in the American administrative system in the districts already occupied, made the task of widening the scope of military government peculiarly complex and difficult. Four types of control were in existence at this time in occupied American territory, to-wit:

1st: The complete territorial system, in the Kreise of Bezirk Treves and in addition in Kreise Adenau and Cochem, Bezirk Coblence;

2nd: A territorial organization under a Division Commander in Ahrweiler;

3rd: Control by tactical units in the bridgehead;

4th: Control by the army in Coblence and suburbs.

The proposed advance would strip the left Rhine bank of nearly all combat units. It would furthermore thin out the troops in the bridgehead, as part of the 4th Division would have to be scattered over the area formerly occupied by the 1st and 2nd Divisions. It was therefore considered advisable to extend the territorial system as widely as possible, at least over the whole area which at that time was occupied by the American Army. In Ahrweiler, such a change would require little alteration in the routine administration of that Kreis. The control of civil affairs had during the first fortnight of June been entirely independent of the 4th Division. Territorial control had then been initiated in view of the expected early return of the 4th Division to the United States. When it was decided to retain that Division temporarily in the Third Army, the Officer in Charge of Civil Affairs at Ahrweiler had again been placed under the orders of the Division Commander. The military police company which had been assigned to the Kreis, when the Office of Civil Affairs had been separated from the Division, was still under the direction of the Officer in Charge of Civil Affairs and was to serve as the permanent garrison of the Kreis.

The situation was less favorable in Mayen, as well as in the bridgehead. The 3rd Division, on its departure, had not only to provide a complete administration for Kreis Mayen but to retain sufficiently experienced personnel with the Division to control civil matters in its new territory. Civil Affairs, during the previous six months, had been in the hands of a single staff officer. His knowledge of local conditions in Mayen was such that his retention in that Kreis was considered by the army authorities as obligatory, even if by so doing the division was placed at a disadvantage. The situation which confronted the Army Commander is a commentary on the failure to provide tables of organization in advance for administering the military government. Although the Office of Civil Affairs had been in existence six months, there never once had been a single surplus officer available, above the requirements of the daily routine, for neither investigation or study of conditions in the advanced area. When the time came for the civil government to expand, there was not at hand the personnel to carry it out in an orderly and efficient manner.

To fill the gap which would be left by the departure of all troops from Mayen, a company of infantry from the 4th Division was ordered to Andernach. This force was considered sufficient to prevent local disorder and maintain the authority of the United States.

The 1st Division occupied, roughly, Kreis Unterwesterwald in the bridgehead, while the 2nd Division had its headquarters at Neuwied with its troops billeted throughout the Kreis. Each Commanding General was ordered to detach sufficient personnel from his division to organize a territorial control unit in its former territory. These officers would thereafter receive the designation of Officer in Charge of Civil Affairs, Kreis Unterwesterwald, and Officer in Charge of Civil Affairs, Kreis Neuwied. Little difficulty was expected from the civil population in this area in view of the constant stream of troops which were expected to pass to and fro, but the presence of contingents of the 4th Division in towns along the Rhine would act as a deterring force to local trouble if any was intended. In order to insure complete tranquility, however, the 299th Military Police Company was placed under the Officer in Charge of Civil Affairs, Kreis Neuwied, and a company of the 4th Division under the Officer in Charge of Civil Affairs, Kreis Unterwesterwald.

The Commanding General of the 4th Division was ordered to dispose his troops so as to secure effectively the passages of the Rhine. He was further warned to pay particular attention to the industrial territory along the right bank of the Rhine. Certain elements of the population of Hönningen, Bendorf and Neuwied were suspected

521572—43——22

of entertaining Spartacist views and it was believed wise to keep them at all times overawed with sufficient military force. No change was contemplated in the administration of the Stadt and Land Kreis Coblence.

The situation in the forward area was recognized as radically different. However much the army authorities may have desired to establish territorial administrations in the zone, such action was deemed impracticable under existing circumstances. To find experienced staffs to administer seventeen Kreise, was an utter impossibility. Territorial control had proven successful and highly desirable in the Kreise in which it had been introduced; but it must be borne in mind that the officers selected by the divisions formerly occupying the districts had been carefully chosen with a view to their fitness for the posts. Furthermore, the majority of them had previously acted as Divisional or Brigade Civil Affairs Officers. To have attempted to institute such territorial units with officers who were as ignorant of the past policy of the American Army in handling civil matters as with the local customs and psychology of the civil population, would probably have spelled failure. If regular officers with long experience in the service had been available, such a step might have been advisable despite the recognized disadvantages, but there were few enough of these officers for even strictly military problems.

It was therefore decided to return to the system in vogue in December 1918, viz.: divisional control of civil matters. The 1st and 2nd Divisions had retained their Civil Affairs Officers and the 3rd Division had several officers available for this work, who had been partly trained by the former Divisional Officer in Charge of Civil Affairs.

Proclamations and regulations prepared in case of further advance.—It was apparent as early as June 11th that the German military authorities were actively pushing the removal of war material from territory immediately adjacent to the allied outposts. On June 17th, the Intelligence Section reported that newspapers in unoccupied Germany were announcing that the great training camp of Sennelager in Westphalia was being evacuated and that war material in the Rhenish industrial district was being moved 150 kilometers back from the Rhine. It was therefore presumed that no resistance would be offered by organized forces during the initial stages of the expected advance. Furthermore, in view of the considerable distance to which the Germans were removing their war material, it did not appear likely that resistance would be offered short of the Elbe line. Therefore, if fighting was to occur, it would probably be confined to encounters with guerillas or armed civilians.

On June 12th, however, the Intelligence Section obtained a copy of a proclamation of the *Oberpräsident* of Hesse-Cassel to the people of his province, which read as follows:

> In the event of an advance of the enemy troops, earnest advice is given to the people of the cities and country not to make an ill-considered flight, but to face the unavoidable in peace and order. The best protection is to remain at home. That is the lesson of the war. A general exodus to the eastern provinces, where there is already insufficient housing and difficult food conditions, would only increase hunger and distress. All public employes have orders to remain at their posts and to support the civil population in this period of difficulty. Let every man therefore go about his accustomed business, avoiding everything which might give the enemy excuse for intervention, and keeping ever in mind the fact that flight will only increase the danger to life and property.

This proclamation contained information, particularly valuable under the circumstances, that the German government intended to leave all officials in evacuated territory, and was urging the civil population to peacefully accept the awaited occupation.

The following two proclamations were received from Allied G. H. Q., about June 18th, to be published by the American Army in every village of the area it proposed to occupy. From the tenor of the proclamations, it is apparent that the French as well as ourselves did not expect serious resistance to further occupation.

(Original translation by French.)

THE COMMANDER IN CHIEF ALLIED ARMIES

ORDER

TO THE CIVIL POPULATION

> The Allies entering into German territory will respect persons and property and will enable the German population to carry on their ordinary business provided that the safety, movements, billets and supplies of the Allied armies are fully guaranteed.
>
> All the personnel of the public administrations must remain at their post, and continue to carry out their duties under the control of the Allied armies.
>
> Any infraction of these orders will be punished in accordance with military regulations.
>
> All persons guilty of any hostility against the Allied armies—whether in firing against Allied troops, or in destroying railways, roads and waterways, or in cutting or damaging

telegraphs or telephones, or in communication with the enemy,—are amenable to military law.

Any house from which civilians have fired on Allied troops, will be burnt.

In all Communes, all arms must be handed over to the Area Commandant one hour after the proclamation of the order.

LE MARÉCHAL DE FRANCE,
Commandant en Chef des Armées Alliés,
Signé: F. FOCH.

(Original translation by French.)

THE COMMANDER IN CHIEF ALLIED ARMIES

ORDER

IN ORDER TO FACILITATE THE WORK OF THE ARMIES AND IN THE INTEREST OF THE CIVIL POPULATION.

All personnel and all material, whether movable or not, of railways, telegraphs and telephones, are requisitioned on the posting of the present order.

Consequently, the personnel will be subject to the military laws and regulations of the Allied Armies. It will continue to fulfill its duties and carry out all instructions issued to it by the Allied authorities entrusted with operating the different means of communication.

These authorities are the following:

The Director of Supplies and Transport of the Allied Armies for all services, in particular:

Railways:	The Inter-Allied Field-Railway Commission and Sub-Commission.
Telegraphs and Telephones:	The Director of Signals in each Army.

All personnel under requisition will be, if necessary, provided with food by means of payable rations supplied by the Allied armies on the following scale:

Flour	320 gr. or Bread 400 gr.
Potatoes	500 gr. or rice (on a basis of 14 gr. of rice for 100 gr. of potatoes.)
Fats	30 gr.
Meat	80 gr.

Any official who leaves his post will be court martialled and dealt with as a soldier guilty of abandoning his post.

LE MARÉCHAL DE FRANCE,
Commandant en Chef des Armées Alliés,
Signé: F. FOCH.

Two paragraphs of the above proclamation are worthy of note. When the Allied Armies advanced to the Rhine in November 1918, no regulations were issued requiring civilian officials to remain at their posts. In view of the lessons learned in the following six months it was now felt wise to make continuance in office obligatory, both to assist the Allied Armies in their administration of occupied areas and to aid the civil population. The other paragraph, dealing with the requisitioning of railways with their personnel, was necessary in view of the lapse of the armistice conditions. That agreement had directed Germany to maintain her railways in occupied territory and place them with full equipment at the disposal of the Allies. With the resumption of hostilities this agreement lapsed and, if measures had not been taken to keep the personnel of the railways at work, the supply of the army would have been endangered.

"Instructions to Troops" were also published by order of the Commanding General, Third Army. It was planned to distribute these instructions in pamphlet form to every officer of the American Army, as soon as it was definitely known that the advance would take place. The purpose of the pamphlet was identical to that of Memorandum No. 4, Third Army, issued in Luxemburg before the troops had begun their march to the Rhine. It was intended by the Higher Command that Americans should conduct themselves in the expected operations in a way befitting the high cause for which they had entered the war. The instructions were further expected to prevent any possible misconception of the attitude which officers and soldiers were expected to adopt when dealing with the enemy's civil population.

In occupying foreign countries, it is manifestly impossible to do away with all crimes and misdemeanors in the invading army. There must always be a criminal soldier element which will belie the good intentions of the High Command. Other groups attach themselves to this element in the excitement which inevitably attends an occupation. Acts of pillage are committed and the reports of such occurrences, magnified to the proportions of atrocities, spread like wildfire from village to village.

It was believed that these instructions would act as a deterring force to any men who might otherwise carelessly or thoughtlessly indulge in such acts. It was realized that no instructions or personal persuasion would have any effect on the criminal element in the ranks and that unfortunate incidents would inevitably occur if discipline was for a moment relaxed.

SECRET.

THIRD AMERICAN ARMY

Instructions to Troops

To be Issued to All Officers

HEADQUARTERS THIRD ARMY
AMERICAN EXPEDITIONARY FORCES
GERMANY

June, 1919.

The following general instructions will govern the conduct of troops and their relations with the civil population in the event of a further advance into Germany.

1. INSTRUCTIONS PERTAINING TO THE CONDUCT OF TROOPS.

1. SAFETY OF TROOPS.—The paramount duty of every Commanding Officer is to insure the safety of the troops under his command.

2. FIRING ON THE ENEMY.—Formed bodies of the enemy's troops will, after the present outpost line is passed, be fired upon unless by their conduct they clearly indicate their intention of surrendering.

3. ENTRANCE OF TOWNS.—Towns and villages will not be fired upon unless resistance is offered. If the enemy offers any opposition to the entry of our troops, such force as may be necessary will be used.

4. PROCLAMATION OF MARSHAL FOCH.—Copies of the proclamation of Marshal Foch will be distributed by the advance guards to German local authorities. By this proclamation, the inhabitants are warned as to the duties and liabilities imposed upon them.

5. BURNING OF HOUSES.—Referring to the order in Marshal Foch's proclamation relative to burning houses, from which our troops may be fired upon, such houses will be burned only by direct order of a Field Officer present on the ground.

6. AIR SERVICE.—Aeroplanes will not fire on enemy ground troops unless such troops are engaged with our own forces, or unless they themselves are fired upon by the enemy.

During the advance, active enemy aircraft will be immediately engaged.

No bombing will be undertaken except under orders from these headquarters.

7. MILITARY NECESSITY.—Military necessity justifies a resort to all measures which are indispensable for securing the end of the war, and which are not forbidden by the modern laws and usages of war.

Military necessity admits of all direct destruction of life or limb of *armed enemies*, and of other persons whose destruction is incidentally *unavoidable* in the armed contests of war;

it allows of the capturing of every armed enemy and of every enemy of importance to the hostile government, or of peculiar danger to the captor; it allows of all destruction of property and obstruction of ways and channels of traffic, travel or communication and of all withholding of sustenance or means of life from the enemy; of the appropriation of whatever the enemy's country affords that is necessary for the subsistence and safety of the army and of such deception as does not involve the breaking of good faith.

Military necessity does not admit of cruelty—that is, the infliction of suffering for the sake of suffering or for revenge, nor of maiming or wounding except to fight, nor of torture to extort confessions; it does not admit of the wanton devastation of a district. It admits of deception, but disclaims acts of perfidy, and, in general, military necessity does not include any act of hostility which makes the return to peace unnecessarily difficult.

II. INSTRUCTIONS PERTAINING TO THE CONDUCT OF THE CIVILIAN POPULATION.

1. MILITARY GOVERNMENT.—The Law of Hostile Occupation will be enforced and any inhabitant, whether soldier or civilian, who offends against the laws will be liable to punishment by military tribunal. Existing rules and regulations regarding Provost Courts and Military Commissions will continue in effect so far as applicable to the new conditions.

Military government is simply military authority exercised in accordance with the laws and usages of war. "Military oppression is not military government; it is an abuse of the power which the law of war confers. As military government is carried on by military force, it is incumbent upon those who administer it to be strictly guided by the principles of justice, honor and humanity—virtues adorning a soldier even more than other men, for the very reason that he possesses the power of his arms against the unarmed."

2. TRIAL.—Where offenses against the laws of war have been committed, trial by military tribunal is necessary in every case, and no offender shall be punished without trial. Examples of such crimes are: abuse of the flag of truce; treacherous concealment by troops of their military character by the use of civilian clothing during battle; pillage and purposeless destruction; treacherous request for quarter; espionage and war treason; damage to ways and means of communication by civilians or soldiers in disguise, and other kindred offenses. Persons found guilty of such offenses may be awarded, by a military tribunal, any punishment except corporal punishment. The death penalty will not be carried into effect without the approval of the Commander in Chief, American Expeditionary Forces.

3. CLOTHING.—A number of the inhabitants are wearing old German uniforms, without distinctive buttons or insignia or rank. Therefore, a decision whether or not men are clothed in uniform will largely depend upon the head-dress worn and

whether or not they are wearing badges or other insignia of rank, and all concerned must carefully differentiate between this class of persons and uniformed government agents such as railroad, postal, telephone or telegraph employes. In this connection, it is to be borne in mind that men in civilian clothing, *part of an organized force*, and openly carrying arms, should be treated as combatants under the provisions of the Hague Convention.

4. NON-INTERFERENCE WITH CIVILIAN INSTITUTIONS. As prescribed in Marshal Foch's proclamation, only such interference with the German local government and institutions will be permitted as the military situation demands.

German officials will be required to retain their posts of duty. A failure on their part to perform their proper functions will result in trial by military tribunal.

The civil authorities will be held responsible that the laws are enforced and that the rules and regulations promulgated by the military authorities are complied with. When the occupation has become stabilized in an area, the existing regulations as announced in General Pershing's proclamation of December 9th, 1918, will be published to the inhabitants.

5. GUARDING RAILWAYS. Until the control of the railroads is taken over by the Inter-Allied Railway Commission, troops will guard stations and other sensitive points. They will prevent the destruction of property and will require the operating personnel to remain at their posts of duty, particularly general and local managers. They will occupy the more important offices of the railway systems and will prevent all movement of German trains unless operated under the control of the Allied or United States Armies.

In the event of the necessity of executing any destruction of tracks, such demolitions as are unavoidable will be of temporary nature.

6. TELEPHONE AND TELEGRAPH LINES AND POST OFFICES. The Signal Corps will assume control of such lines as are necessary for military purposes.

Important postal, telephone and telegraph offices will be occupied and guarded and the operating personnel required to remain at their posts of duty. All civil communication on telephone and telegraph lines will be stopped, and German operators will not be allowed to use their telephone or telegraph instruments except in the presence of a Signal Corps representative.

Neither telephone nor telegraph lines will be destroyed unless the military situation demands such destruction.

BY COMMAND OF LIEUTENANT GENERAL LIGGETT:

MALIN CRAIG,
Chief of Staff.

OFFICIAL:
J. C. MONTGOMERY,
Col. Gen. Staff,
A. C. of S., G-3.

American orders and regulations were so well known by the civil population of occupied territory that it was not considered necessary to issue further instructions in the rear area. It was, however, intended that the Oberpräsident of the Rhine Province should, on the resumptiton of hostilities, warn the population that existing orders and regulations remained in force.

Recapitulation.—On June 22nd the German cabinet, headed by Scheidemann, which had taken a strong stand against signing the peace-terms, was deposed by the National Assembly of Weimar. This action signified that Germany had made up her mind to accept the inevitable and would sign in short order. On June 23rd the Peace Conference at Paris was officially notified that Germany was ready to sign and, on June 28th, the actual ceremony took place at Versailles. On June 23rd, the Allied and Army Commands revoked all orders which had been issued to govern the proposed advance. The American divisions, which had moved forward to the demarcation line of the bridgehead, were withdrawn to their former billeting areas, and control of Civil Affairs reverted to its former status.

It is believed that, if the forward movement had occurred, the problem of civil government would have been better handled than during the original advance to the Rhine. The unexpected termination of the war in the autumn of 1918 and the consequent occupation of German territory had not been foreseen until a relatively late date. For this reason no adequate preparations had been made to institute a civil government in the territory to be occupied. No corps of specially trained officers existed to handle civil matters and, in consequence, each American policy had to be developed bit by bit, with the inevitable mistakes and failures which must ever follow in the wake of lack of organization and inexperience.

In June 1919 however, plans for the extension of the area of military government had been taken up far enough in advance to carefully weigh various methods and policies and to determine which would best suit the altered conditions brought about by the renewal of hostilities.

Furthermore, by June there was a considerable staff of officers and men, already trained in the duties of civil administrators. These officers and men would have quickly and smoothly introduced our policies and regulations into the new territory, thus avoiding the confusion and vagueness which existed in December 1918.

CHAPTER NO. 17

COMPARISON OF THE ALLIED STAFF ORGANIZATIONS FOR CONDUCT OF MILITARY GOVERNMENT

General considerations; Staff Organization in Tenth French Army for Military Government; Staff Organization in British Second Army for Military Government; Staff Organization in Belgian Army for Military Government; Staff Organization in German Army for Military Government of Belgium; Summary.

I. General considerations.—During the whole period of the military occupation of Germany, advantage was taken from time to time, through the courtesy of staff officers of the French, British and Belgian armies, to study the different systems evolved to meet the problems confronting their armies in the matter of military government. Whether any of them had mature plans prepared in advance for solving these problems, is not clear, but it is believed that they like ourselves evolved the machinery required from day to day. The conclusions drawn from this comparison are therefore interesting as showing the various methods followed in arriving at the same results. It must be constantly borne in mind in this connection that, under the terms of the armistice, the Allied commands were required to exercise their authority through regularly elected or appointed German officials. They were not, of course, required under the general rules of international law to retain a hostile or inefficient official in office, and all the armies fully exercised the right to remove from office and even to try by military courts any officials who failed to discharge their duties to the satisfaction of the Allied authorities. Comparatively speaking, however, few officials were replaced, and as the German bureaucracy was a carefully selected, trained body of officials with reasonable prospects of life tenure of their positions, supervision of their work was a far simpler matter than actual government by Allied officers would have been. It is dangerous, however, for any army to occupy enemy territory without a thorough organization of its military government. The American occupied territory lay between the British and French zones, and we therefore had greater opportunity to become familiar with their methods than we did with those of the Belgian army, whose territory at no time touched ours. The British, with their larger experience in colonial administration, realized more fully than either the French or ourselves, the absolute necessity of assigning an ample number of trained staff officers to this duty. We had had, for fifteen years, a General Staff system in name only, and we were compelled to learn General Staff principles as applied to *operations* after the war began. As we had adopted the French General Staff system practically in its entirety, we might reasonably have expected to evolve a General Staff system for the military government of occupied territory based upon the French system. As a matter of fact, at no time during the American occupation did we apply the principles of General Staff control to our military government, as did the French, but gradually approached the British system without however appreciating as fully as did they the vital necessity for a *staff system* to deal with the vast number of intricate problems of government incident to a hostile occupation—problems which form no part of the course of study at any of our military schools. The gradual evolution which took place, and which tended toward the adoption of the British system, may be considered therefore as the result of a common system of thought inherent in the two armies as a result of a long period of common history in the evolution of principles of government. This was naturally most noticeable in the administration of justice, since both were familiar with the general principles of criminal law originating in common law which differs greatly from the civil law as respects the rights of the accused. Here again, however, the lack of trained officers for civil duty in the American army led our provost courts to proceed along lines of military discipline alone, and to assess penalties wholly disproportionate to the offenses committed, if viewed from the standpoint of the civil population.

II. Staff organization in Tenth French Army for military government.—A general outline of the staff organization for conducting civil affairs in the zone occupied by the Tenth French Army was as follows:

1. A special section of the General Staff (**Bureau des Affaires Civiles**) was created for all such matters. In this the French flexibility of mind showed itself, since their General Staff system, upon which our own is modeled, revolves around *Operations*, everything else being very properly subordinated thereto. When the period of operations was over, however, they promptly realized the necessity of creating a special section of the General Staff to deal with this totally different phase of work, for the very simple reason that to parcel it out among the various General Staff sections and services would have loaded them down with duties and would have prevented their giving their whole attention to keeping the army instantly prepared to renew, if necessary, operations against the enemy. There would also have existed no assurance that the officers who had been selected, with other objects in mind, would have special aptitude for this class of work. Military occupation of an enemy's territory can

take place only after the defeat of his armies in the field, and history will judge the occupying army thereafter by its success in administering the occupied territory. In the present enlightened state of international public opinion, no army can afford to ignore this phase of staff organization. Being a member of the General Staff, the Chief of the Bureau of Civil Affairs met in daily conference with the chiefs of the other sections of the General Staff, thus insuring harmonious action and a common policy.

The original zone of occupation of the Tenth French Army embraced four different political organizations, viz.: Bezirk Wiesbaden in the province of Hesse-Nassau; the province of Rhenish Hesse, and a portion of each of the Bezirke of Coblence and Treves in the Rhine Province.

(a) For each of these political subdivisions, the French commander appointed a Chief Administrator (*Administrateur Supérieur*), at Wiesbaden, Mayence, Kreuznach and Saarbrücken, respectively. These administrators served directly under the Chief of the Bureau des Affaires Civiles;

(b) Each Kreis had an administrator who served under the *Administrateur Supérieur* of the Bezirk or Province;

(c) Liaison officers were placed at Coblence and Treves with the Regierungspräsidenten of those Bezirke, who acted at the same time as liaison officers with the American Headquarters in those cities.

All of the foregoing officials met at Mayence one day each week for a conference with the Commanding General of the Tenth French Army. Orders within the occupied territory concerning purely civil matters were issued by the Civil Administrators; those concerning purely military ones by division or other commanders. In case of conflict of policy between the two, the matter was referred to the Army Commander. This system differed from the British system of having all civil officers serve on the staff of the Military Commander. The French system, however, had the great merit of territorial organization, thereby permitting any reassignment of troops without disturbing the Civil Administration.

The organization and duties of the Bureau des Affaires Civiles may be summarized as follows:

First Section—Control of German Administration:

Public administration; justice, religions; finance; education; German employes; German public officials; legislation; jurisprudence; tribunal for "Simple Police" (Provost Courts); requisitions; damages; restitution; censorship; archives; civil affairs bulletin.

Second Section—Organization and Officials:

Political affairs; public works; railways; post; telegraph; telephone; navigation; fisheries.

Third Section—Civilian Food Supply:

General organization; receipt of foodstuffs; allotment to civil population.

Fourth Section—Economic:

Financial questions; industrial service; control of factories; commercial service; statistics.

Fifth Section—Miscellaneous:

Press and propaganda; publication of "Le Rhin Illustré"; connections with French and local press; telegraph agency; French performances in local theaters; military libraries; theaters for troops.

Embraced in the foregoing duties of the French civil affairs organization, are many matters for which no provision was ever made in the American zone. The French Economic Section concerned itself with stimulating French trade in Germany, which was a matter of vital importance to France in view of the depreciation of the franc in New York and London. None of the duties performed by the Fifth Section of the bureau had a counterpart in the American zone, as we indulged in no propaganda of any sort, whether by papers or theatrical performances. In particular we never in any way interfered with German budgets and therefore had no organization to look after such financial matters, whereas the French supervised them very minutely.

The following matters were assigned by the French to staff departments, viz.

(a) Sanitation and Public Health was managed by the medical service of the army. We exercised a very close control over German sanitation in order to protect our troops, who were more closely billeted on the people than in any other zone, and as this required control of a large number of German officials, civil sanitation was supervised by an officer of the Medical Corps assigned to civil duty and serving with the civil administration. This was in order that control of such officials might be centralized in one office which followed a uniform policy. This system worked well.

(b) *Public Works and Public Utilities.*—The French assigned the supervision of these activities to the army engineers. In the American zone, they were at first supervised by the Office of Civil Affairs at Advanced G. H. Q., but in the course of time came to be handled by the Chief Engineer of the Army. Neither system worked to the best advantage: the first, because the army had ultimately to do the work anyway; the second, because it required one more American officer to give orders to German officials and prevented centralization of authority in dealing with them. The principle followed in the case of Sanitation and Public Health would have worked much better, viz.: to assign an Engineer officer to civil duty to supervise these activities.

(c) *Elections.*—The policy regarding these was determined in the French army by the *Bureau des Affaires Civiles*, the Intelligence Department of the army furnishing any secret data required. At first, the policy regarding elections in the American zone was handled by the Second Section of the General Staff, but such supervision of elections as was undertaken later was done by the Office of Civil Affairs.

(d) Circulation of private individuals was controlled in the French zone by the Intelligence Department of the army; that of public officials, by the Bureau of Civil Affairs. In the American zone, the rules regarding circulation were laid down by the Office of Civil Affairs at Advanced G. H. Q., actual control being exercised by the Second Section of the General Staff of the Army. This system worked well but, as noted elsewhere in this report, circulation from the American viewpoint was unnecessarily restricted in order to conform to a common policy in all the occupied territories.

(e) Censorship in the French army was entirely controlled by the Intelligence Department. With us it was handled like Circulation, viz.: supervised by the Second Section of the Army General Staff, under rules laid down by the Office of Civil Affairs at Advanced G. H. Q. The system worked well, except that it involved considerable supervision of the German postal department which should, like other departments, have been supervised as to administrative matters by the same office which controlled other departments of the government.

The strong features of the French organization appear to be:

(a) Recognition of the Army Commander as the head of both military and civil adminstration;

(b) Creation of a special section of the General Staff to deal with civil matters, thus leaving the military staff free to center its attention on operations;

(c) Control of Civil Affairs through the bureau thus created;

(d) Organization of civil administration on territorial lines.

Its weak features appear to be:

(a) Failure to assign the civil staff officers to the staffs of corps, division and other commanders commanding in their respective territories, thus leaving an opportunity for friction, unless the powers of the corps and division commanders over the civil population were very carefully defined;

(b) Failure to fully realize the vast range of civil administration, resulting in the assignment of a personnel too limited for this work.

In concluding this survey of their staff organization, it should be added that the administration of the French, though firm, showed commendable restraint, in view of the powerful incentives which might properly have moved them to resort to harsh measures. Considering the severity of German administration of the occupied districts of France during the war, the conduct of French administration in Germany (except for some unnecessary interference in German internal affairs, having in view what they conceived to be France's urgent necessity for guarantees for her future security) is remarkable for the absence of retaliatory measures.

III. Staff organization in Second British Army for military government.—The British military government, perhaps because of their greater experience, put into effect from the beginning an organization which more nearly embraced all the fundamental principles of military government than did that of any other army. The spirit of British institutions lends itself easily to the adaptation of new methods to new situations, since their unwritten constitution requires great flexibility of mind in this respect. Republics have modeled their institutions largely upon the American Constitution which has the tendency, as all students of constitutional law realize, to attempt to arrest development by cast-iron formulae. Only the genius of our Supreme Court under the original inspiration of Chief Justice Marshall has permitted us to break away from this tendency. Through this development, we have come to ignore many constitutional prescriptions and have introduced many others. The same tendency toward rigidity in staff organization will prevent real development of General Staff policy unless the principles followed so successfully by the Supreme Court in interpreting our fundamental law are followed.

The first principle applied by the British army was that the Army Commander was the supreme authority. British G. H. Q. did not attempt to interfere in the administration of military government, concerning itself solely with the larger questions of policy emanating from Marshal Foch's Headquarters. The British army did, however, make what would appear to be an error in designating their Chief Civil Administrator as Military Governor. Whether this caused any confusion in British military minds is not clear, but it certainly did in the minds of Germans, since they looked upon the Military Governor as the supreme authority, which he was not, at least after the first few months. If there was any confusion as to divided authority among the British staff at first, it was quickly removed, for the strong feature of British staff administration is its spirit of co-operation. This is required by the nature of their loosely-knit staff organization. Here again the silent influence of institutions molds military thought the same as it does political thought.

The second principle applied by the British army was to create another special staff department to meet the new conditions following active operations. The British General Staff (G), being the evolution of minds long accustomed to think in terms of economics, concerns itself only with training (T), intelligence (I) and operations (O). Supply and transportation (Q) and administration (A) were, in theory at least, coordinate with the General Staff. In practice however the Chief of the General Staff ordinarily controls all staff departments. None of these divisions, all of which are naturally subordinated to operations, was equipped to handle civil administration. A new staff division (C) was therefore created for the conduct of Civil Affairs. A subdivision of this organization—referred to as (P)—concerned itself with matters of policy rather than of administration.

The third principle used by the British was that the Area Commandants controlled absolutely the civil administration in their respective areas. In this they differed from the French practice. The British system is believed much the better and is certainly better adapted to American military thought. The principle that subordinate commanders should have a trained civil staff (C) was carried out however, thus freeing the commander and the military staff from the burden of civil administration. This principle was carried down to the lowest units, until the Town Commandant (Town Major) became the supervisor of the local Burgomaster.

The fourth principle involved in the British organization was that army commands were made co-extensive with the political subdivisions of occupied territory. This is a serious defect, as it subordinates the responsibilities of the operations section to the necessities of civil administration. It is the application of the territorial, as distinguished from the tactical, system of military organization which has done so much to prevent the development in America of real tactical commands. The true principle here is believed to be to assign civil administrators permanently according to the political organization of the occupied territory, and then provide for their automatically becoming staff officers of the commander whose unit occupies the area. In this way only can the error be avoided of subordinating the strategical, tactical and training dispositions of troops to the necessities of civil administration. Even though the military government of an occupied territory is ordinarily the main function of the occupying troops (after an armistice or capitulation), nevertheless an operations or training section ought not to be thus hampered.

As previously noted, one branch of the civil staff concerned itself with matters of policy as distinct from those of administration. No particular advantage can be seen from this division, as the two are closely related. Thus P. C. handled censorship matters; P. F., food and coal; P. I., industry; P. P., politics and propaganda; P. S., signal security. P. I. was subdivided into dyes, chemical, engineering, finance, labor, imports and exports branches. In this, the close relation between the army staff organization and economic matters designed to assist British trade is evident. A bureau of information for business men, where German exporting and importing houses could get in touch with those from other countries, was maintained in the Military Governor's office, and its facilities were courteously extended to Allied citizens. Cologne, being the largest industrial city in occupied territories, was the logical center for such information.

British military government embraced more of the fundamental principles of staff organization for a military government than did that of any other army, viz.: centralization of authority in one commander; creation of a special staff section to handle matters of civil administration; assignment of such staff officers to the staffs of subordinate commanders; control through a territorial organization conforming to the political subdivisions of the occupied area; and assignment of sufficient personnel to perform the duties required of the military government. The title of Military Governor was admitted to be misleading, as the supreme authority was in reality the Army Commander. By reason of his broad experience, the senior British officer is by nature and training an excellent civil administrator. He is broad-minded, essentially fair and just, always dignified but withal firm. The great dignity with which the British Military Governor surrounded his administration naturally impressed itself upon the German official who is very fond of this sort of government, while the fairness and liberality with which the German population was treated did infinitely more than the clever British propaganda to bring about a *rapprochement* between the British and the German peoples. Considering the bitterness of the feeling between them during the war, this was a notable achievement. The philosophy of the magnanimity of the victor was never better demonstrated.

IV. Staff organization in Belgian Army for military government.—The Rhine, province north of Cologne is, along with Westphalia adjoining it, the industrial center of modern Germany.

The Belgian army had no bridgehead on the right bank of the Rhine, all of its troops being on the left bank. This thickly populated industrial section was a difficult one to control during the occupation on account of the prevailing labor unrest. To assist the Belgian army in this control, Marshal Foch assigned the French 33rd Corps to the Belgian army, with its Headquarters at Aix-la-Chapelle.

As in the French zone, there was no confusion in the Belgian area as to who was the real head of the military government—this authority was the Commanding General of the Army. The Belgians thus applied correctly the first principle of military government.

In the Belgian army, the first section of the General Staff controls operations; the second, intelligence; the third, personnel and administration; and the fourth, transportation and supply. In order to provide for the necessities of military government they added three more sections: (a) *Direction du Contrôle Administratif;* (b) *Justice Militaire* and (c) *Section Economique.*

The section in charge of administrative control was the real directing spirit of the military government. Its head was designated *Directeur du Contrôle Administratif* and acted directly under the Chief of Staff, just as did the first four sections of the General Staff. The Belgian army, however, was the only one to add more than one section to their staff on account of the military government. While the addition of one section for civil affairs is fundamentally necessary, no reason is seen why this one could not be made to perform all of the additional duties required by military government. From the American viewpoint, therefore, the addition of special sections dealing only with military justice and economic matters was a mistake in organization.

The Belgian zone comprised parts of two Bezirke, viz.: Aix-la-Chapelle and Düsseldorf. These Bezirke were made the heads of administrative districts, the French Corps Commander at Aix-la-Chapelle being in charge of that Bezirk, while the Belgian Commander for Bezirk Düsseldorf was at Crefeld. On the staff of each of these commanders was a *Controleur en Chef* who was at the head of the civil administration. Similarly, on the staff of cavalry and infantry division commanders there was a *controleur* to handle civil affairs. A civil administrator was also assigned to each Kreis. These various administrators were charged with the supervision of all administrative, financial and judicial matters in their respective areas. The Belgians thus followed the German territorial organization in supervising the work of civil administration. However, by attempting to control it through their tactical organization they found that the corps areas fitted into the German organization better than did the divisional areas, thereby emphasizing the necessity for keeping the civil administration independent of tactical requirements. Reports by corps and division commanders were required once every two months, but it does not appear that conferences were held periodically on civil matters. In the French Tenth Army all administrators met at Headquarters once a week for conferences. The same rule obtained after the first few months in the American zone, except that, on account of the large area, the Officer in Charge of Civil Affairs met the officers for Bezirk Treves (usually at the city of Treves) once a week, and those for Bezirk Coblence at Army Headquarters in Coblence on another day, also once a week. Uniformity of administration was assured in this way.

Special phases of the military government were assigned by Belgian G. H. Q. to General Staff sections as follows:

(a) Circulation, censorship and police matters were assigned to the Second (Intelligence) Bureau;

(b) Requisitions to the Third (Administrative) Bureau;

(c) Civilian food supply to the Fourth (Supply) Bureau.

As might be expected, the Belgian methods of military government were closely related to the French, since the two peoples have a common language and to a large extent common institutions, particularly in jurisprudence. For the same reason the American and British methods were much alike.

V. Staff organization in German Army for military government of Belgium.—Complete data are not available as to the methods followed by the German army in Belgium and, on account of the many criticisms of their administration, German officials have been timid about discussing it. Through the courtesy of various German staff officers connected with that government, however, a general outline of the system followed by them has been procured and is inserted here for purposes of comparison with the various systems followed by the Allies. With the relative effects of the various systems of administration we are not here concerned, the purpose being to compare those systems. Their results depend not only upon the methods of application adopted by the occupying forces, but upon the attitude of the civil populations concerned toward the occupying armies, both of which powerfully affect the measure of results attained.

The German genius for organization is shown by the fact that they adopted, so far as is known, practically all of the principles heretofore noted as requisite to a correct form of military government. The Governor-General was a military officer (*Generaloberst*) but he had on his staff trained civilian administrators who were drawn from the German bureaucracy at home. These men were selected not only for their ability in civil administration but also for their perfect knowledge of French. This principle was carried down to the lowest grades. The Belgians were the only Allied forces who created special civilian administrators, giving them military authority but without military rank. These civilian administrators on the staff of various army commanders down to the *arrondissments* not only supervised the administration of affairs by Belgian officials, who remained in office, but also attended to various activities caused by the fact of occupation, such as laying out German field-cemeteries, identification of the dead, control of requisitions, civilian food relief and social welfare.

The German government generalship of Belgium was territorially organized, its jurisdiction extending over the provinces of Luxemburg, Limburg, Liege, Namur, Brabant, Hainaut and East Flanders. Each province was governed by a Lieutenant General assisted by a Civil President of the grade of *Landrat* or *Regierungsrat* in Germany; the arrondissments were governed by Major Generals or Colonels assisted by a Civil Commissioner of the grade of *Regierungsassessor* at home. This system was general except in the zone of the line of communications directly behind the fighting armies, but even here the *Etappeninspektor* had a trained civil official with him to handle civil affairs.

It will be seen from the foregoing that the German system ought to have functioned perfectly. We are as yet too close to the facts to judge their results in the calm light of history, but when the real truth is known, if serious mistakes are found to have occurred, the errors will perhaps be found to exist in the antipathy of the civil population toward the occupying forces and in the unnecessary harshness on the part of the German administrators in their effort to maintain strict control of the civil population between their zones of operations and their bases in Germany.

VI. Summary.—Summarizing the principles which can be deducted from a comparison of the American administration (described in Chapter No. 4) and the British, French, Belgian and German administrations outlined herein, we find:

First principle: Undivided and Centralized Control:

(a) American; At first military and civil authority divided between Advanced G. H. Q. and Third Army; civil control gradually transferred to army and finally taken over completely on abolition of Advanced G. H. Q.

(b) British: Undivided control but confused by designation of Chief Civil Administrator as Military Governor.

(c) French: Undivided control; Army Commander supreme.

(d) Belgian: Same.

(e) German: Same.

Second Principle: Creation of separate staff department for civil administration.

(a) American: Assigned to an "Officer in Charge of Civil Affairs" but his powers never defined and his relation to General Staff undefined; this officer at G. H. Q. at first designated "Chief of Staff for Civil Affairs" but designation later revoked.

(b) British: Separate staff department (C) created, coordinate with General Staff (G), Supply (Q) and Administration (A).

(c) French: Separate section of General Staff created: *Bureau des Affaires Civiles.*

(d) Belgian: Three separate staff departments created, coordinate with General Staff, viz.: *Direction du Contrôle Administrative, Justice Militaire* and *Section Economique.*

(e) German: Separate staff section created, functioning independently of other staff sections and presided over by trained civil official with civil title, having military authority in civil administration but without military rank.

Third Principle: Civil administrators of lower political subdivisions serving on staffs of corresponding military commanders.

(a) American: Detailed on corps and division and, where required, on brigade staffs.

(b) British: Same.

(c) French: Civil administrators independent of tactical commanders.

(d) Belgian: Civil administrators on staff of subordinate commands.

(e) German: Same.

Fourth Principle: Civil administrators selected for special qualifications and personnel sufficient for work required.

(a) American: In principle, selected for special qualifications; in practice, not. Insufficient in number.

(b) British: Apparently specially selected and sufficient in number.

(c) French: Apparently specially selected but insufficient in number.

(d) Belgian: Specially selected; probably insufficient in number.

(e) German: Specially selected from trained bureaucracy and apparently sufficient in number.

Fifth Principle: Civil administration corresponding to political subdivisions of territory.

(a) American: In beginning confused with tactical considerations; later made to correspond with territorial requirements.

(b) British: Tactical requirements subordinated to necessity of territorial organization for civil administration.

(c) French: Tactical requirements unhampered by territorial organization for civil administration.

(d) Belgian: Tactical requirements subordinated to civil administration; worked well in corps areas, not so well in division areas.

(e) German: Administration wholly territorial.

CHAPTER NO. 18

THE CLOSING PERIOD OF MILITARY GOVERNMENT

General Considerations; The Luxemburg Commission; The Inter-Allied Rhineland Commission; The Origin of the Principle of Government in the Rhineland by a Civil Commission; The Rhineland Agreement and its Limitation of Powers of the Army; Situation Caused in American Zone by Coming into Force of the Peace Treaty; Agreement reached by High Commission and Army Commander as to the Status of the American Zone after Coming into Effect of Peace Treaty; Practical Application of Agreement Reached by Commanding General with High Commission as to the Latter's Status in American Zone.

Reference.—The Luxemburg Commission, -------- Chapter No. 3.

General considerations.—The period of the supreme authority of the Allied armies in the Rhineland virtually came to a close on January 10, 1920, with the declaration of peace between the Allied Powers and Germany. In the final acts which put the peace treaty into effect America was not represented, the United States Senate on that date not having ratified the treaty. Although in a legal and technical sense the Commanding General, American Forces in Germany, still continued to retain the powers vested in him by the armistice, he deemed it advisable in the interests of unity among the allies to put the ordinances of the High Commission (promulgated in accordance with the terms of the Rhineland Convention) in effect in the American zone. This arrangement proved virtually the only way out of the situation in which the American Army found itself on the declaration of peace. The Allied governments, through their representatives on the Council of Ambassadors at Paris, had at first taken the position that the High Commission must be recognized in all its powers everywhere throughout the occupied Rhineland. This viewpoint was not held either by the Army Commander or by the government at Washington. Negotiations ensued between the High Commission and the Army Commander, which resulted in the latter agreeing to publish the Ordinances of the High Commission in the American zone, as military orders, although technically retaining all the powers of a Military Commander granted him by the terms of the Armistice. The Army thereafter exercised the powers vested in it by the armistice terms in very few instances only, the High Commission recognizing the authority of the Army Commander in the American zone. All Ordinances and decisions of the Commission were being transmitted to the Army Commander and by him published to the civil population as orders of the military authorities.

This agreement, which came into force on January 10, 1920, terminated to all intents and purposes the period of military government in the Rhineland. Technically, military government was still supposed to exist in the American zone, but as all important questions of general policy falling within the scope of the powers of the High Commission were decided by that body, only a few matters remained to be decided by the Army Commander under the terms of the armistice.

The termination of military government is usually an involved and difficult matter. Though in a majority of historic occupations termination of the powers of military government has gone hand in hand with evacuation by the army of the territory in which the powers had been exercised, there are precedents to the contrary. Thus, in the Philippine Islands in 1900 and 1901 the powers, which during the insurrection had been vested in the army, were gradually turned over to a civil commission. The circumstances surrounding the transfer of authority in the Philippines were not dissimilar to the factors influencing the creation of a supreme civil body in the Rhineland. In both cases the government believed it inadvisable to continue in force indefinitely a strictly military government. In the Rhineland, as was the case in the Philippines, it was believed wiser to place the civil government under the guidance of civilians rather than of soldiers. Military government finds its *raison d'être* in the state of war, and from this very fact is unsuited to a condition of peace. There were, however, a number of circumstances which make it impossible to carry the comparison between 1900–01 and 1920 to any great length. The High Commission was created not for the purpose of governing the Rhineland, but for looking after the safety, maintenance and requirements of the army—an intermediary body interposed between the troops and the civil population. The Civil Commission in the Philippines was an American body, while the I. A. R. H. C. was composed of four Commissioners equal in rank and representing four distinct nations—France, Belgium, Great Britain and the United States. There were other factors leading to its establishment which did not exist in the Philippines. In the peace of Versailles, the United States had agreed to retain a military force in the Rhineland as a symbol to France that its army and wealth would ward off any future attack by Germany. Though the garrison at Coblence was to be merely nominal, it yet insured American participation in the peace settlement on a basis of equality with our Allies. The President, however, in return for such participation, insisted that the government of occupied Germany be restored to the Germans and that a civil commission be appointed to look after the interests of the armies.

The High Commission came into existence formally on January 10, 1920. To a certain extent, however, it was merely the successor of the Inter-Allied Rhineland Commission, whose authority in the economic field had been supreme since April 26, 1919, and which in turn had developed out of Marshal Foch's Luxemburg Commission. Though the charters and functions of these several commissions were radically different, the personnel directing their activities remained much the same and their policies were for that reason markedly similar. In fact, for several months before the coming into force of the peace treaty, the I. A. R. C. had been mainly preoccupied with preparing for its transformation into the High Commission. This preparation had consisted in drafting the Ordinances and Instructions which were to govern the Rhineland when the new *régime* should be instituted. The statement therefore that the High Commission was the virtual successor of the Economic Commission, existant during the armistice period, is in many respects true though the connection was not intended when the Rhineland Convention was drawn up. The transfer of authority from the army to the High Commission can therefore be better understood if we trace the gradual infiltration of civilian influence into the military government; how little by little many of the powers at first exercised by the army were withdrawn from its competence and transferred to one or the other of the Civil Commissions.

The Luxemburg commission.—The reasons governing establishment by the High Command of a uniform Allied control in the zones of all armies are fully discussed in the Chapter No. 3 of this work.

The Rhineland being economically, if not politically, a unit, it was realized that the armies would have to co-ordinate their efforts if the problems of unemployment and consequent radical agitation were to be solved. Nor had the belief, prevalent at this time, been branded yet as a myth that Germany was in a position to regain quickly her former trade and so outstrip the Allies victorious on the field of battle. If the armies had reached the conclusion that the industries of occupied territory ought to be resumed sufficiently to absorb the unemployed, they still had no intention of permitting the enemy to increase his military strength thereby. Export and import trade was to be hedged round with such restrictions that the Allies alone could utilize the wealth of natural resources in the Rhineland in case of resumption of hostilities.

The Luxemburg Commission was organized on January 6, 1919, by direction of Marshal Foch and established its headquarters at Luxemburg. Its official title was the Inter-Allied Economic Commission, but the name by which it became generally known was derived from the city in which its meetings were held. The commission was supposed to be an adjunct of the Office of Comptroller General of Occupied Territory, an office exclusively French and created by Marshal Foch in his capacity as Commander in Chief of the Allied Armies.

M. Tirard, who later became the President of the High Commission, was appointed by Marshal Foch to be the first Comptroller General, and with him were associated a number of other comptrollers for certain specific matters such as finance, food, etc. The authority of the comptroller was supposed to extend into both Allied and French zones, Marshal Foch's intention being to regulate himself all economic matters within occupied territory. The Luxemburg Commission was supposed to subordinate its functions to those of the Comptroller General, but as the Comptroller's office was exclusively French and the Committee was Interallied differences between the two were inevitable. As a matter of fact, the Allied armies never fully recognized the authority of the Comptroller General and the Luxemburg Commission gradually came to assume a more and more independent position. The Commission was composed of representatives of the French, Belgian, English and American armies, the chairman being the French General Maugas.

The Commission, during its brief existence and despite the handicap of the varying policies of the several Allies, did achieve a certain success. Branch Economic Sections were set up at Aix-la-Chapelle, München-Gladbach, Treves, Cologne, Mayence and Ludwigshafen, the internal organization of each branch being modelled however by the Army involved irrespective of the others. The larger part of the work, of both the branch sections and the central commission, revolved around the regulation of imports and exports. A marked tendency to moderate the former embargo on traffic with unoccupied Germany soon developed simultaneously in all zones. The Luxemburg Commission gave authority to the branch Economic Sections, in the case of certain special articles, to approve and to disapprove applications for export or import. In the case of certain other articles, ultimate decision was reserved by the main Commission, the application being forwarded to it by the Economic Sections.

Shortly after the organization of this Inter-Allied Commission, a committee with somewhat similar functions was set up in Paris under the auspices of the Peace Conference. This Committee was known as the "Committee for the Left Bank of the Rhine", and functioned directly under the Supreme Economic Council. Its Chairman was M. Seydoux, representing France, and the American representative was Mr. Sheldon Whitehouse, councillor of the American Embassy at Paris. As the Luxemburg Commission did not function harmoniously in all matters, its troubles were passed on to this Sub-Committee of the Peace Conference. Part of the difficulty seems to have originated in French complaints that the American Army was not respecting the decisions of the Luxemburg Commission. In consequence, Mr. Wallace Day, one of the American representatives on the Supreme Economic Council, was transferred to the Luxemburg Commission in the hope that he could adjust the difficulties which had arisen. The representative of the American Army was thereupon withdrawn, and the Army was therefore represented on the Luxemburg Commission solely by Mr. Day.

The Commission had never been able to achieve any positive results other than in the regulation of imports and exports. Financial matters were wholly outside its jurisdiction, presumably being dealt with by the French Comptroller of Finance, but in reality, by each army within its respective sphere. Fuel and food problems were also partly handled by the armies and partly by other international commissions, such as the Inter-Allied Coal and Hoover Commissions. Much difficulty constantly arose from the lack of co-ordination of policies of these bodies. Great Britain, shortly after the arival of Mr. Day in Luxemburg, had also replaced her military representative by a civilian. Both delegates quickly realized that the difficulties confronting the Commission had arisen from its lack of authority, which had prevented it from exercising the influence over conditions on the left bank of the Rhine which was contemplated at the time of its organization.

The Supreme Economic Council, having been notified of the conclusions of the British and American members, thereupon organized the Inter-Allied Rhineland Commission, on April 21, 1919, vesting it at the same time with full authority to co-ordinate the economic policies of the various Allies.

The inter-allied Rhineland commission.—The charter of the new commission was promulgated by the Supreme Economic Council in its sitting of April 21, 1919, at the Quai d'Orsay, Paris. It read as follows:

1. That an Inter-Allied Commission, consisting of four Commissioners, one from each Ally concerned with the administration of the occupied territories, should, together with an Italian liaison officer, be set up with full authority to co-ordinate the administration of the four Army Commands on all economic, industrial and food questions, in accordance with the policy laid down from time to time by the Supreme Council.
2. That orders would be issued under the authority of the Supreme War Council to the Army Commands in the various areas, that directions given by the Commission shall be uniformly executed throughout the whole area.

The powers granted the Commission were greatly in excess of those formerly exercised by the Luxemburg Commission. The decision to lodge authority in the new Commission for co-ordination of the food policies of the various armies, was in itself a long step towards a unified control of occupied territory. The influence which the Commission actually came to exercise during the occupation was due, however, as much to the high calibre of the individuals selected by the governments to represent them on the Commission as to the charter itself. The member appointed by France, who later became its President, was M. Tirard a former French civil governor of Morocco and later the Comptroller of occupied territory under Marshal Foch. Sir Harold Stuart, an administrator with long experience in the Indian civil service, was sent as representative of Great Britain. Mr. Pierrepont B. Noyes and M. Transconster, both successful manufacturers, were appointed as members of the Commission by the American and Belgian governments, respectively.

At the same time that the Supreme Economic Council promulgated the charter of the new Commission, it took steps to see that decisions taken under the powers therein granted should be respected by the Armies. In consequence, the Commanding General, Third Army, received a letter from General Pershing, dated May 6th, directing that the American Army assist the new Commission in every way. There was, of course, no desire on the part of the Army to obstruct the Commission, and as a matter of fact, it was largely due to the cordial reception accorded it by Lieutenant General Liggett that Coblence was made its headquarters.

The Commission had temporarily established its offices at Luxemburg, in order to smooth over the transition between the *régimes* of the two commissions. It was however realized that inasmuch as Luxemburg was a neutral country, it would be both more fitting and convenient if headquarters were established in a Rhenish city. After an inspection of the facilities available in the larger towns in the area of each army, it was decided to move to Coblence. This town was not only the most important governmental city in occupied territories, but the American Commander was entirely agreeable to having the Commission take up its headquarters in the same city, despite the tax which this move would place on billeting facilities. The transfer took place during the third week of May, the offices being established in the "*Gerichtsgebäude*" (Court House). On September 3rd, it having been determined in the meantime that the Rhineland Commission would eventually become the High Commission, its offices were transferred to the *Oberpräsidium*—the seat of government of the Rhine Province—a much larger and more suitable building than the *Gerichtsgebäude*. The accommodations of the High Commission were arranged for and requisitioned by the Army which undertook the work in an entire spirit of co-operation despite the facts that Coblence was already over-billeted and the arrival of the High Commission meant a large additional amount of work for the Army.

The charter of the I. A. R. C. specifically authorized it to coordinate the administration of all economic, industrial and food questions in the various zones. This involved an inspection of all laws and regulations put into effect on these matters by the German or Prussian government, and a signification of approval or disapproval for each specific law. The announcement of the decision was communicated directly by the Commission to the Army Commanders who took the necessary steps to communicate it to the civil officials concerned. Decrees establishing new taxes formed the larger part of the new German laws and they were almost without exception approved by the Commission.

The dyestuff industry, one of the most important in Germany, had been so shut in with regulations, first by the Supreme Economic Council and later by the peace treaty, that a large number of problems pertaining to it alone were also constantly being referred to the Commission for solution. A special sub-committee dealing entirely with this industry had to be established eventually. Matters pertaining to this industry were however of small interest to the American Army as there were no dye factories in its area.

Under the power granted the Commission over financial matters, was promulgated a decision on December 19 stating that Allied banks which had been established in Germany should not be subject to civil judicial tribunals for violating the German laws of so-called "fiscal evasion" if the transactions had been made with the explicit or implicit consent of the military authorities. This decision, too, did not affect the American Army.

A decision of October 28, however, concerned the American Army primarily. The United States had, since about March 19, been disposing in occupied territory of a large proportion of its war material. It did not seem proper to permit this property to be taxed, even after its sale to Germans. Therefore the Army Commander had forbidden the German government to levy customs and other duties on the civilian purchasers. The decision in question ruled that thereafter German purchasers should pay the regular customs duty then in effect, thereby revoking previous army orders. However, the larger quantity of the Army's surplus stores had been sold by October and the decision did not have the importance it would have had if published some months earlier.

Probably the most important decision of the Commission was that of November 15, which authorized the German customs officials to take up their duties on the Western frontier. Although America and England had never raised any objection to Germany's continuing to collect customs duties along her Western frontier, the French and Belgian armies had taken a different attitude, thereby creating what was commonly referred to as the "Gap in the West" through which the merchandise of Europe flowed into the interior of Germany duty-free. This at the same time permitted Germany to be stripped of merchandise without payment of export duties. Such exports were greatly stimulated by the depreciation of the mark, thus permitting of greatly increased purchasing power of the currency of Allied countries. Such a situation naturally gave the German government, whose financial position was already grave, cause for concern. The decision of the I. A. R. C. was therefore greeted by all civil officials with expressions of pleasure and was undoubtedly an act of wisdom in the interest of the Allies as well as those of Germany, inasmuch as the lack of a customs frontier undoubtedly would have diminished the possibility of Germany's payments for reparations.

The Rhineland Commission also served to communicate decisions of the Supreme Council to the armies. The agreement of the Allies with Germany to permit her to levy customs on a gold basis was transmitted in this manner to the armies on January 8, 1920. This was another wise measure which tended to bolster up the disorder into which German finances had fallen.

Another numerous group of decisions published by the Commission pertained to food. One of the most important of these, of date of November 13, took steps to prevent the transportation of foodstuffs from Germany over the Western frontier by Belgian, French and Dutch merchants, whom the depreciation of the mark enabled to buy foodstuffs in Germany cheaper than at home. This was a matter of grave importance, as Germany had far too little for her own needs. All German civil officials were by the terms of the decision instructed to carry out their own laws under pain of removal from office, but were also assured of the support of the Allied armies in the performance of their duties. Another instruction, of November 25, endeavored to lay down a set of rules to guide the armies in their requisitions and purchases of foodstuffs. This was a very difficult matter, as the customs and policies of the armies proved to be in some cases diametrically opposed. However, certain general rules were therein promulgated by the Commission, which undoubtedly served to lighten the task of the civil food officials.

The dissolution of the Inter-Allied Coal Commission on October 12, 1919, placed additional work on the Commission. A special subcommittee was at once formed to deal with the many problems in connection therewith, but before any real constructive work could be begun the Supreme Council on November 12 turned back the responsibility for coal production and distribution to the German government.

There had been appointed at Coblence, alongside of the future High Commission, a representative of the German government, termed *Reichskommissar*, who after peace had been declared was to assume a superior position to all other civil officials within occupied territories. To the *Reichskommissar* was now attached a representative of the German Coal Commissioner in Berlin, with authority to direct the distribution of coal within occupied territories. The Rhineland Commission was supposed to retain thereafter only a loose supervision over distribution of coal to civilian industries, contenting itself with taking necessary measures to secure the amounts required by the armies and public utilities. On December 27, it approved the settlement of a strike in the brown coal district near Cologne, by the terms of which the miners obtained a seven-hour day.

The more important work of the Commission was of a preparative nature. It put much study into the drafting of the future code of laws, to be called "ordinances." In this work, the American Commissioner kept constantly in touch with the Army Commander, hearing his views and guiding his policy in many cases in the light of the experience gained by the Army. The Rhineland Commission passed out of existence on January 10, 1920, the day on which peace was declared, and the High Commission assumed the position of supreme representation of the Allied governments in occupied territories. As the United States had not at that time ratified the peace treaty, the American Commissioner, with the consent of the Allied governments, was authorized by the State Department to attend the sessions of the High Commission as the American representative.

Origin of the principle of government in the Rhineland by a civil commission.—It is now generally known in Europe and America that the conception of governing the Rhineland by a civil commission, after declaration of peace, originated in American circles. The President's proposition to that effect was based on a letter of Mr. Noyes, American representative on the Inter-Allied Rhineland Commission, and embodied certain definite recommendations. The principles of government therein formulated were accepted by the President as his own, and were not only presented by him to the Council of Four but agreed to by the representatives of the Allies. The plan as outlined in the letter was then referred to a sub-commission which drew up the Rhineland Convention.

The letter in question was written at Coblence, where Mr. Noyes was at that time representing the United States on the Inter-Allied Rhineland Commission. The views which he held and presented to the President were substantially those of a large number of officers of the American army who for some months had had an opportunity of closely following events in the Rhineland. The close connection between Dr. Dorten's fiasco in attempting to foment revolution in the Rhineland, and Mr. Noyes' letter to the President, had considerable weight in shaping future policy. The French command of the Tenth Army had approached Lieutenant General Liggett on May 21st, asking for American assistance in promoting a Rhenish republic, while Mr. Noyes's letter to the President was sent on May 27, only a few days later. There was reason to believe that Dr. Dorten's "revolution" had the backing of the French military party which at the time undoubtedly had other aims in prospect than a Rhenish state within the German confederation. The President not having acceded to the French conception of the Rhine as France's political frontier, the Army was bound by its instructions and the terms of the armistice to respect the *status quo* in occupied territories. American and British policies were so radically opposed to any scheme involving the forcible separation of the Rhine Province from Germany that a very critical situation might have developed if a supreme authority were not created in which the Allies could exchange views and take action as a unit.

There seemed to be two possible forms which such a supreme authority might assume, viz.: an Inter-Allied Military Commission, in which the Allied armies stood on an even basis; or, a civil body in which the governments were similarly represented. Mr. Noyes, after a thorough discussion of the question with American officers, decided to recommend the latter plan as his solution of the problem. His reasons, in outline, are stated in his letter to the President, as follows:

American Commission to
Negotiate Peace PARIS, May 27, 1919

To the HONORABLE WOODROW WILSON,
President of the United States of America;
11, PLACE DES ETATS-UNIS, PARIS.

Dear Sir:—

After a month spent in the Rhineland as American Commissioner, I feel there is danger that a disastrous mistake will be made. The "Convention" for the government of these territories, as drafted by the military representatives of the Supreme War Council on May eleventh, is more brutal, I believe, than even its authors desire upon second thought. It provides for unendurable oppression of six million people during a period of years.

This "Convention" is not likely to be adopted without great modification. What alarms me, however, is that none of the revisions of this document which I have seen, recognize that its basic principle is bad—that the quartering of an enemy army in a country as its master in time of peace and the billeting of troops on the civil population will insure hatred and ultimate disaster.

I have discussed this matter at length with the American Commanders of the Army of Occupation; men who have seen military occupation at close range for six months. These officers emphatically indorse the above statements. They say that an occupying army, even one with the best intentions, is guilty of outrages and that mutual irritation, in spite of every effort to the contrary grows apace. Force and more force must inevitably be the history of such occupation long continued.

Forgetting the apparent ambitions of the French and possibly overlooking political limitations, I have sketched below a plan which seems to me the maximum for military domination in the Rhineland after the signing of peace. Our Army Commanders and others who have studied the subject on the ground agree with this programme:

Skeleton Plan

I. As few troops as possible, concentrated in barracks or reserve areas, with no "billeting", excepting possibly for officers:

II. Complete self-government for the territory, with the exception below:

III. A Civil Commission with powers:

a. To make regulations or change old ones whenever German law or actions

(1) threaten the carrying out of treaty terms, or

(2) threaten the comfort or security of troops.

b. To authorize the army to take control under martial law, either in danger spots or throughout the territory, whenever conditions seem to the Commission to make this necessary.

Very truly yours,

(Signed) PIERREPONT B. NOYES,
American Delegate,
Inter-Allied Rhineland Commission.

There was still another reason, which while influencing Mr. Noyes to propose the above plan, was not stated in the letter to the President. The economic condition of Germany, although at the time of writing no worse than it had been during the winter, showed no sign of improvement. More important than Germany's losses of territory, which the negotiations at Versailles foreshadowed, were the severe restrictions which it was believed would be placed on both the finances and industry of the land. The Rhineland was its most important province in wealth and population. If Germany was ever to get on her feet and meet her obligations she must have a free hand in restoring her economic condition.

The deliberations of the committee of the peace conference, to which Mr. Noyes's proposition, approved by the Council of Four, was referred, are narrated at some length in a so-called "White Book" of the French government. The Rhineland Agreement, signed by the representatives of the German and the Allied governments at Versailles on June 28, 1919, was a separate and distinct pact from the peace treaty, being an agreement between America, Great Britain, France and Belgium on the one hand, and Germany on the other.

The Rhineland agreement and its limitation of powers of the army.—The armies of occupation had so different a position under the Rhineland Agreement than under the Armistice that a brief comparison should be made here of their powers during each period. It was largely because of the erection in the American zone of a supreme

Allied authority of an international character that the position of the army, with the coming into force of the peace treaty, became such a delicate matter. Around this situation hinged the negotiations between the Commanding General and the Commission as to the validity of the ordinances of the High Commission and their effectiveness in the American zone.

The Armistice of November 11, 1918 provided for the occupation of the left bank of the Rhine and the administration by the armies of occupation of the areas or bridgeheads to which they were separately assigned. It was provided in substance in Article 5 of the Armistice that no person should be prosecuted for his participation in military measures previous to the signing of the Armistice.

International law places upon the Commanding General the responsibility broadly speaking of preserving order, punishing crime and protecting lives and property within the territorial limits of his command. His power in the premises is as great as his responsibility. The Armistice in no sense checked, or refused to the military forces, any of the powers usually and ordinarily exercised by an invading army, except as above noted. A reading of the Armistice clearly shows that each army of occupation was to act as the representative of its respective government in the conduct of the military operations with which it was charged. There was nothing in the Armistice removing from the Commanding General (with the exceptions noted) any of the authority expressly or by inference vested in him by international law or usage.

Provision was made in the treaty for occupation of German territory west of the Rhine during a period of fifteen years, such occupation, upon conditions being broken, to be continued as long as necessary. If, however, the treaty provisions were complied with before the expiration of the fifteen year period, the troops of occupation were to be withdrawn. (Articles 428–431.) The German government agreed to maintain the troops of occupation in the same manner as during the Armistice. (Article 249.)

The source of power of the Inter-Allied Rhineland High Commission is found in the agreement of June 28th, 1919, between the United States of America, Belgium, Great Britain and France, of the one part, and Germany of the other part. Such agreement was entered into by the high contracting parties acting under powers conferred upon them by their respective governments and authorized in Article 432 of the Treaty of Peace, which provides: "All matters relating to the occupation and not provided for by the present treaty shall be regulated by subsequent agreements, which Germany hereby undertakes to observe."

Under the terms of the Rhineland Agreement, the High Commission is made the supreme representative of the Allied and Associated

Powers within occupied territories, except in so far as the treaty otherwise provides. Specifically, the High Commission was given power to issue ordinances for the maintenance, safety and requirements of the Allied and associated forces. It could further, in case necessity arose, suspend the civil administration in the Provinces, Bezirke and Kreise, and in addition take any other steps necessary in order to adapt the civil administration of the German government to the needs and circumstances of military occupation. After publication, the ordinances were to have the force of law and be recognized by the German civil authorities as well as by the Allied and associated military authorities. The High Commission was also given the power to declare a state of seige in the whole or in any part of the territory under its jurisdiction. The personnel of the High Commission was to be exempt from German direct taxes. It was to negotiate with the German government for suitable premises for its housing, and all its costs and expenses were to become a charge against Germany. Its supplies were to be transported free of charge and free of all import duty, and its members to be granted diplomatic privileges and immunities.

While the High Commission was given authority to legislate for the security, maintenance and requirements of the armed forces and whenever necessary to declare a state of siege in the whole or any part of occupied territories, the military authorities were empowered, when public order is disturbed or threatened, to take such temporary measures as may be necessary to put down insurrection or to restore order, and to requisition in kind and to demand services in the manner laid down by The Hague Convention of 1907. It was further provided in the Rhineland Agreement that the military authorities retain exclusive jurisdiction over the members of their respective forces and persons accompanying them, to whom the general officers commanding the armies of occupation had issued revocable passes. The same applied to any person employed by or in the service of the respective occupying armies. Persons who committed offenses against the persons or property of the armed forces of the Allied or Associated Powers could be made amenable to military jurisdiction.

Situation caused in American zone by coming into force of peace treaty.—When the Rhineland Agreement was being drawn up in Paris, the eventuality of non-ratification of the peace treaty by any one of the belligerent powers had unquestionably not been foreseen. France, England, Italy and Belgium ratified the treaty and the Agreement within a reasonable time, but desiring American co-operation in the plebiscite areas and on the numerous commissions, put off a formal declaration of peace from week to week. When, in the latter part of December, the divergence of opinion within the United States Senate seemed to be so great that early ratification became improbable, the Allies decided to put the Treaty and Agreement into effect without further delay. On January 10, 1920, the Allies and Germany accordingly signed the treaty by which peace was brought again to Europe.

It is hardly necessary to remark that both the American Forces in Germany and the American Commissioner on the Inter-Allied Rhineland Commission now found themselves from a legal point of view in an extremely peculiar position. The agreement was essentially a product of American thought. The principles enunciated therein had originated with Mr. Noyes, the American Commissioner, himself. Peace had now come, and the Allies were bound by their pact with Germany to put the Rhineland Agreement into force. America, however, was still at war; her armies stood on German soil by virtue of the Armistice Agreement of November 11, 1918, and could certainly not technically recognize a pact which their own government seemed unwilling to ratify. Under the agreement, the territory occupied by the Americans was entitled equally with the other Allied zones to the mitigation of strict military government. It may well be questioned whether the Allied governments had the legal right to put the Convention into effect, as the pact expressly provided that America, Belgium, Great Britain and France should all be represented on the High Commission. Non-ratification by the United States Senate had of course made it impossible for the American Commissioner to attend the sessions in any but an informal capacity. Any objection by Germany to government of occupied territory by the High Commission would have been difficult to refute legally. However, knowing that the American point of view would receive as much consideration in the sittings of the Commission as if she had been formally represented, Germany raised no objections. The Ordinances of the High Commission lifted many restrictions formerly in force by orders of the various armies.

The position of the American zone was the stumbling block to amicable agreement among all parties. Major General Allen had received instructions from the War Department in no way to anticipate ratification of the peace treaty by the Senate. An out-and-out recognition of the authority of the Rhineland Commission within the American zone would have unquestionably been a deliberate violation of these instructions.

The announcement of the approaching declaration of peace at Paris served to arouse both Commission and military authorities to the gravity of the situation which would thereby be brought about. Negotiations were begun in the hope that some informal arrangement might be concluded, by which the ordinances of the Commission could be put into force in the American zone as orders of the Commanding General. These negotiations were still in progress when the Supreme Council at Paris endeavored to take matters into its own hands. In its sitting of

December 30th, a resolution was passed, which announced that the jurisdiction of the Rhineland High Commission would, from the day of the coming into force of the peace treaty, be extended to all Rhenish territories including those occupied by the American troops. Mr. Wallace, the American Ambassador to France who was present informally at the sitting, stated that he would refer the resolution to Washington for instructions.

This action of the Supreme Council at once reacted on the French, Belgian and British Commissioners. Those governments having stated their position so strongly, their Commissioners felt that they could not shift their standpoint by entering into an agreement in the nature of a compromise. General Allen now found himself in a very embarrassing position. Desiring above all things to maintain the Allied unity, yet he could take no step towards an understanding with the Commission without disobeying orders of the American government. In his anxiety to find a way out of the *impasse*, he had even referred the legal aspect of the situation to a committee of his staff officers, hoping that the legal difficulties in the path of such a settlement were not insurmountable. This committee, however, after a careful study of the ordinances of the High Commission, stated that even the putting them into effect as military orders would clearly constitute an act anticipating ratification. The General determined that there must be some solution and decided to recommend this very course to the War Department, and request necessary authorization to carry it out. On January 2, 1920, he telegraphed the American Ambassador at Paris, stating his intentions and asking whether they had the approval of the State Department. A further telegram was dispatched to the Adjutant General at Washington on January 5, asking for similar specific authorization. The American Commissioner, Mr. Noyes, also wired to the Secretary of State, outlining the compromise which the American army would endeavor to have the High Commission accept. Mr. Noyes strongly urged that the State Department assent to this compromise.

Peace was declared at Paris on January 10, unfortunately before replies to these telegrams could be received. The army and High Commission found themselves confronted with a situation demanding an immediate solution, which could not be delayed until specific authority arrived.

Agreement reached by high commission and army commander as to the status of the American Zone after coming into effect of peace treaty.—The High Commission was at this time composed of M. Tirard, representing France, Sir Harold Stuart, Great Britain, and M. Rolland Jacquemyn, Belgium, Mr. Noyes being present at its sittings in an informal capacity. It intended as its first act to publish two important documents throughout occupied territory. The first of these was a proclamation announcing the assumption of supreme authority in the Rhineland by the High Commission. The second was the "Ordinances", i. e., a code of laws which was to replace automatically the existing orders of the various armies. Any agreement reached between the Army Commander and the High Commission naturally revolved around the promulgation of these two documents. After some preliminary discussion, the Army Commander formally proposed that the ordinances of the High Commission be published in the American area as military orders. The proclamation, on the other hand, was to be published as it stood, General Allen believing it incompatible with his instructions to place his name under a statement that supreme authority in the Rhineland now lay with the High Commission. This proposition was informally agreed to by M. Tirard, the President of the High Commission, on the 11th, and was confirmed the next day in a letter in which M. Tirard in the name of his colleague expressed his appreciation of the spirit of conciliation shown by the Commanding General. A good deal of uncertainty arose soon after the publication of these documents as to whether the Army or the Commission was in control. The contradiction between the Ordinances and the Proclamation was so striking as not to be overlooked, and requests for explanations reached the Office of Civil Affairs from both the *Oberpräsidium* and the *Regierung*. They were informed that decisions would be rendered from time to time as actual difficulties arose, and not on hypothetical questions.

Answers to the telegrams previously sent the State and War Departments reached Army Headquarters on January 13. The answer of the Adjutant General was very brief, merely advising General Allen that the War Department approved his recommendations to put the Ordinances into effect as American military orders. The State Department's answer was also on the whole favorable to the compromise adopted, but stipulated that measures taken by the Army Commander must not conflict with the terms of the Armistice.

The agreement made by M. Tirard and the Commanding General on January 11 was intended to be a mere temporary solution of the difficulties. It was contemplated that further negotiations would ensue when a reply to Ambassador Wallace's telegram had been received from Washington. This reply had now been received, but it gave no indication that America would make any further official recognition of the High Commission. The Supreme Council, confronted by an insurmountable obstacle in carrying out its resolution of December 30, finally receded from the position which it had assumed, and soon after recognized the authority of the American Army under its war powers.

Practical application of agreement reached by Commanding General with high commission as to the latter's status in

American Zone.—Recognition by the Supreme Council that America's position in occupied territories was limited solely by the terms of the Armistice, immeasurably simplified the relation between the High Commission and the Army. The Commanding General had published the Ordinances of the High Commissions as military orders and now intended to carry them out as far as was practicable. Despite the admission of the Supreme Council, he now saw no reason for altering this agreement. With certain small exceptions, there did not seem to be any reason why the High Commission should not at once take over the duties to which it had been assigned by the Rhineland Agreement and which it had assumed by virtue of the publication of the Ordinances. The High Commission understood perfectly that the General was still the supreme authority within the American zone, and that his action in transferring certain powers to the High Commission was merely an act of courtesy, revocable at any time if emergency required. General Allen published a circular letter to his command on January 14, urging all officers and men to enter into the spirit of the *modus vivendi* which had been adopted on his initiative, and to conduct themselves as if the Senate had already ratified the Rhineland Agreement.

There were still certain practical difficulties facing the Army in its purpose to abide strictly by the Ordinances of the Commission. The Commanding General, after all, was alone responsible for the security of the Army. Under the Rhineland Agreement, this responsibility, except in cases of emergencies, was to devolve upon the High Commission. It was very questionable how far the responsibility of the Army Commander for public order could be transferred to other shoulders. It also seemed probable that the Army might have to take certain measures not provided for in the Ordinances, in order to satisfy the requirements of security. There were other practical difficulties, largely centering around the respective duties of the O. C. C. A.'s and the representatives of the High Commission in the Kreise, and it was not clear how many of the duties formerly performed by the O. C. C. A.'s could now be abandoned or turned over to these representatives, without prejudicing the maintenance of public order.

The day after the *modus vivendi* had been agreed to, negotiations were begun between the parties there concerned, looking to a satisfactory solution of these difficulties. A number of verbal conferences were held between Mr. Noyes and the Officer in Charge of Civil Affairs, and as a result thereof the Commanding General on January 17 formally put in writing the steps he would take to conform the administration of the American zone with that of the rest of the occupied territories. Very few deviations were made from the reading of the Ordinances. Article 16 of Ordinance III prescribed that notice of political meetings alone should thereafter be reported in advance to the local representatives of the High Commission. General Allen's letter of January 17, stated that he felt it inadvisable to carry out this Ordinance literally, as the larger part of the political unrest in the American zone had been stirred up in industrial meetings. Notice of these meetings had till then been given forty-eight hours in advance by the civil officials to the military authorities. General Allen felt that the removal of all control over such meetings was highly inadvisable and therefore stood in this matter on powers granted him by the Armistice.

In the matter of inspection of German prisons, it was provided that the inspections should be made by the representative of the Commission and the O. C. C. A. together. Suits by civilians against the military were not to be permitted until peace was ratified. Saluting was still demanded of all civilians in uniform. On the whole, however, the modifications made in the Ordinances were few, and it was expected that the High Commission would assume a position in the American zone equivalent to its position in the other areas. On January 31, 1920, the orders of the Army in existence during the period of military government were revised by Paragraph 58 of the Civil Affairs Bulletins. A careful study of the previous orders had been carried out and the new regulations were entirely in the spirit of the Ordinances.

January 10, 1920, the day on which peace was declared, marks the virtual close of military government. The period necessary for readjustment still required a large amount of work but instances of disagreement were very few and were always amicably adjusted.

The Army, whose commander had proposed the *modus vivendi*, fully lived up to the spirit of the agreement, once it had been accepted by the High Commission.

Thus one of the most critical phases of the relations of America to the Allied powers, one which in other hands might well have been disastrous to unity of action, passed into history. To the forbearance and farsightedness of the Commanding General of the American Forces in Germany, the world today is indebted in a large measure for the solution of a matter vital to the welfare of the Allies and Germany.

The period of American Military Government over that corner of Germany bordering the Rhine and Moselle, an area which has witnessed the conquering armies of so many nationalities come and go, has now passed into history. As to the justness and humanity of the rule of the American Army during this period, the verdict must to a large extent be moulded by the historians and students of tomorrow. The Army today awaits that verdict, confident of the justness of its deeds and of the humanity of its purpose. For certain it is that the Army of the United States has its roots deep in the soil of the American people and that the spirit of its service on the Rhine has only reflected the spirit of service to humanity which animated the nation throughout the World War.

To each officer and man, whether regular or national, who did his part to demonstrate the spiritual force of Democracy, America may well be grateful and to the leaders of the Army, in whose hands devolved the interpretation of the spirit of America, the nation should be especially grateful—Major General J .T. Dickman, Lieut. General Hunter Liggett, their Chief of Staff, Brigadier General Malin Craig and Major General Henry T. Allen and his staff.

Made in the USA
Monee, IL
04 November 2026

67387532R00213